SOLAR-ELECTRICS
RESEARCH and DEVELOPMENT

by

Robert L. Bailey

Associate Professor
Department of Electrical Engineering
University of Florida, Gainesville

230 Collingwood, P.O. Box 1425, Ann Arbor, Michigan 48106

Library of Congress Catalog Card Number 79-54871
ISBN 0-250-40346-3

Manufactured in the United States of America

CONTENTS

PREFACE

This Work Is About the important future new field of solar-electricity generation and the integration of its subareas into a coherent body of knowledge. These are now in varying stages of Research and Development (R&D). Thus the R&D in solar-electrics of the 1970's is in anticipation of likely unfolding developments of the 1980's. Solar-electrics are defined as the set of advanced[1] technologies for obtaining electricity from inexhaustible solar energy. They are a part of the larger field of 'Alternate Energy Sources'. Solar-electrics presently includes the advanced high growth sub-fields of solar-thermal, photovoltaic, wind, solar-water, biological, energy storage, systems, and novel solar-electrics. Only photovoltaics, wind, energy storage, and systems of solar-electrics are treated in this book, however. The past and present art of these are assembled in a unified way of particular interest to power systems engineers and others. Time and funding did not permit reviewing the remaining sub-fields. These are planned for a later textbook.

Solar Heating vs Solar-Electrics Most of us are aware of the growing world-wide interest in utilizing solar energy in sensible and practical ways. The sun's energy is generally recognized, in the present energy conscious world, as the only source of inexhaustible 'energy income'. In trying to utilize it, various forms of solar heating and cooling now enjoy a large and growing following because such practical uses are highly developed. Solar heating is technically comparatively simple. A growing number of commercial firms are entering solar heating, and this part of solar utilization is becoming widely applied and growing in economic importance.

In contrast, the generation of electricity from solar energy is, as an overall technology, younger than solar heating. Also solar-electrics are--many think--technically more complex and therefore presently less developed in terms of large-scale commercial markets. Presently, for example, widespread use of solar cells is minuscule compared to the square meters of solar heating collectors installed in the United States. The result is that total solarly generated electric energy is presently infinitesimal compared with the approximately 1916 billions[2] of KWHe[3] of bulk electric power produced in the United States in 1975 by conventional electric generation plants (fossil + nuclear + hydro).

Future Solar-Electric Industry In the macroscopic sense solar-electric generation is a new industry in its infancy. The technology is now mostly in R&D stages with only a relatively few commercial products offered--e.g. solar cells and a few windmills--for specialized small-scale applications. Solar-electric products do not yet appear in the Sears catalog!

---

[1] 'Advanced' throughout the literature means still in some stage of R&D. A further connotation is that the technology cited is not widely commercially used compared to other existing competing technologies which are widely and commonly used. The term 'Advanced' is typically loosely used.

[2] "Factor Affecting the Electric Power Supply, 1980-85. Executive Summary and Recommendations," Federal Power Commission, Bureau of Power, Dec. 1976. Available from National Technical Information Service (NTIS), Springfield, VA 22161 as PB-264 760.

[3] The e indicates energy in electric form as distinguished from thermal, which would be indicated by t.

Solar-electrics--in terms of a large-scale economically important industry--then becomes a vision of the future. Such a large-scale industry is not now in situ; but it is in the very nature of R&D, in an overall sense, to reveal what realistically could be done--perhaps on a large-scale--by bringing science and engineering together in positive creative ways to synthesize new devices and systems. The vision of a future large-scale solar-electric industry is founded on solid reality, as we shall see herein. It could eventually be a multi-billion dollar/yr. industry. To actively engage in such R&D--clarifying dreams of the future and making them realities--requires that we be optimistic, at least in the long run. Finally, as a careful study of history shows, the grand scheme of events seems more for visionaries and innovators than for pessimists or those with no plan.

Readers outside active solar-electric R&D areas must therefore understand that the devices and systems to be presented herein as products of R&D are not "finalized, fully ready for commercialization." Substantial engineering design work beyond R&D is a must before commercialization of specific solar-electric devices or systems can occur.

The viewpoint here is that a state-of-the-art survey of solar-electrics in the 1970's is in anticipation of developments likely to occur in the 1980's.

Unfortunately, there will be many who will judge this book as if solar-electric technology is fully ready for commercialization. Others will undoubtedly feel it is premature, considering the overall development of solar-electrics.

The Need For An Integrated One Volume Solar-Electric Work is brought on by the rapid R&D growth of solar-electrics since 1972. A well-executed comprehensive single volume on solar-electrics could tend to promote orderly growth of the field. The need for a one volume solar-electric book has also been created by the influx of many newcomers to the field. Newcomers--both in and out of active R&D--legitimately need to quickly and authoritatively be appraised of the prior art before trying to extend it.

Then the government's need to help integrate this already diverging sophisticated advanced technology is pressing.

There is also a strong interest in solar-electrics among engineering students--at least at the University of Florida. Here I taught in 1975 what may have been the first course in U.S. higher education devoted exclusively to solar-electric engineering. It was repeated in 1976, 1977, and 1979. We had a clear need for a single volume study guide which integrated solar-electrics. The course formed the structure from which the present research evolved. It later led to this book.

Users Of This Book will probably be active solar-electric researchers, R&D managers, electric utility planning engineers and managers, engineering faculty, future engineering students, engineers outside the solar-electric field, the resourceful do-it-yourselvers throughout the land, and others peripherally interested in solar-electrics, the latter--chiefly motivated by reducing their ever-rising electric bills.

Government The Catalyst For Solar-Electrics Much of the interest in and recent growth of solar-electrics in the United States can be attributed to the important catalytic effect of the Department of Energy (DOE) and its predecessors, the Energy Research and Development Adminstration (ERDA) and the National Science Foundation's older Research Applied to the National Needs (RANN) programs. Other government agencies, as NASA, also helped catalyze solar-electrics.

The genesis of solar-electric growth was with the 1972 work of the NSF/NASA Solar Energy Panel. It was composed of a group of 44 of the nation's experts in solar energy. The Panel's report[1] was the nation's first coherent solar energy R&D plan. The plan was adopted and became the principal solar program for RANN, ERDA, and now--somewhat modified--for DOE.

Integral with that larger U.S. solar R&D plan was important activity to be undertaken in various solar-electric areas. Congress subsequently funded solar R&D at high levels each year, a portion going to solar-electrics R&D.

The impetus and major progress in the 1970's growth of solar-electrics has been largely fueled by federal funds; but federal monies also had an important catalytic effect in attracting corporate and private R&D funding to the advanced solar-electric area.

We should not lose the perspective that before solar-electrics could be 'sold' to decision makers as a viable technology it was first necessary to sell the nation an integrated R&D program on general solar energy utilization. This effort is still underway. Had the general solar sales effort not been moderately successful, undoubtedly solar-electrics would not be a growth industry today.

The Scope of this document encompasses an examination of the more important known means to obtain useful electric energy and power from solar energy. We deliberately limit to solar-electrics, and little of the important broader solar field will be injected. For example, as useful as solar hot water heating is, it generally does not result in electricity as a practical output and is therefore only peripherally relevant to the scope of this document.

The scope is further limited chiefly to the United States situation. We are not attempting here to exhaustively survey solar-electrics on a worldwide basis, though some progress in other countries will be injected at appropriate places.

The viewpoint is principally terrestrial. We are seeking means of achieving practical useful solar-electric devices and systems on earth as opposed to doing so in space.

Overall, we want to survey advanced solar-electric systems technology for factually what has been done, to integrate in sensible ways the various subdisciplines within solar-electrics, and to prognosticate the probable future of solar-electrics as we all too slowly transition toward inexhaustible solar energy utilization.

Because of the relative short research contract (only 14 months) and the limited personnel (Principal Investigator full time and limited part-time help), the depth and completeness of the survey was further limited. Time and funding simply did not permit examining solar-thermal-electrics, solar-water-electrics, biological-electrics, and novel solar-electrics.

Perspective Sought All of us seek perspective. Perspective and the creation of a logical structure is particularly needed in the growth phases of any new technology. Solar-electrics is no exception. We are already seeing much narrow vision; e.g. the solar thermal people who consider their Power Towers the only way to generate electricity solarly, the solid state applied physicists who know little of Power Towers but know their solar cells are the way to go, the wind

[1] "An Assessment of Solar Energy As a National Energy Resource," Dec. 1972. Available from NTIS as PB-221-659.

people who know little of solar cells but see wind power utilization as the only sensible solar-electrics generation means; and then there are the bio-people who understand little about windmills or solid state physics of solar cells but who see utilizing various biological means as the best route for generating electricity from solar energy. The list can easily be extended. The situation is not unlike examining the skin of a large pachyderm with a microscope, failing to see him macroscopically.

We seek to break out of tunnel vision and to see the perspective of the whole elephant of solar-electrics. One perspective, for example, clearly is that, in all likelihood, each solar-electric method will be in competition with the others for the customer, either directly or via delivering useful electric power to the utility grid. Also there is the perspective that as a nation we shall need electric energy from any source--probably of multiple kinds--in the future. There clearly will be many systems integration problems. To see these and other perspectives and acquire a coherence for the whole of solar-electrics we necessarily must travel sans some details, leaving these for the interested reader in the cited references.

It was felt that such perspectives could best be seen from the vantage point within a university where presumably objectivity reigns.

The Research Method principally used for this state-of-the-art survey was selective sampling. In 1966 I began accumulating a specialized solar-electric library. To it numerous research reports were added on this contract, as were solar-electric books and conference proceedings. In toto, the collection was a representative sampling of solar-electrics as of the present time. Several conferences relevant to solar-electrics were also attended to further insure information currency.

We must realize that selective sampling of a field is inherently not enhaustive. The acquisition of all the government research reports alone would have far exceeded both fiscal and time budgets for this research.

Because of the nature of sampling we, inherently, cannot include 'what was done last week by __________'. The perspective is more like 'this was roughly the state-of-the-art as of about the mid 1970's', i.e. in a few years time frame--and even that can not be very precise because of the dynamics of many people in multiple organizations working on innumerable complex problems each difficult to summarize as of a single time increment, Δt.

An integral part of the sampling process was knowing, talking with, corresponding with, and otherwise interacting with many, but not all, of the people actively involved in solar-electric R&D.

The conclusion is that selective sampling by a knowledgeable person is the most practical modus operandi for a state-of-the-art survey.

Acquiring Authoritative Solar-Electric Information was a nontrivial problem in this research. Though solar-electrics, in toto, is yet young compared to established technologies, e.g. turbo-generators, already the literature is growing faster than an individual can assimilate.

One of the most helpful entrée sources to the field was Energy[1]. This government publication identifies, classifies, and abstracts government sponsored research reports on energy. A subsection deals with solar energy from which specific reports on solar-

[1] One of the Weekly Government Abstracts. Available from NTIS as a subscription.

electrics were selected based on a general knowledge of the field. The reports were ordered from the National Technical Information Service (NTIS), studied, digested, and their content appropriately woven into this volume. There were many detailed operational problems, however, in acquiring such reports.

An alternate source limited to the solar field is the government's Solar Energy Update.[1]

Many computer print-outs of literature searches on various solar-electrics subfields have been made by various agencies--some more helpful than others. These are cited where appropriate.

Numerous conference proceedings were also accumulated and studied. The December 5-7, 1977 Miami International Conference on Alternative Energy Sources[2] was particularly helpful. Much useful information on solar-electrics is contained in the various Intersociety Energy Conversion Engineering Conference (IECEC) Proceedings[3]. These annual conferences are devoted exclusively to advanced energy conversion. The International Solar Energy Society's (ISES) annual conferences are a repository[4] of much state-of-the-art work.

The Electric Power Research Institute (EPRI) issues timely research reports[5] containing some relevant solar-electric R&D art.

A limited number of books on subfields of solar-electrics are starting to appear, and these were acquired, studied, and appropriately cited.

Another source of information was subscriptions to several prominent energy newsletters[6]. Various R&D and professional magazines and newspapers were also scanned for solar-electric content.

Direct correspondence with other researchers was a potent and helpful means of acquiring some needed information for this survey.

---

[1] Originates from U.S. DOE, Oak Ridge, Tenn. 37830. Subscribe through NTIS.

[2] Proceedings of Condensed Papers, Miami International Conference on Alternative Energy Sources, 5-7 December 1977, Miami Beach, Florida. Copies available from School of Continuing Studies, Conference Services, University of Miami, P.O. Box 248005, Coral Gables, Florida 33124.

[3] These Proceedings are published by a different engineering society each year. The August 28 - September 2, 1977 12th Intersociety Energy Conversion Engineering Conference Proceedings was published by the American Nuclear Society, 555 N. Kensington Avenue, LaGrange Park, Ill. 60525.

[4] Proceedings are available through American Section of ISES, Inc., c/o American Technological University, Killeen, Texas 76541.

[5] Their Publication List is available from Research Reports Center, P.O. Box 10090, Palo Alto, CA 94303.

[6] Energy Digest, Scope Publications, Falls Church, VA 22041. Also Solar Energy Intelligence Report, Business Publishers, Inc., P.O. Box 1067, Silver Springs, MD 20910, and Energy Insider, C460, U.S. Department of Energy, Washington, D.C. 20545.

In summary, technical information from any source was accumulated for this research which dealt with any facet of solar-electrics. References are cited throughout where the reader can find more detailed information.

Serious solar-electric researchers, development, and design people are encouraged to build their own library of information around their own interests and involvements. Again, both the initial acquisition difficulty and the ongoing nature of solar-electric information collection must be emphasized.

Interdisciplinary Newcomers will quickly sense that solar-electrics is inherently a strongly interdisciplinary field. Electrical engineering, mechanical engineering, thermodynamics, fluid mechanics, solid state and general physics, chemistry and chemical engineering, instrumentation, aerodynamics, hydrology, biological systems, environmental engineering, social systems and impacts, and invention are a few disciplines one encounters. No matter how well educated we may be in one discipline, we encounter exciting learning opportunities in such interdisciplinary areas. The interdisciplinary view is absolutely necessary for making progress in solar-electrics.

Style I have tried to make the book easily readable, interesting, accurate, and challenging. The writing style chosen is casual, unstilted, and has a minimum of mathematical tediousness unless absolutely necessary. Hopefully throughout we have treated most--though undoubtedly not all--of the principal fundamentals of the various subfields one needs to know for making further progress.

The Units used are mixed, simply because such a mixture is the present state-of-the-art; but System Internationale (SI) units are used where possible and convenient. Many published source data used are given in the older English units. Converting all such data to SI could not be achieved within the short time of this contract. Realistically any contemporary working in solar-electrics must be facile in several unit systems until the technology eventually gets converted to SI units.

Current Barriers To Progress and Future R&D Problems for that subfield of solar-electrics, as I see them, are simply listed in the latter part of each chapter. These should allay fears of shallow thinkers that 'everything has already been done,' identify bonafide technical and nontechnical topics needing additional thought and work by competent people for progress, help DOE and other government agencies and corporations formulate clear future programs, and stimulate and challenge the younger generation by clearly and cleanly laying out the problems[1] needing additional creative work. This entire solar-electric field in its initial growth phase is so rich for the application of creative minds[2]! Many specific R&D, Design, and commercialization opportunities are identified. Others undoubtedly exist. Little attempt has been made to justify any single future R&D problem. Such justification is best left to those who choose to pick up the problem(s).

---

[1]Engineering students chronically complain that the older generation "keeps the real problems covered up where we can't see them!"

[2]Bailey, Robert L., Disciplined Creativity for Engineers, [Ann Arbor, Michigan: Ann Arbor Science Publishers, Inc., 1978].

Author Accountability Opinions, findings, conclusions, recommendations, or forecasts expressed in this publication are mine--for which I alone accept full responsibility. They may not reflect the views of the research sponsors (the University of Florida, the United States Department of Energy, Florida Power Corporation) or the publisher.

Gainesville, Florida

Robert L. Bailey

Acknowledgements

Dr. Robert L. Sullivan's encouragement and assistance in initiating this research is gratefully acknowledged. Interactions with him throughout this research were most helpful, particularly in the systems area.

The decisiveness of Mr. Ken Klein, Division of Electric Energy Systems, U.S. Department of Energy, and Mr. Ned Spake, Florida Power Corporation, made this research possible through their sponsorship. The University of Florida's Faculty Development Program contributed significantly to the project funding via a sabbatical leave for the Principal Investigator. The content of this book--with a few minor subsequent changes--was created by the author under research Contract EC-78-S-05-5624 for the United States Department of Energy and Florida Power Corporation. The final report was number ORO-5624-4, spanning the October 1, 1977-November 15, 1978 period.

Helpful suggestions and guidance were made by Mr. Phillip Overholt, Department of Energy, who managed the contract. Mr. Larry Rodriguez, Florida Power Corporation, in reviewing and critiquing the numerous detailed manuscripts, saliently contributed to the finished work.

Drs. Sheng S. Li and J. K. Watson made many inputs to the total work with helpful discussions. They made the research a pleasurable adventure.

Mr. Robert E. Sparkman, Electrical Engineering graduate student, assisted in the initial information retrieval and preliminary organization.

Stimulating preparatory discussions occurred with Dr. Johnny R. Johnson, Electrical Engineering Department, Louisiana State University, during his 1976 sabbatical at the University of Florida.

Numerous solar-electric researchers cited throughout kindly responded to requests for copies of their technical papers and thereby contributed to the work. Many publishers, cited throughout, kindly granted permission to use selected copyrighted illustrations.

The following Electrical Engineering graduate and undergraduate students endured the birthpangs of the Principal Investigator's first 1975-1977 solar-electric engineering course at the University of Florida and enthusiastically and materially assisted with the initial collection and organization of solar-electric research materials: Frank L. Bellissimo, John R. Blyth, Clark H. Briggs, Charles E. Crutchfield III, Charles W. Curry, Jr., Danny Lee Dees, Chester R. Gangaware, Ronnie D. Gillespie, Joseph R. Gracia, Richard Harrison, James A. Heaney, Fernando Irizarri, Robert T. Jones, Walter J. Jones, Willie H. Jones, James A. Kimball, Stephen A. Kuras, Kenneth Lee, Carl A. Letter, Mary Claudia Maire, Michael J. Marsden, James K. Massey, Kirk Matchett, L. Chas. McHenry III, Johnnie C. Mears, Jr., Darrell D. Moyer, Marshall Dwight Odum, William F. Pope, Lorenz H. Porzig, Hans Björn Püttgen, David A. Ramirez, Thomas Reedy, Harold G. Robertson, David C. Rush, Jane Terry Scarbrough, Eric R. Schultz, Randy Sell, John A. Sloane, Roddie S. Sorrell, Kenneth B. Stackhouse, Leroy Tucker, Ronald E. Wheeler, Wayne Whitley, Robert Stanford Winter, and Daniel L. Zucker.

Majorie L. Summers, Marjorie A. Niblack, Jeanne Trudeau, and Illeya Hamwi contributed strongly by preparing the graphics. Mr. John L. Knaub and his assistants performed the photographic work for the illustrations.

Mr. Gary A. Moore greatly assisted in the onerous task of proofreading the entire book manuscript and of insuring uniformity among the many symbols, abbreviations, and word usages.

Mr. Michael Frye assisted with the Index preparation.

Miss Edwina Huggins, Miss Shirley Sweeting, and my wife, Betty Bailey, faithfully, patiently, and accurately transformed my manuscript into readable typed versions for the original DOE research report. Miss Stacey Garvin assisted with the final book manuscript.

ROBERT L. BAILEY has been involved in solar-electrics since 1968 when he did research in advanced photovoltaics for NASA Goddard Space Flight Center. In 1966 he introduced the first courses in Direct Energy Conversion Engineering at the University of Florida--courses he continues to teach and which contain some solar-electric converters.

An Associate Professor of Electrical Engineering at the University of Florida and member of the Graduate Faculty, the author in 1975 introduced the first course in Solar Electric Engineering at the University of Florida. It is believed to have been the first such course in the nation. Solar-Electrics started with that teaching experience, growing into a U.S. Department of Energy research contract in 1977-78 from which it evolved to the present book.

He is a member of the Electrical Engineering Department's Electric Energy Engineering group.

This is Professor Bailey's second book. His Disciplined Creativity For Engineers was published by Ann Arbor Science Publishers in 1978.

The author, has worked in a wide range of engineering management, and research activities. His 11 years with General Electric and 19 years at University of Florida have included many creations in research, development, design and manufacturing. He was a consultant and leader in the 1972 NSF/NASA Solar Energy Panel which formulated the nation's first coherent solar energy R&D program for the 1972-2020 period. That program launched the present solar boom and is now being implemented by DOE, other agencies, and companies. He is inventor of the Electromagnetic Wave Energy Converter which he has researched for the past ten years.

Professor Bailey received his BEE from Auburn University and MSE from the University of Florida where he also did additional graduate work. He is a member of the Institute of Electrical and Electronic Engineers, International Solar Energy Society, American Society of Engineering Education, and several honor societies.

Professor Bailey enjoys fine woodworking as a hobby.

CHAPTER 1

INTRODUCTION TO SOLAR-ELECTRICS

"Of the possible worlds we might choose to build,
a solar-powered one appears most inviting."
Denis Hayes [1]

- World energy situation.
- United States energy situation.
- Solar: Our major inexhaustible energy income.
- Electricity: A vital energy form.
- Solar-Electrics: A part of larger solar and electric technologies.
- Dispersed vs Centralized utilization
- Multiple conversions: A recurring principle.

Introduction

Solar energy utilization is a part of the larger field of alternate energy sources. Solar-Electrics is the sub-set of advanced technologies for obtaining electricity solarly. It belongs to both solar and electric science and engineering.

Since our goal in this work is to survey solar-electrics technology in the mid 1970's and to integrate this state-of-the-art in a sensible way, we introduce the subject in this chapter.

We first obtain some perspectives on the broad complex energy situation into which the world and the United States are headed. It is in this larger context and macroscopic background that the solar-electrics field emerges in the mid 1970's. Experience indicates that many currently working in specialty fields of Solar-Electrics are somewhat deficient in their own minds about integrating these larger important issues. So we treat them here first.

Out of our collective insatiable needs for more and more energy to sustain civilization arise important questions: Why is our need for energy increasing? What are the principal forcing functions? Should we continue using up our precious fossil fuel supplies? Is there an energy crisis? Now? Later? When? What are the key facts? Where will the needed future energy come from? Are renewable energy sources capable of meeting the mind-boggling energy demands of the future? If so, what form(s) (heat, electricity, chemical, biological, and others) should our principal renewable energy form take? If a major energy form is to be electricity, shouldn't we examine the fundamental merits of energy in electric form? What are they? Shouldn't we have some perspectives on the present electric energy industry, against which any renewable energy source must compete? Then if one asserts solar-electrics is a viable inexhaustible energy technology, how is the field structured so we get a feel for it? Finally, should this emerging technology--new compared to conventional fossil fuel electric plants--be deployed across the nation on a dispersed or a centralized basis and what are the merits of each? These, of course, suggest many others.

We seek at least partial answers to the above important questions in this Chap. before plunging into the details of solar-electrics in later chapters. As we progress through this first Chap., bear in mind we are compacting enormously complex and important issues. We will generally see here only small but significant parts in our attempt to secure manageable perspectives for issues and industries so large they tax our mental abilities to cope with them.

The reader interested in more details will find them in the references; but be forewarned of the enormous complexity of the lifeblood of our world and nation--energy.

1-1 World Energy Situation

Importance of Population Our search for perspective begins with examining the world's population--its past and probable future. The world's energy situation, as we shall see, is directly influenced by the number of humans existing at any one time on the earth.

Austin and Brewer, in an interesting and significant paper [2], examined the world population problem and its many complex implications. They nicely summarized existing theories of population growth, starting with Malthus in 1798, and appraised the validity of each theory. Their key result is reproduced in Fig. 1-1.

In Fig. 1-1 the world population data points believed to be most accurate are plotted, recognizing the best data on population are from about 1900 forward. Curve-fitting techniques were used. They argue that the exponential law[1] holds only for a small population with a constant fertility rate[2] and that the law must be modified to accurately predict human growth. One modification was advanced by Foerster et. al. [3] in 1958. They proposed the coalition law concept. It is based on the postulate that since we humans possess an effective communication system we are able to form coalitions, as urbanization, industrialization, and others, until all elements are strongly linked. Foerster's reasoning lead to the coalition law plotted in Fig. 1-1. While the mathematics, including least squares curve fitting, need not bother us here, their equation leads to the prediction of a "Doom Time" about 2027 A.D. when the predicted world population would reach infinity. Note that the coalition law nicely fits the actual population up to 1965 when the world's population was about 3.3 billion.

Austin and Brewer say that though the coalition law will probably remain valid until about 1990, the reasoning is flawed, as there are limits to the birth and death rates.[3] Hence the fertility rate cannot grow indefinitely. There is also the obvious problem of environmental resistance entering, for

---

[1] A clear exposition of the mathematics for an exponential function, including relating the doubling time to the yearly growth rate, will be found in Krenz [79, p14-18].

[2] Fertility rate = r = b - d, where b = birth rate and d = death rate.

[3] Interestingly, note that World War II--with all its slaughter--had negligible effect on the world's population as seen in Fig. 1.

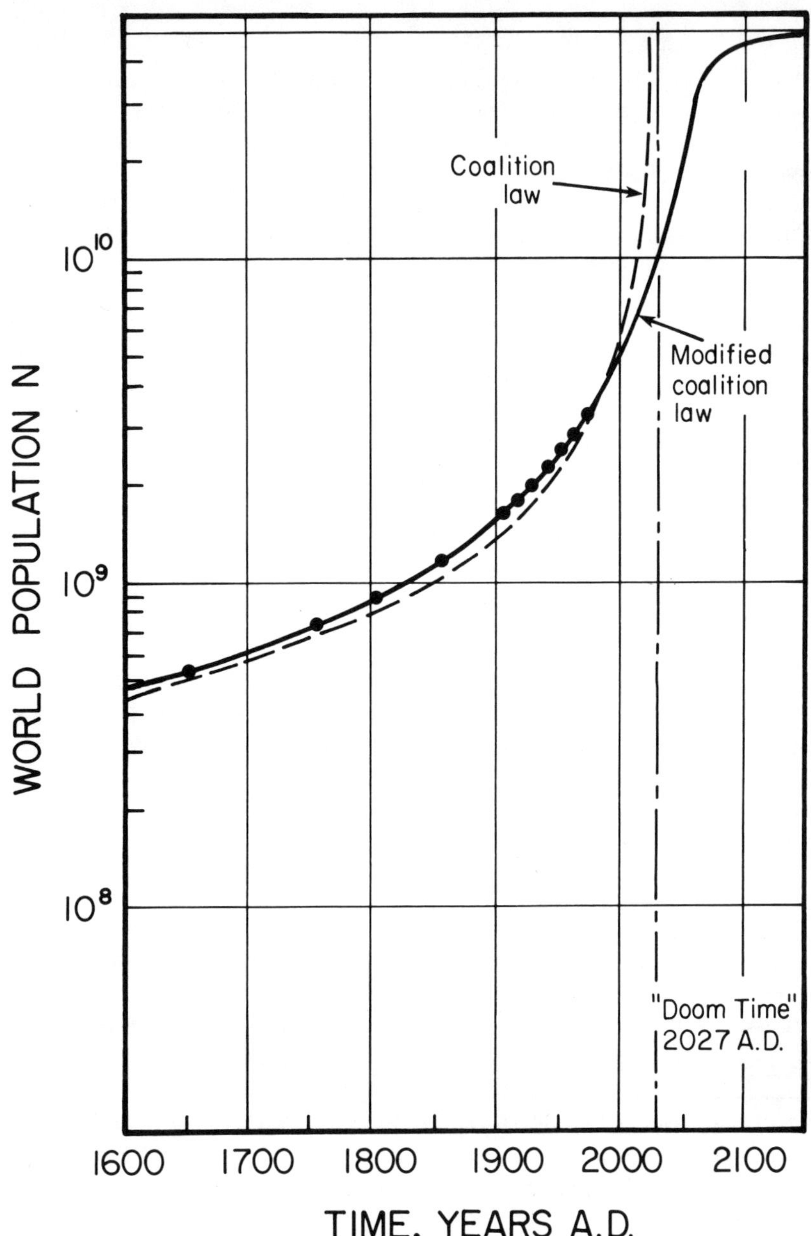

FIG 1-1 World population since 1600 A.D. from Austin and Brewer [2]. Copyright © by the Institute of Electrical and Electronics Engineers, Inc. Reprinted by permission, from IEEE SPECTRUM, Vol. 7, No. 12, December 1970, pp. 43-54.

this finite earth cannot support an infinite population. Finally, they develop a modified coalition law, also plotted in Fig.1-1, that envisions an ultimate earth population of about 50 billion people--about 15 times the present. Their limit is based on the available arable land, temperature, and water limitations. No one, of course, knows this limiting figure precisely, and what it is likely to be is currently being argued.

It is clear from Fig.1-1 that factually we are living in a very dynamic time of the world's history--a time when worldwide catastrophe could lie shortly before us in the areas of population explosion, energy, pollution, other environmental effects, and food supplies unless appropriate worldwide control measures are implemented. There are many other implications--social, political, and technical--all beyond our scope here.[1] The principal implication here is regarding energy.

It is intuitively clear that more people imply higher total energy needs for the world, even if each person continues to use the same energy amount per year as others preceding him. Unfortunately, the world's energy situation is further aggravated by the historical fact that since World War II the total energy used per capita has increased at a linear rate [2 , Fig. 3]. We shall speak more of this in Sec. 1-2.

The world's energy situation might be loosely analogized to the scene of a finite oasis lake with people drinking water therefrom; but the scene is not in steady state. Instead the number of drinkers from the lake is rapidly increasing, while simultaneously the drink each takes each time is linearly increasing. Obviously the lake will run dry eventually. Disaster lies ahead for such a community unless some restraining forces are brought on the scene or a switch made to use their water income[2] instead of their water capital.[3] The decision could well be the most important the community members ever make.

<u>The Fossil Fuel Delta Function</u> A different perspective on the world's energy situation is in Fig. 1-2. Here we imagine summing the entire world's fossil fuels consumed each year for all the world's work. That gives a point on the curve. Repetition at other times traces out the curve up to the present. The curve's shape beyond the present is somewhat conjectural, though many known facts tend to verify that the curve will be something like that shown. The resulting curve has been called the 'fossil fuel delta function'.[4]

---

[1] We repeatedly find throughout the population and energy literature concerns over possible major social upheaval, revolution, economic collapse and depressions, unrest and general instability throughout the world unless the dual problems of population and energy are controlled in rational ways.

[2] The energy analog is 'energy income'. Energy income is energy available on a more or less continuous basis.

[3] The energy analog is 'energy capital'. Energy capital is stored energy, usually accomplished over millenia. All fossil fuels are energy capital. They come from the sun's conversion. Some prefer the term 'energy savings' instead.

[4] Because of its loose similarity to a mathematical delta function which has the properties of occurring at a single time, reaching infinity, having zero width, and an area under the curve of 1.0. The analogy, obviously, is not mathematically rigorous--the situation with all analogies to some degree.

Note the abscissa which is a compressed time scale.

From Fig. 1-2 we learn that humanity's total fossil fuel energy use up to the present era has been (1) gradually increasing, and (2) the energy used per year was orders of magnitude lower than we are now using. Further thinking reveals that the area under the curve up to any time t represents the total fossil energy used up to that time. An extension of this reasoning to the lower right part of the curve would quantify all the world's fossil fuel used, i.e. at that time the world's fossil fuel supply would be completely exhausted. We are presently on the left side of the curve. This implies fossil fuel use will continue to increase until eventually it peaks. At or near the peak men would become aware--many for the first time--of the finiteness of the world's fossil fuel supply. They in wisdom would therefore start to use less in order to make the supply last as long as possible, i.e. to effectively widen the curve.

The curve shown in Fig. 1-2, while admittedly being an oversimplification, has broad philosophic, social, political, technical, and other implications. It also potentially represents a tragedy, for were we to stand at 3000 A.D. and look back, the fossil fuel 'energy pimple', as it has also been called, would have long since vanished. Historians there looking back would see in perspective the transient nature of the fossil fuel era while simultaneously expressing amazement that we of our era burned up the fossil fuel--our

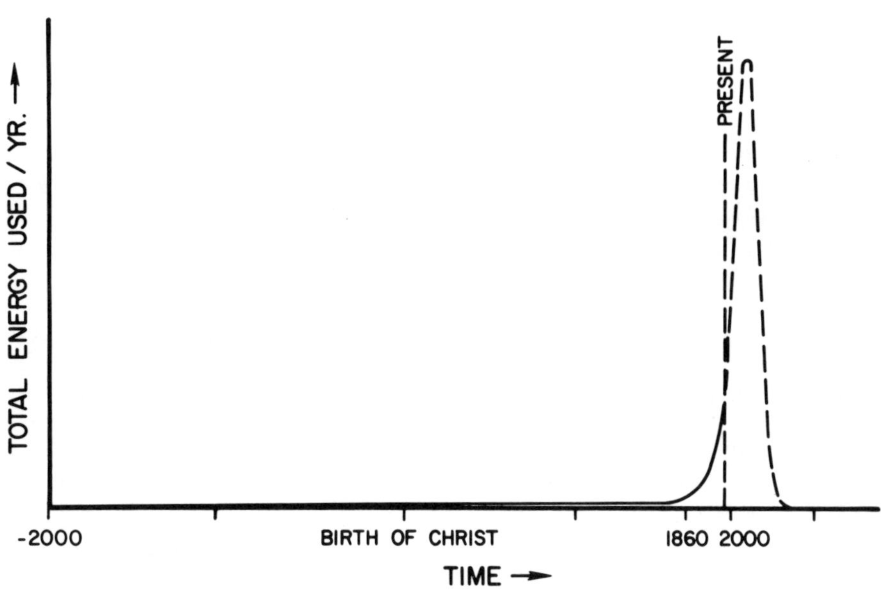

FIG. 1-2 The 'fossil fuel delta function' also known as the 'energy pimple'.

irreplaceable energy capital--instead of using it for more lasting purposes. They may also express amazement at our civilization's failure to live off its energy income instead of its energy capital. Therein lies the tragedy.

Hopefully we collectively shall be wise enough to prevent such a tragic plight from occurring to the world. But to prevent the potential tragedy we must become unhooked from the notion that we have to use fossil fuels alone to supply the world's energy needs. Such a notion comes hard for us, for the present generation has been raised in a time of abundance of cheap fossil fuel energy and a seemingly limitless supply. Fortunately, more and more people and nations are coming to realize the finiteness of the world's fossil fuel energy supplies and of the necessity to turn to other inexhaustible energy supplies--before it is too late.

The tragedy of tragedies would be for humanity to arrive at the right side of the curve with all fossil fuels exhausted and no energy left to construct apparatus for utilizing renewable energy resources. Civilization would perish.

We leave Figure 1-2 by observing that, in spite of our present closeness to the common widespread use of fossil fuels, their use appears relatively short in macroscopic perspective. It is entirely a question of what time increment window Δt, we want to look through. Finally, the perspective cast by such a simple figure causes all thoughtful people of the world to be seriously concerned about the world's energy future--a future that most certainly will touch every person.

An Expanded Look Lest it be thought by some short-range thinkers that Fig. 1-2 is all qualitative philosophic nonsense, we now take an exploded quantitative view of the curve on the left side and around the peak except that now we look only at oil, presently the world's most widely used fossil fuel. We start by examining Fig. 1-3 which plots the world's energy demand in barrels of oil equivalent vs time. Before discussing Fig. 1-3 we first summarize how it was generated.

Early in the 1970's Wilson [4] at MIT "believed that the world may be moving steadily and with little apparent concern toward a new and massive energy crisis." She set up a Workshop on Alternative Energy Strategies composed of 88 energy knowledgeable people selected from 15 industrialized countries around the world that used 80% of the world's energy in 1972. The participants met over a 28 month period between October 1974 and February 1977 and focussed principally on the world's long-range energy issues. In an impressive and significant manner their work explored the world's probable energy future. They examined various sensible energy scenarios and clearly laid out their assumptions[1] for each.

---

[1] The problem of assumptions runs all through the many studies of energy. Unfortunately, not all authors lay out their assumptions in a crisp clear manner. The researcher new to the energy area is forewarned that many assumptions are in fact, difficult or impossible to ascertain from the literature.

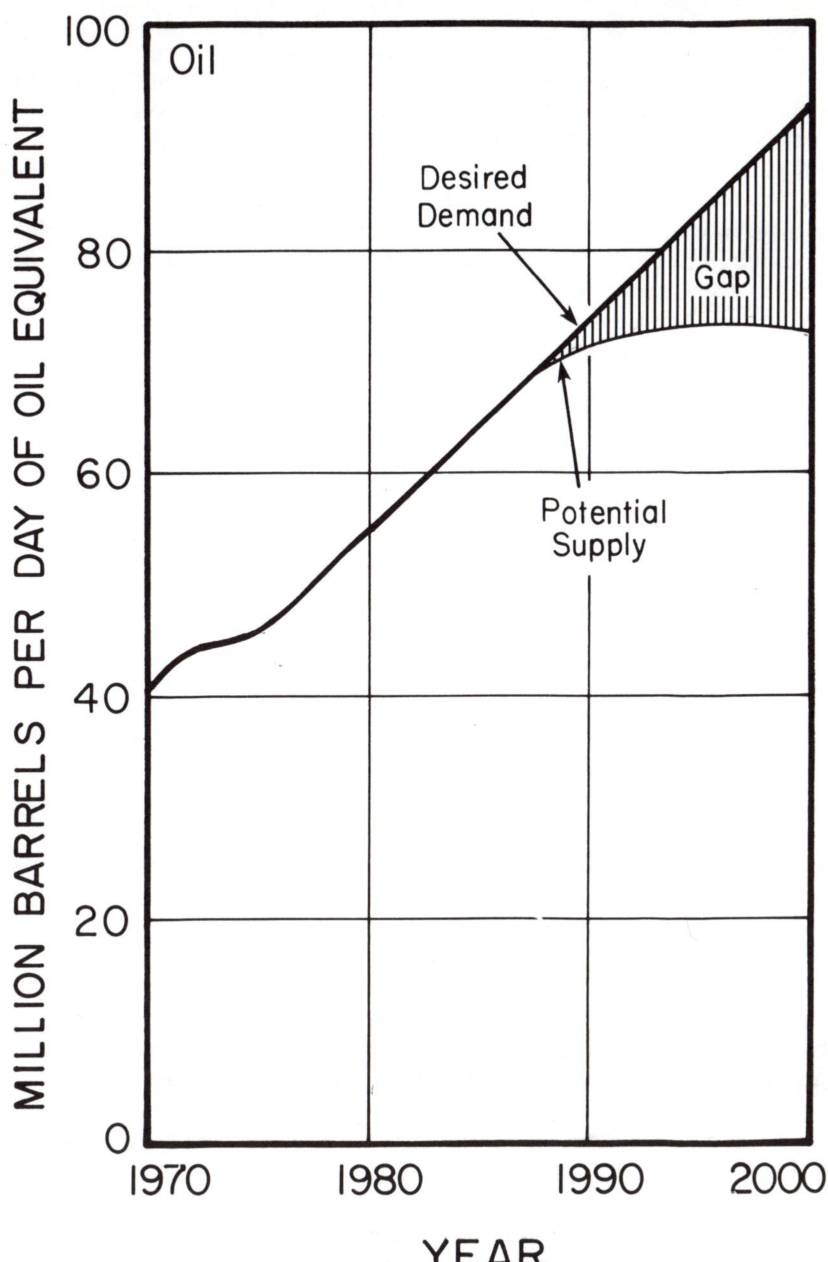

FIG. 1-3 Energy demand and supply in the world outside communist areas. From Energy: Global Prospects 1985-2000 by Carroll L. Wilson [4], Copyright © 1977, McGraw-Hill Book Co., Inc.

Finally after careful study they arrived at many detailed conclusions--too numerous to include here--which the interested reader will find in their report.

Fig. 1-3 is based on their assumptions of a high economic growth rate, a rising energy price, vigorous government responses to the world's energy plight, and coal rather than nuclear power as the replacement fuel. Note that the MIT group indicates the world's energy supply and demand are in approximate balance now and will probably continue so until about 1988, though the year is not precise. After that time "the supply of oil will fail to meet increasing demand before the year 2000, most probably between 1985 and 1995." It is clear to those researchers, who have thoroughly and objectively studied the world's energy problems, that there will be a widening gap between energy demand and supply for the world after about 1985. They conclude that "the change from a world economy dominated by oil must start now" and that "--it requires the will to mobilize finance, labor, research and ingenuity with a common purpose never before attained in time of peace; and it requires it now."

Hayes' book [5] explores the transition to a post petroleum world and reveals helpful perspectives on the world's complex energy plight. Usmani's address [6] exhibits a worldwide viewpoint and quantitatively examines alternative energy sources the world is now considering. There are many other talks, books, reports, and studies--too numerous to mention here--which have been published describing the world's energy situation. If one sums them up, he arrives at the approximate picture being presented here.

We shall say more about the finiteness of the world's oil supply in the latter part of this section.

Energy And GNP An entirely different facet of the world's energy problems can be seen from Fig. 1-4. Here we normalize total energy use to a per capita basis, E_T, and inquire how E_T might be related to productivity per capita as measured by the Gross National Product (GNP) for many of the world's countries. The resulting wide data variability is evident in Fig. 1-4. A rough trend line is indicated.

We see from Fig. 1-4 that the developed countries are high users per capita of energy while the developing nations are low users per capita. The developed countries also make better use of their energy resources by producing more per capita of the world's goods and services.

Curves of this nature are sometimes used as a moral indictment against the United States with about 6 percent of the world's population but using about 35 percent of the world's energy resources. Some authors fail to point out, however, that with that high energy use the United States produces more per capita than any other country in the world, as is evident from Fig. 1-4. Nevertheless there are valid moral points, which are being aired worldwide, about a more equitable distribution of the world's energy resources.

The final conclusion from Fig. 1-4 is that energy and productivity are related, though not linearly. Stated another way, at the personal level the high energy users tend to also be the highest producers. At lower energy use levels small increases in energy use result in marked increases in productivity while at higher energy use levels the productivity increase is less marked. Some saturation or limiting phenomena tend to come into prominence as one moves up the curve.

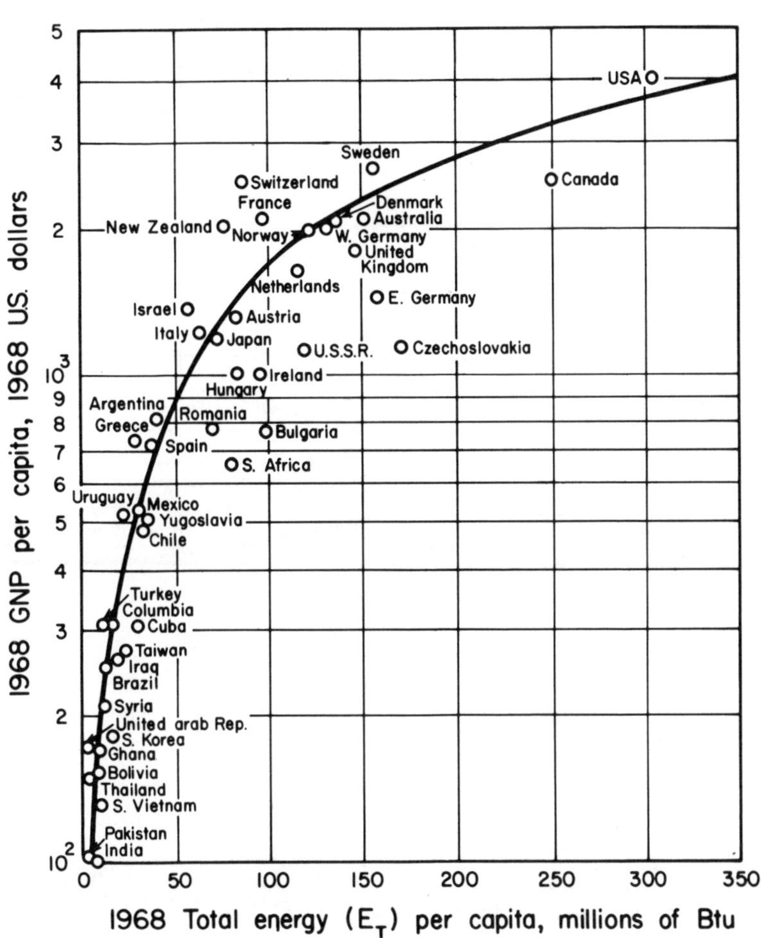

FIG. 1-4 Gross national product per capita vs total energy consumption per capita, 1968. From Hill [7].

The Concept of Limits in the world's per capita total energy use has been explored by Felix [8] up into the 24th century. His result is plotted in Fig. 1-5. Note that the ordinate is a logarithmic scale which has the effect of greatly compressing curves. According to Felix, the exponential energy use curve per capita on which the 'limits to growth' theory rests is in error and has mislead many people to support the zero growth theory. He favors the tapered growth model as being more a realistic future pattern for the world. It avoids the errors of what he calls the 'exponential growth

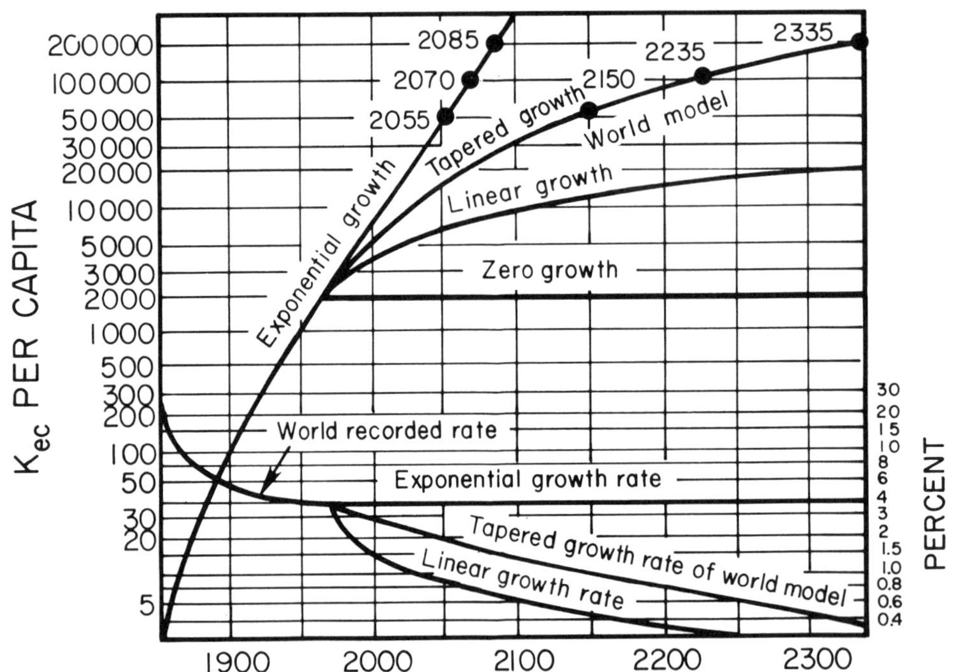

FIG. 1-5 World's personal energy use for several growth models. The ordinate, K_{ec}, is in terms of kilograms per capita energy equivalent of coal. The bottom curves show corresponding growth rates (right scale). Source: Felix [8]. Copyright © 1973 by The Institute of Electrical and Electronics Engineers, Inc. Reprinted by permission, from IEEE SPECTRUM, Vol. 10, No. 9, September 1973, pp 63-68.

syndrome' and "provides for extra decades--and even centuries-- of technological advances." But Felix's implied assumption that there will be an adequate supply of fossil fuel energy until the 24th century and plenty of time to develop energy alternatives must be weighted against the bleaker estimated availability of world oil curves which we now present.

Future World Oil Availability We have already indicated that the world's total use of energy is growing at alarming rates--so high, in fact that the term "Energy Crisis" permeates the energy literature. Central in this concern shared by the advanced thinkers of virtually all nations is the availability of oil for, at the present anyway, oil is the major fossil fuel of the world.

Of the increasing difficulty of finding and successfully extracting oil over the last few decades there can be little serious doubt. Within the past decade or so it has become increasingly necessary to move off land and extract oil from the more hostile places on the earth's surface. Technology has been created for the difficult feat of offshore drilling. Now offshore oil well rigs dot the seas of the world--some in very hostile environments such as Europe's North Sea. Offshore oil wells are rapidly being deployed around the world.

Then man has moved into the hostile Arctic and developed the field on Alaska's North slope;[1] subsequently the Alaskan pipeline--man's largest engineering job--was created in the mid 1970's against great difficulties to transport the oil. Finally super tankers--with all their environmental problems of oil spills--have been created to bring the world the oil it demands when it wants it. All this technology was unknown a generation ago. It has been created relatively recently--all in the name of energy.

We have earlier argued that the world's use of fossil fuels in aggregate will eventually be exhausted for all practical purposes.[2] The supplies are finite and cannot last forever at the fantastic rates we are currently using up this precious 'energy capital'. Hot debates have occurred over when exhaustion will happen,[3] as that determines how much time we have left to develop alternate energy sources.

Some feel for the estimated production of oil may be had by ferreting out the predicted curves put forth by some of the world's knowledgeable people in the oil area--people who have perspective and can see decades and centuries instead of only into next year. One such man is M. King Hubbert, a research geophysicist with the U. S. Geological survey. Hubbert's forecast for the world's oil production is shown in Fig. 1-6. Note that the general curve shape is similar to our earlier Fig. 1-2 for all fossil fuels.

According to Hubbert, the majority of the world's oil supply will be gone by about 2025 or 2050--only about one generation away--and "from the standpoint of human history the epoch of the fossil fuels will be quite brief." He also forecasts a similar curve for coal which would not exhuast until about 2400-2500 A.D. because of its plentifulness.

We earlier introduced Wilson's MIT workshop group formed of international energy experts. On a smaller time scale than Hubbert, their forecast of oil production outside of Communist areas is shown in Fig. 1-7. Note the

---

[1] One recent result of the Prudhoe Bay, Alaska development is the problem of how to get the natural gas from there to where it is needed in the United States. A 4800 mile Alcan Pipeline is to be built through Canada. It will be the largest privately-financed project ever undertaken that will cost more than the Apollo II moon walk, the Golden Gate Bridge, the Alaskan pipeline and the movie, Star Wars, combined [9].

[2] Actually the likelihood of complete fossil fuel exhaustion of the world is a fiction. As we go deeper and deeper in the earth's surface to extract energy there will eventually occur some depth at which the energy recoverable from the fuel exactly equals the energy required to extract it. To go lower would use more energy than would be extracted, i.e. it becomes an energy losing proposition. Thus there will always be some fossil fuels left.

[3] One can find in the energy literature of the pre-World War II era predictions that the oil supply will be exhausted in a few decades. So far it hasn't happened, though Fig. 1-6 indicates it is imminent.

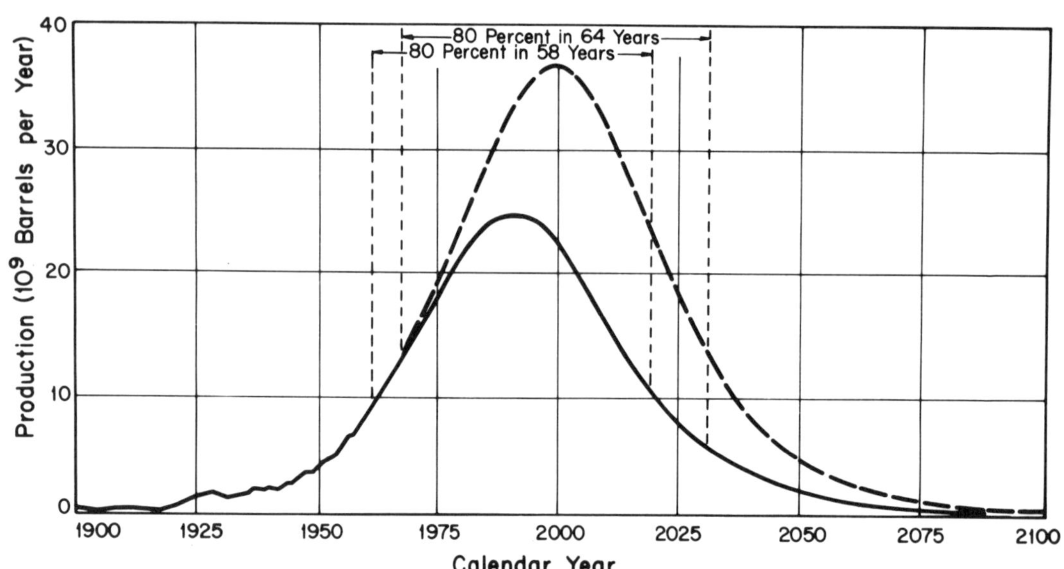

FIG. 1-6 World oil production. The dashed curve is Ryman's estimate of 2,100 x 10^9 barrels. The solid curve is Hubbert's estimate of 1,350 x 10^9 barrels. From "The Energy Resources of the Earth", by M. King Hubbert [10]. Copyright © 1971 by Scientific American, Inc. All rights reserved.

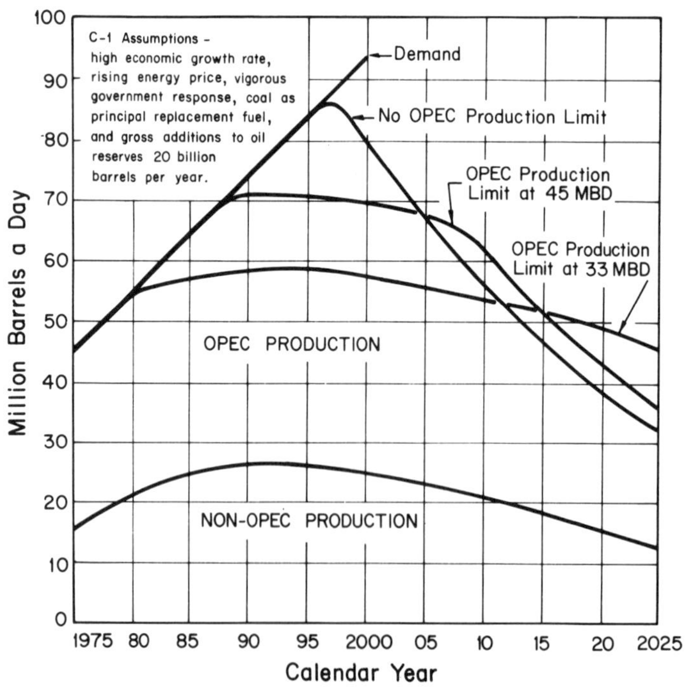

FIG. 1-7 Oil production for the world outside communist areas according to Wilson et. al. [4, p 20]. Note the MIT group's clear statement of their assumptions. By permission of McGraw-Hill Book Co.

general agreement of the curve shape with Hubbert's.

The above curves all underscore the importance of the world's energy problems. They also indicate that fortunately we have some small time at our disposal for coping with the ominous problems of developing **alternative** energy sources; but time is short. We earlier indicated how tragic it would be for civilization to leave our grandchildren a world exhausted of precious fossil fuels.

We close this section by challenging the reader to acquire some personal feel for the enormity of 50 million barrels of oil a day; then consider what the quality of life on our world would be like without it.

Finally we've not explored here where the world's oil comes from, where it will likely come from in the future, environmental aspects, the morality of irreversibly scarring the earth's face, the politics of energy[1] and many, many other complex facets of the world's energy problems. Remember, we must travel sans details to see some of the larger perspectives before arriving at our principal concern, Solar-Electrics.

We shall obtain other perspectives on the larger world energy problem by reviewing the U. S. situation in the next section.

1-2 United States Energy Situation

Our Total Energy Needs The 'Energy Crisis' in the U. S. has been so well publicized as to require only a brief recounting here of a few salient facts for us to achieve our desired perspectives. As previously with the world, we start by examining our total energy needs from all sources.

Gaucher, a former Scientific Planner with Texaco, in a significant integrative paper [11] has clearly gotten at some of the key facts we need. One of his graphs is reproduced in Fig. 1-8. Here energy in units of 10^{15} BTU's[2] is plotted vs time (the curve labeled Total Energy) over an

---

[1] Mangone [12] examines the energy policies of various countries in an overview fashion. He concludes that: "Few issues will be so important to world politics in the future as the national use of energy resources --."

[2] Convenient energy units for the U. S. total are in terms of "Quads". A quad is a quadrillion BTU = 10^{15} BTU. Thring [13] has pointed out that there are different meanings for the energy units in Europe and America:

	Europe	America
Billion	10^{12}	10^{9}
Trillion	10^{18}	10^{12}
Quadrillion	10^{24}	10^{15}
n tillion	10^{6n}	10^{3n+3}

A Quad staggers our imagination, and we have a difficult time acquiring a feel for such enormous energy units. Bueche [14] helps us out. He says, "A Quad is about 50 million tons of coal. It would be a string of railroad cars stretching from New York to Alaska. Alternatively it is about 7 1/2 billion gallons of gasoline--enough to run 10 million automobiles for a year. Finally a Quad is about the total energy used to run a city of 1 million people for 3 years."

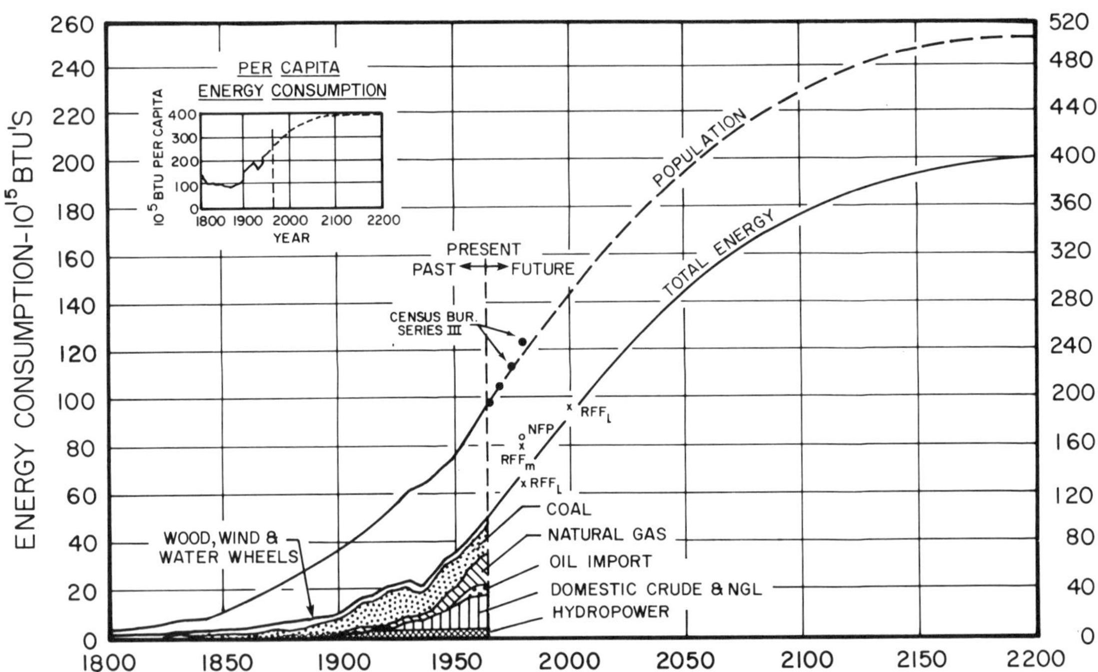

FIG. 1-8 Energy consumption in the United States, past, present and future. Reprinted with permission from Solar Energy, Vol. 9, No. 3, L. P. Gaucher, "Energy Sources of the Future for the United States, Copyright © 1965, Pergamon Press, Ltd. [11].

appreciable time period--400 years. Note he is forecasting a gradually increasing total energy need for the country with a rate of increase that slowly diminishes. Note also that from the time of his survey--about 1965--we were using about 50 Quads of energy/yr. His forecast indicates the total need may go to about 200 Quads, a 4 fold increase over roughly the next 200 years. Some later forecasters think the 200 Quad level will be reached by the year 2000 and continue increasing. As expected, there is wide variability on such predictions, depending on the assumptions one makes.

The effects of the depression and the acceleration caused by World War II are evident from the notch in the total energy curve.

We also observe from the total energy curve that coal and petroleum products--all fossil fuels--have historically been our chief energy sources. We further observe from Fig. 1-8 that the population and total energy curves track, as expected. Naturally, if the population were in some socially acceptable way contrained, then the future U. S. energy burden could be correspondingly eased. The impact of the population on the energy needs of any given country is a recurring theme throughout energy literature.

Gaucher sketches for us in the upper left-hand corner of Fig. 1-8 the energy per capita/year. Note that this has been increasing since about 1900. It is thus apparent that in the United States our per capita energy use has been on the rise for a long time, and it was not suddenly precipitated by the Arab Oil embargo of late 1973.[1]

It is interesting to contrast Gaucher's U. S. data with Austin and Brewer's [2] data for the world where they showed the energy used per capita has been linearly increasing only since World War II.

In summary, it is clear that the United States' energy situation is aggravated by both a growing population and a rising energy use per capita. Here again the situation is not unlike the oasis lake analogy we used for the world in Sec. 1-1. The dual needs for population growth restraints and personal energy conservation are apparent from these facts. Gaucher concludes with, "What we think of as big business today--the 400 million ton per year coal business, the million barrel per day oil business, the 14 to 15 trillion cubic feet per year natural gas business and the 900 billion KWH electric power business--will be dwarfed by the corresponding businesses of the future."

Gaucher went on to forecast the probable composition of energy supplies for the future, factoring in all developments--political, social, economic, and technological--which probably will impact the mix of energy sources. His result is shown in Fig. 1-9. From it we can see that:

- Fossil fuels will continue to exist and be used, though--in aggregation--at lower fractional parts of the total.
- The percent of energy supplied by any one fuel type changes with time--some decrease while others increase.
- The absolute amount, in Quads, of any one fuel may markedly increase; yet its fractional part of the total needed may decrease because the total is growing at such a high rate.
- Hydropower has historically played a rather small role in the total energy picture and will probably continue so, according to Gaucher.
- He foresees the absolute necessity of solar energy or some other yet undiscovered energy source as being essential in the long-range future if we expect to meet the rising bulk energy needs of the country.

Energy And GNP for the United States have been plotted by Sailor [15] as shown in Fig. 1-10. Note that his energy used curve is similar to Gauchers. We also note that the GNP curve tends to track the total energy curve, a not unexpected result in view of relationships between energy and GNP of Fig. 1-4

---

[1] The reader will find it interesting to speculate on why the U. S. per capita energy use curve has risen and why the forecast is to continue rising.

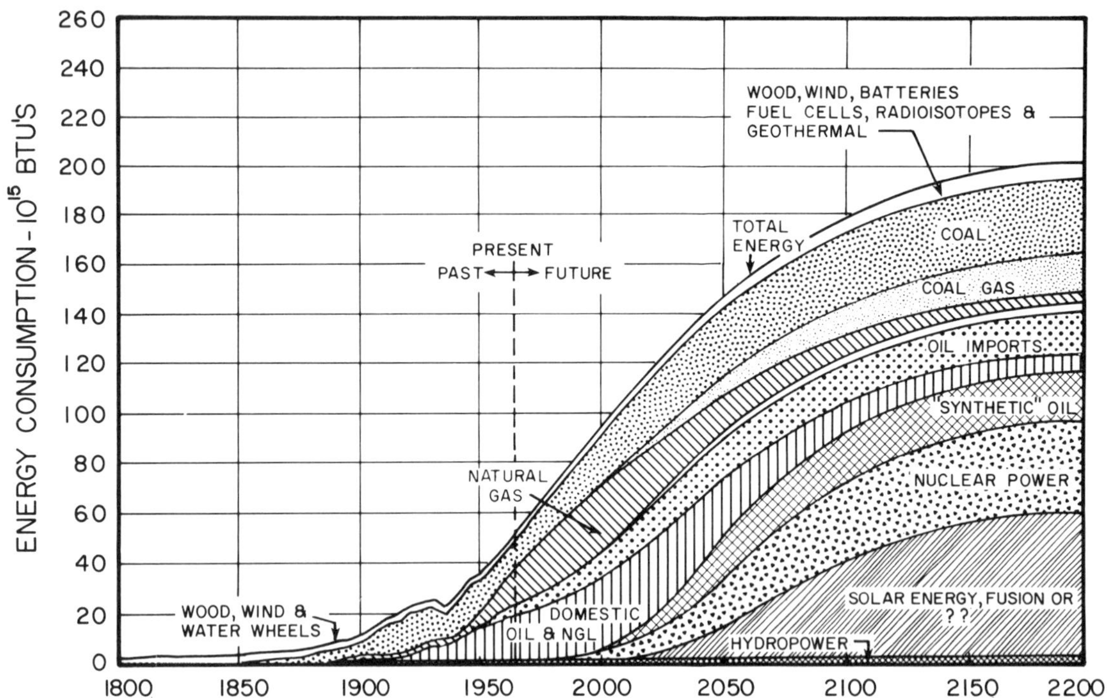

FIG. 1-9 Energy consumption in the United States with the future given in detail. Reprinted with permission from Solar Energy, Vol. 9, No. 3, L. P. Gaucher, "Energy Sources of the Future for the United States, Copyright © 1965, Pergamon Press, Ltd. [11].

for the world. Again we see that high energy use in the United States tends to result in high production of useful goods and services as measured by the GNP. Figure 1-4 established that the United States makes better use of its energy resources--in terms of GNP/capita--than any other country in the world--all in spite of the recognized energy wastes permeating the production-distribution-use system. A curve similar to Fig. 1-10 is in reference [16 p 9-12] along with other data relating energy and GNP.

Another way of looking at the relationship between productivity and energy use has been advanced by Felix [17] whose plot is shown in Fig. 1-11.

He plotted the energy used per dollar of GNP (we'd want it low) versus the GNP per capita (we'd want this one high) for various countries based on Ramsdell's data [18] from the United Nations. His sketched-in curve in Fig. 1-11 shows the countries with the lowest energy use in relation to GNP. By Felix's criterion the U.S. ranks the most favorable in having the least energy used per dollar of GNP for the highest GNP per capita, again a result we saw in connection with Fig. 1-4.

How America uses its energy is important to know in our search for energy perspectives. Unfortunately there are many newcomers to the energy scene who don't have clear perspectives and fall into the error of extolling some

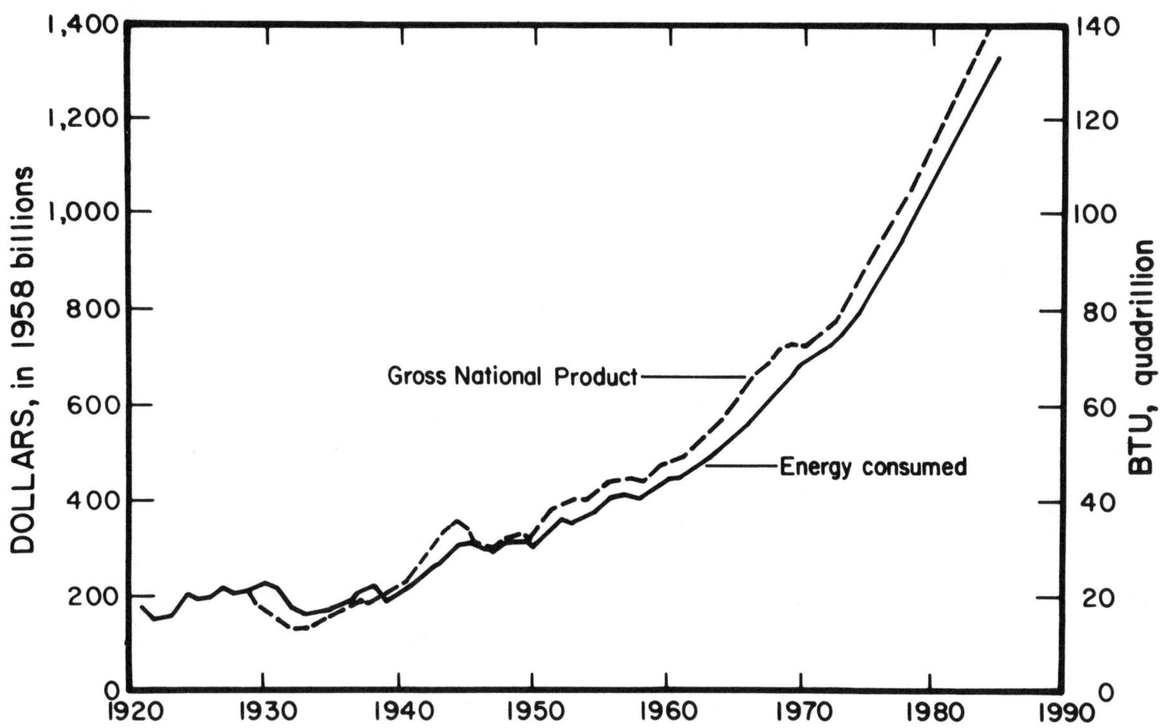

FIG. 1-10 Energy Consumption and gross national product, 1920-1970 with projections to 1990.
Source: Sailor [15].

small energy-using area[1] as the area which will solve all the nation's energy problems--if we'd just work on it. The picture is much more complex, but here we'll try to cut through the mazes of energy data and facts to see a few major energy use points.

We first inquire "What major sectors are the large energy users in America? How much energy is used now for each? What do the foreseeable future needs look like for each sector?"

The answers are in Fig. 1-12 which nicely summarizes this whole complex energy use area by major sectors.

---

[1] "Unplug all your electric toothbrushes!" is a ridiculous overworked example.

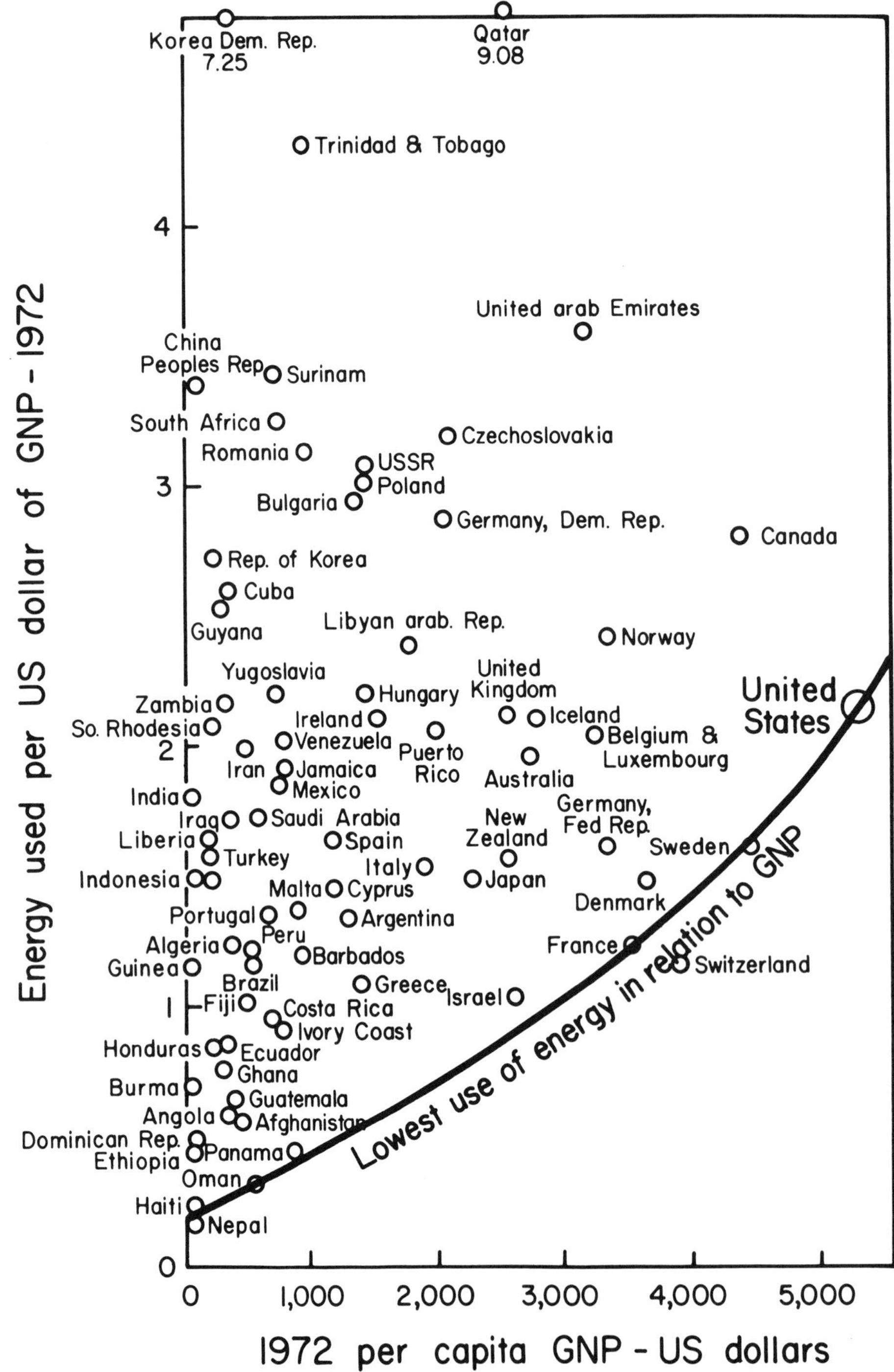

FIG. 1-11 Relationship between energy used and GNP on a per capita normalized basis for various countries. From Felix [17]. Original data from Ramsdell [18]. Used by permission of *Electrical World*.

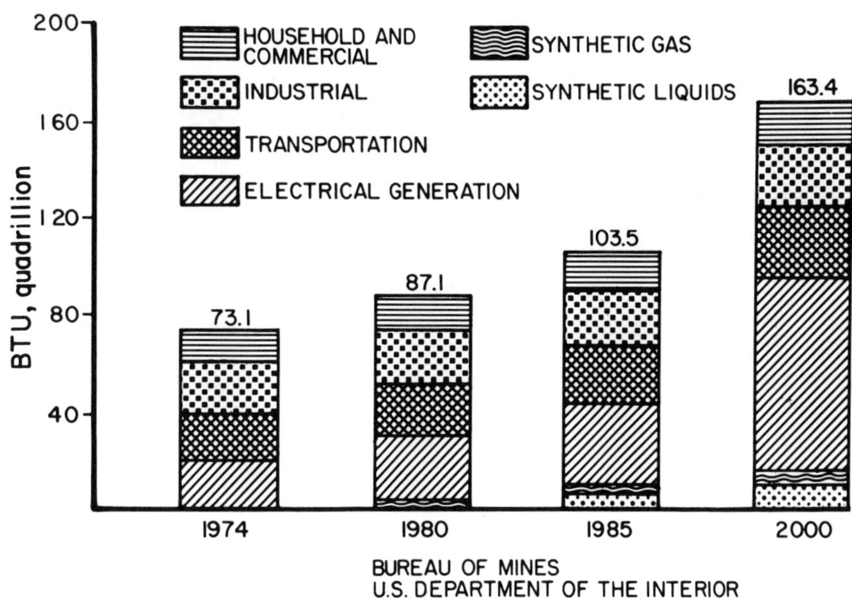

FIG. 1-12 United States energy consumption by sector, 1974-2000. Source: Dupree [19].

It is immediately evident that there is no single dominating sector which if eliminated or worked on so as to significantly reduce its energy use would essentially solve all the nation's energy problems;[1] rather the situation is more like a smearing out of our energy needs into several major sectors. It is clear for 1974, however, that energy used in the industrial sector and energy used to run the entire electrical industry are about equal and are both large energy users--over 20 Quads each. The transportation sector follows. The household and commercial sector is the smallest user of Quads.

Fig. 1-12 while also showing the growing total energy needed to run the country, additionally forecasts enlargement of the electric sector until it reaches a dominant position by 2000. According to Dupree's data, the energy resources needed to supply the electrical sector will approximately quadruple over the 1974 energy used by 2000. Whether this change actually occurs depends on the assumptions and decisions made in the immediate future by the nation's energy planners.

---

[1] There repeatedly arise those on the contemporary scene, however who erroneously hold out such false hopes. Many nuclear power advocates border on this error. Electrical engineers sometimes also fall into this error by implying most of the nation's energy needs are for electrical energy. We shouldn't lose sight of the fact that, as important as electrical energy is, the United States survived a long time without electrical energy in bulk!

Finally note Fig. 1-12 envisions the arrival of synthetic fuels on the national scene the latter part of the century.

We should mention that forecasts as Fig. 1-12, including the total numbers and the sector distributions, are subjects of continuous lively top level discussions by various U. S. governmental agencies and energy planners. Such curves have a way of being rapidly outdated as new assumptions and data arrive on the energy scene.

Another way of finding where America uses its energy is shown in Fig. 1-13. Immediately we note the four major sectors of the previous figure. We also note the complex flow arrangements from energy resource to ultimate users--and remember this is a simplification. Actual energy flows for the nation are much more complex. A not so subtle, but to some disturbing point, is that in 1970 of all the energy resources fed into the nation only about half of it exits as useful energy. The other half is rejected energy.[1] Hill's forecast for 1985 is for this situation to deteriorate rather than improve.

We thus find a major part of the energy used in America is rejected, accomplishing no useful purpose. In many cases it is in fact environmentally harmful. Clearly a significant and obvious part of the United States' energy problem centers around minimizing wasted energy and thereby saving energy resources needed for the front end of the nation's energy system. Major national conservation of energy efforts have been made because of the above. We shall likely see a great deal more such activity which, in toto, has the potential for short-range energy savings to the nation. More fuel-efficient autos, better house insulation, energy efficient home thermostat settings, and creation among people of a conserving attitude are some of the present beneficial efforts in this area. Many believe that conservation offers the nation the greatest savings per dollar invested.

Coupled to conservation is a growing awareness among the people of the need to start developing and using unconventional alternate energy sources, as solar energy. It is clear, though, that conservation, to have any national impact, must be a massive effort undertaken by millions. The conservation of a few individuals--no matter how dedicated--will have negligible effect.

A somewhat more personal view of the nation's use of energy may be seen in Table 1-1. We immediately see that the greatest users of bulk energy are industry, transportation, and homes and offices in descending order. Conversely, according to the perspective of Table 1-1, if everyone in America shut off all lights in homes and offices, provided nothing else is done simultaneously, it would have an insignificant effect on the total energy required to run the country, probably to the surprise of many. Table 1-1 further suggests the diffuseness of our energy uses, i.e. as a nation

---

[1]A conventional electric power plant is a classical example. There about 2 out of every 3 energy units put into it is rejected as heat. Only about 1 energy unit is useful in electric form.

U.S. ENERGY FLOW
1970

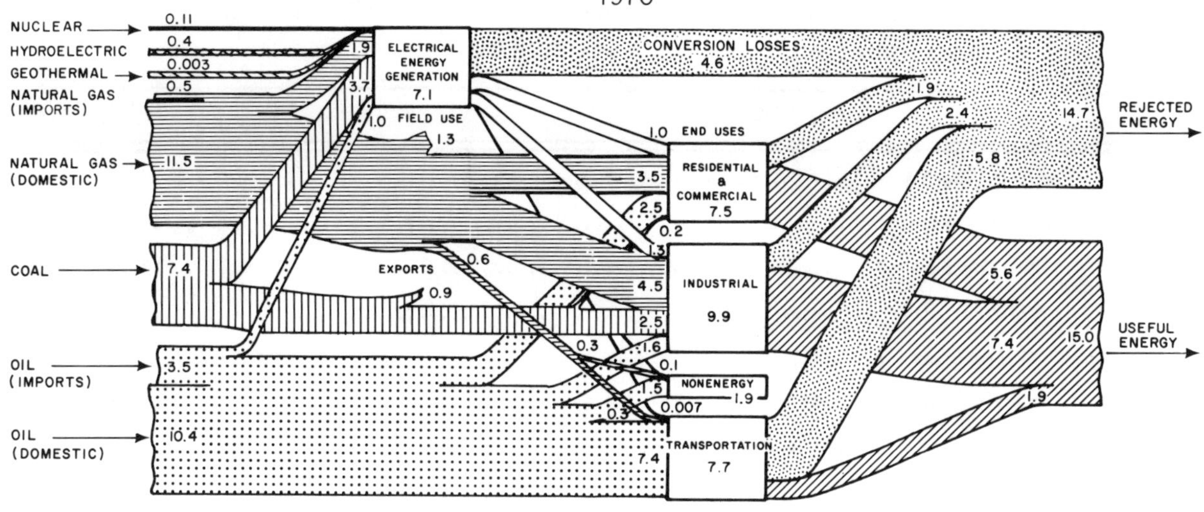

(UNITS MILLION BBLS/DAY OIL EQUIVALENT)

U. S. ENERGY FLOW
1985

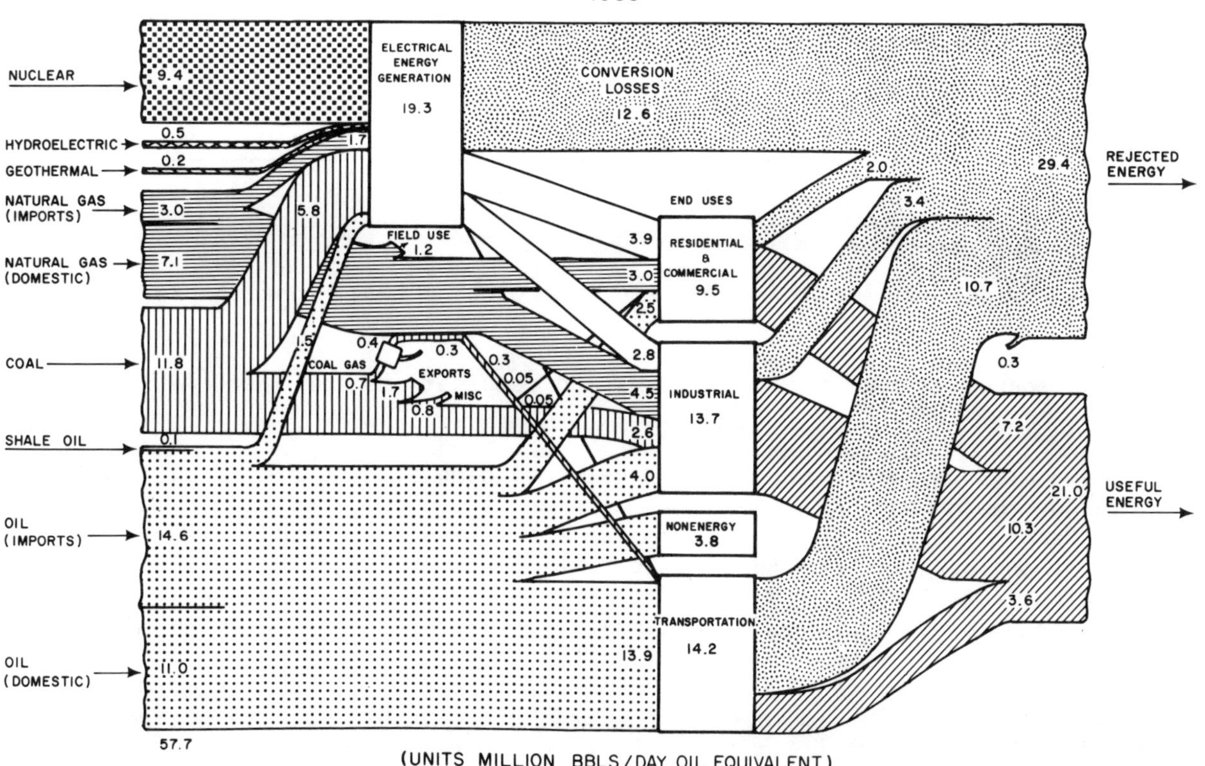

FIG. 1-13 U. S. Energy flow for 1970 and 1985. From Hill [20]. (Source uncertain Believed originally from a government report.)

TABLE 1-1 Where America's Energy Goes

	Percentage of All Energy Used in U. S.
Running Industry	37.3%
Powering Transportation	24.8%
Driving cars	13.2%
Driving trucks and buses	5.5%
Flying planes	3.2%
Driving farm and other off-road vehicles	1.2%
Fueling ships and boats	1.0%
Fueling trains	0.7%
Heating Homes and Offices	17.9%
Providing Raw Materials for Chemicals, Plastics	5.5%
Heating Water for Homes and Offices	4.0%
Air-Conditioning Homes and Offices	2.5%
Refrigerating Food	2.2%
Lighting Homes and Offices	1.5%
Cooking Food	1.3%
Other Uses	3.0%

Source: Office of Science and Technology; Chase Manhattan Bank

we use energy in many different forms and combinations for a multiplicity of tasks.

We close by observing that extensive ongoing studies are being made by both government and private sectors of America's energy use patterns. It is a facet of the larger energy problem not likely to vanish. Note also that if one persists through the plethora of energy literature he probably can find facts indicating his desired energy use perspective. An example might be a large manufacturer contemplating entering the solar home heating market. He might want to know if all the nation's homes could be solarly heated what percentage of the nation's total energy use could be eliminated. Table 1-1 would tell him in this case, and he'd have his desired perspective; but in the more general sense he'd have to enter the complex area of energy flows as cited in government literature. Reference [21] introduces the subject and contains a wealth of detailed useful information on energy uses and flows in the United States.

Sources of America's Energy We naturally inquire as to what the principal facts are for what energy sources the nation has actually used in the past by way of seeking our future energy perspectives. Historically we have used several different kinds of energy sources. There are well documented detailed histories surrounding each type of energy resource. We have neither the space or time here for such. Instead what we need here is an overview. We find it in Fig. 1-14. Gaucher has given us some perspectives on the various principal energy sources used in the United States over a wide time span. He plots total energy used, say, for wood, wind, and waterwheels combined as a function of time. Repeating the procedure for various other energy sources gives us the total figure and some sense of perspective on the nation's current energy sources.

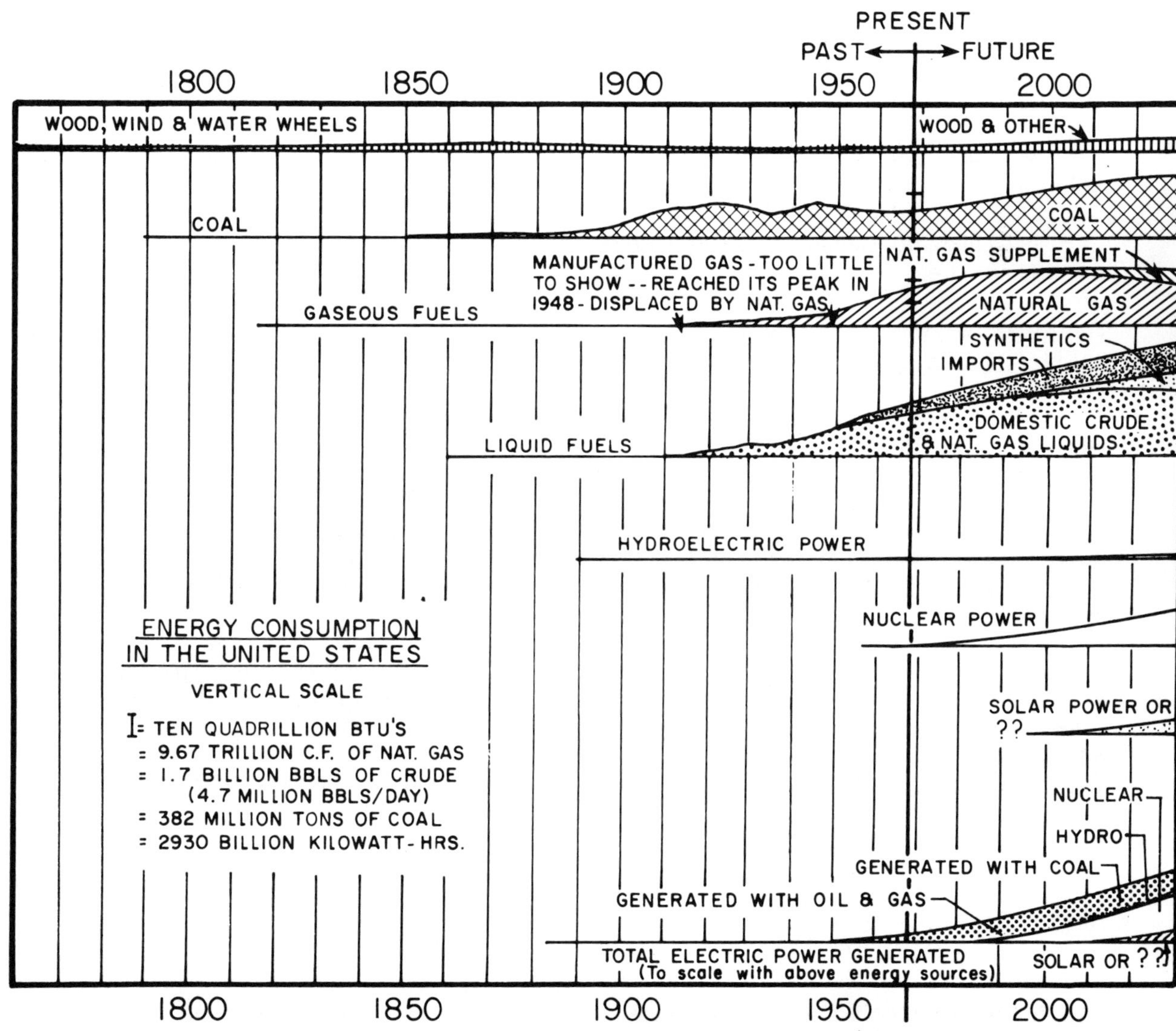

FIG. 1-14 Chronological development of the nation's energy sources. From Gaucher [22].

First we note that wood. wind and waterwheels have been used since the nation was formed. Coal got off to increasing use in the 1870's as steam power became important. Gas came next, but it wasn't until 1930 when it started being piped long distances that it came into prominence. Liquid fuel use climbed rapidly with Kettering's invention of the auto self-starter in 1919. Hydroelectric power went commercial about 1890. Then over half a century passed until nuclear power arrived. Thus, in toto, historically a new energy source has appeared on the American scene every 30 years or so.

We should also observe, by way of perspective, that several of the currently popular alternate energy sources, e.g. wood, wind, waterwheels, and hydropower have in fact been used for a long time and are not "new", as many newcomers to the advanced energy area initially think.

Gaucher indicates that if history is any guide we should expect a new energy source--solar power or some other undefined useful energy source--to appear on the national scene around 2000. Note also that whatever the new energy source of the future may be, the use of fossil fuels--in various mixes--will probably continue into the foreseeable future.[1]

A different way of showing the rise and fall of the nation's principal energy sources and their various mixes on a percentage basis of the total used is shown in Fig. 1-15. Note that "it has taken some sixty years from the point at which a transition to a new energy resource was first discernible until that resource, in turn, reached its peak use and began to decline relative to other sources [23]." In light of the historical evidence of Figs. 1-14 and 15 we might well ask "Can the nation afford another 60 year wait for the next fuel type to appear?" The question is nonrhetorical and all the more urgent in light of our treatment of the larger world energy situation in Sec. 1-1. The increasing number of electric energy blackouts and brownouts are giving us a feel for living in a society where the power demand occasionally exceeds the supply and circuit breakers are opened to relieve the strain. The 'energy crisis' poses many major problems for America's utility industry [81].

We earlier stressed the importance of oil as currently the major energy source for the United States. What does the supply picture for oil[2] look like for the nation? Fig. 1-16 answers. Here we see the total number of barrels of oil per day (in millions) needed to run the country plotted vs time. The graph shows both domestically produced oil and imported oil, the two chief sources.

---

[1] But let us not forget Hubbert's curve of Fig. 1-6 which showed most of the oil gone by about 2025!

[2] An impressive compilation of facts on oil and other energy alternatives is summarized in reference [24] for the reader caring to go deeper into the energy alternatives for the nation. Also Weaver's article [25] gives helpful perspectives on the nation's energy alternatives.

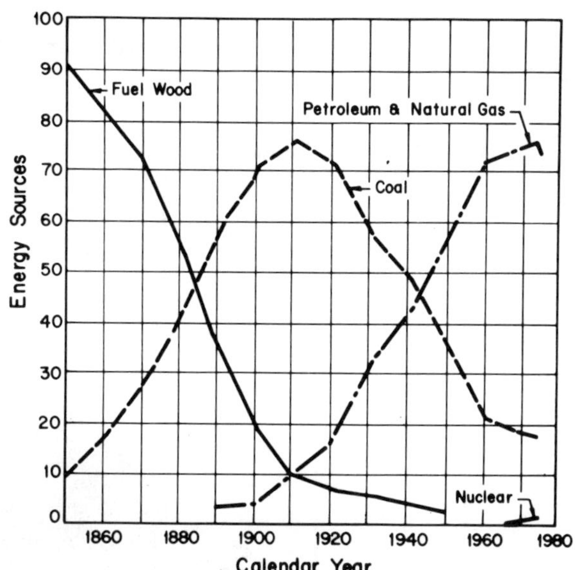

FIG. 1-15 U. S. energy sources and their patterns.
Source: [23].

Since total demand obviously exceeds domestic supply, the U. S. turned to importing oil. In the early years shown the availability of inexpensive imported oil served as a disincentive to domestic production. The combination of declining domestic production and rising demand led to an increasing dependence on imported oil. This situation continued until the Arab oil embargo[1] of late 1973-74 which had the singular effect of clearly demonstrating the vulnerability of the nation to supply disruptions[2] and oil price increases.[3] The oil imports have also caused an increasing outflow of U. S. dollars to oil supplying countries, principally in the Middle East. In 1976, for example, the U. S. spent about $27 billion on oil imports alone. Thus sale of the single commodity, oil--because of the enormity of it--we find seriously upsetting

---

[1] An interesting summary of this situation is in [80, p81].

[2] It is interesting to observe we learned the same lesson in World War II when foreign oil supplies from many sources were severly disrupted or cut off. We also learned similar lessons about the nation's rubber supply. Do we <u>really</u> learn from history?

[3] Pre 1973 oil cost $2-3/barrel. Following the embargo, imported oil rocketed to the $11-$13 per- barrel range.

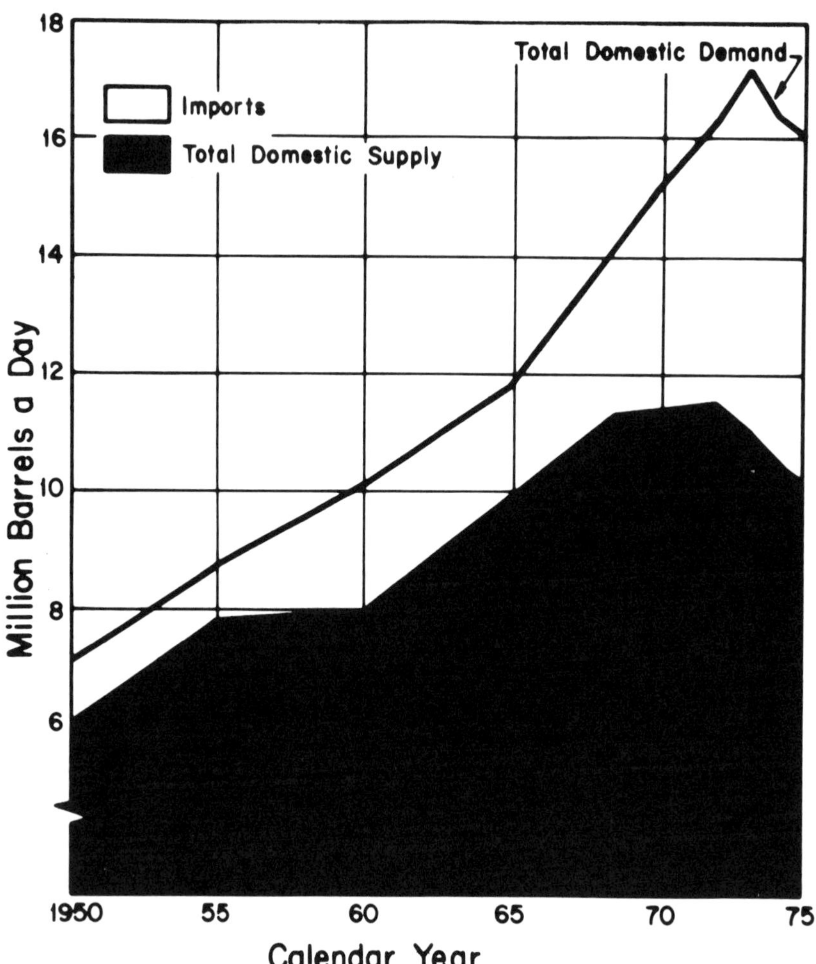

FIG. 1-16 United States domestic and imported oil. Source: [26].

the U. S. trade balance, a situation with many important social and political overtones beyond our scope here.

Sources [25] and [26] have many other facts on oil and other energy resources and energy alternatives from which the busy reader may get a more detailed view than we can include here. A by-product of source [26] adapted for regional energy planning is in [27].

<u>Where Will The Nation's Future Energy Come From?</u> Another perspective on United States' use of fossil and other fuels is in Fig. 1-17. Here we see plotted the years remaining (from about 1975) for various energy sources.

Observe that of the principal fossil fuel alternatives open to the nation, we've chosen to use up the ones in least supply, petroleum and natural gas. The towering energy source of coal has been largely ignored until in recent years. Now we are embarked on a great national effort to, once again (see Fig. 1-15), utilize our coal energy resources. But pursuing this option is fraught with many difficulties: air pollution, rebuilding the nation's railroads, strip mining and reclamation, black lung disease, adequate compensation

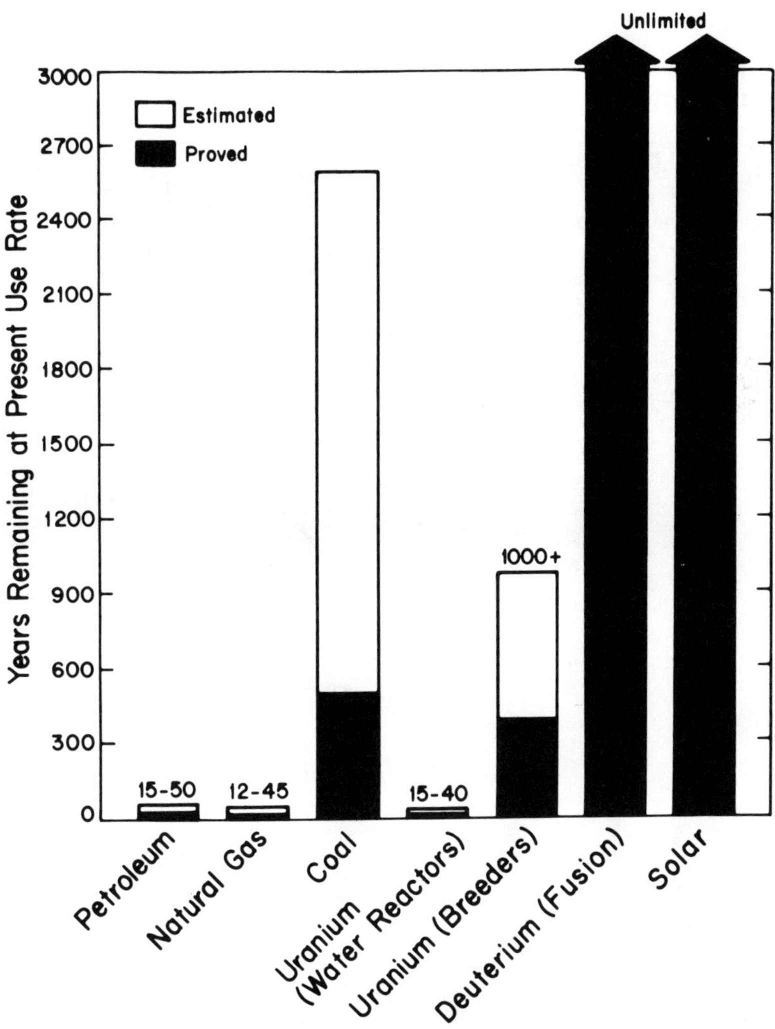

FIG. 1-17 United States energy resources.
Source: Hill [20].

for coal miners, paralyzing strikes, and many other issues somewhat characteristic of the far reaching social and political impacts implicit in any energy option.[1]

Fig. 1-17 also impresses us with the unlimited number of years available via the energy sources for fusion and solar power. We shall say more about solar power in Sec. 1-3. Observe in passing that, amazingly, we as a nation have largely ignored these potentialities in favor of exploiting and exhausting the shorter-range energy finite supply sources. Recalling our perspective for the world from Fig. 1-2, one can only hope 500 years from now when historians look back on us they shall exhibit grace for us in our largely ignoring the obvious.

We should not leave Fig. 1-17 without saying in the interest of objectivity that the numbers for the estimates vary--some widely--throughout the energy literature, depending on the assumptions again. One can also have lively discussions about what is meant in Fig. 1-17 by "proved", particularly for fusion which is well-known to not yet be a reality for any appreciable time even in the laboratory.

In addressing the serious problem of sources of energy for the future, particularly with regard to energy imports, ERDA's 1975 proposed national energy plan [23] contains a set of six scenarios identified in Fig. 1-18. Note that here net imports are plotted in quads, since natural gas is not normally

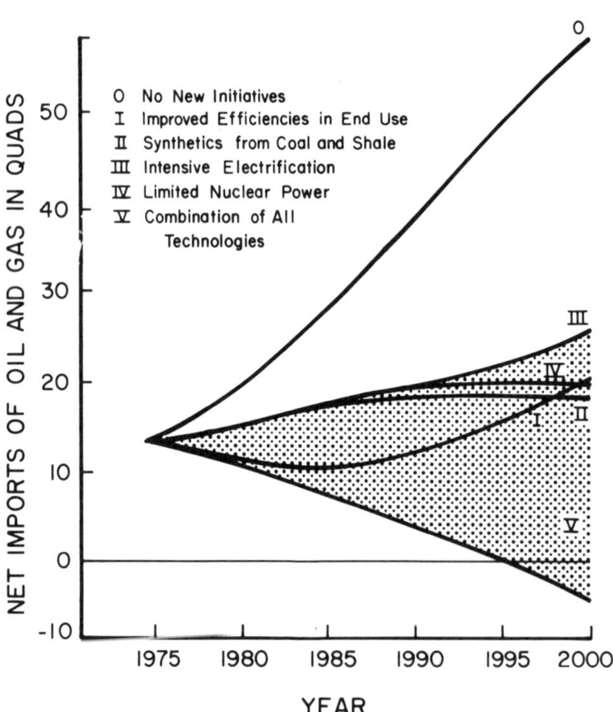

FIG. 1-18 United States future imports of oil and gas. Source: [23].

[1] Felix [8], Rose [28], Dance [29], Palz [30 , pp24-32], and Friedlander [31], among others, examine the options.

sold in barrels. ERDA felt that all scenarios are unacceptable individually, as they show increasing imports. Only a combination of all technologies, scenario V, can make importing fuel a matter of choice.

The extent to which the United States is able to reverse its current oil import trends and achieve national energy independence[1] remains to be seen.[2] It is an important national problem which is not likely to vanish in the near term.

Transition To Inexhaustible Energy We close this sec. by observing that ERDA's 1975 proposed national energy plan[3] recommended, among others, the long-term (past 2000) priority "To pursue vigorously those candidate technologies which will permit the use of essentially inexhaustible resources." It went on to specifically recommend solar-electric technologies of wind, thermal, photovoltaics, and ocean thermal gradient, be pursued and that the unresolved questions in technical, economic,[4] environmental, and social areas be answered by vigorous development efforts.

Since that solar-electric technology recommended by ERDA, now DOE, is founded on solar energy as one viable energy alternative for the nation's future energy needs, we now leave the larger more general energy area and examine the nature and characteristics of solar energy in Sec. 1-3 before later addressing specific solar-electric conversion methods.

1-3 Solar--Our Major Inexhaustible Energy Income

In Sec. 1-1 we introduced and defined the terms energy capital and energy income. In both Secs. 1-1 and 1-2 we made the point that the world and the nation are living off their energy capital, the accumulations of fossil energy

---

[1] Felix [32], among many others, has explored the energy independence. idea. He concludes the U. S. can be energy independent by 1985 by utilizing nuclear power and coal.

[2] Indications are that production from the Alaskan North slope, region which began producing in mid 1977, has already caused the trend in U. S. oil imports to significantly decrease instead of increasing as depicted in Fig. 1-16 [33]. Thus increases in domestic supply can reduce oil imports provided total demand stays about constant.

[3] It is popularly supposed the United States has no comprehensive energy plan. Such a plan, in fact, exists and was proposed by ERDA in 1975 to the Congress and President. An earlier plan was proposed in 1971 by President Nixon [34]. Also an active government program was mounted in 1977 to secure feedback directly from the people towards a national energy plan [35] culminated by President Carter's energy plan of spring 1977.

[4] Farrelly [36], among others, has explored the impact oil price increases may have on advanced energy technologies.

created over millions of years. As every investor knows, it is not a prudent thing to live off one's capital very long, for there will surely come a day of reckoning when the capital is exhausted. Instead wisdom indicates the desirability of living off one's income.

The sun is the only energy income we have to the earth[1]. "The sun is the greatest source of heat and light that man has ever known" [37]. From it originally all our fossil fuels were created which are solar energy that has been stored over millions of years. So it is a natural thing, in the light of the finiteness of our fossil fuels and the certainness of their exhaustion if we continue using them as shown by the various charts in Secs. 1-1 and 1-2, for us to go back to the original source of the earth's energy supply, the sun.

Man has recognized the importance of the sun from the earliest times. Many cultures have worshiped the sun as the God of Life.[2]

The ancient Egyptians and their cult of the sun god Ra immediately come to mind.[3] From ancient Greek mythology came the sun god Helios who was sometimes referred to as the god of fertility and creator of harvests. Some current solar terminology springs from that source, e.g. Heliostat, pyrheliometer, and others.

Rau [37, p16] tells us that "nobody in all that time seems to have hit upon the idea of using the sun as a source of power and energy." He goes on to point out [p43] that the only exception was Archimedes who in 212BC succeeded in setting fire to the sails of Roman warships attacking Syracuse by using a "burning mirror."[4] Almost 1800 years passed before anyone turned toward seriously utilizing the sun's energy. The fascinating story of some of the more modern pioneers in solar energy utilization is is Rau [37, Ch. 7], Daniels [38, Ch. 2], Kemper [39], and other sources.

---

[1] Some think the energy in the seas perhaps is harvestable via fusion; but this is stored energy and not energy income. Its potential is undoubtedly huge.

[2] A charming summary is in Rau's [37] Chap. 4: "Solar Myths and Solar Power."

[3] One frieze recovered by archaeologists [see Palz 30, pii] depicts the sun disk with individual rays coming down to Egyptians. Each ray had a hand on its end. Thus the ancients saw the sun symbolically as offering man helping hands. Perhaps they were closer to the truth than we'd thought. The symbology becomes all the more powerful as an age of solar energy utilization is unfolding before us.

[4] It is interesting that this experiment was successfully recreated in modern times by Ioannis Sakkas to demonstrate this achievement of Archimedes [40, p72]. "With help from the Greek Navy he set fire to a 150 foot row boat at Piraeus by using 70 mirrors of 25 sq. ft. area each" [41]. This basic concept is essentially the same as that currently being researched in modern "power towers".

The Amount of Solar Energy arriving on the earth is vital to know if we expect to utilize solar energy to supply any significant portion of the enormous amount of energy needed to run the modern world. The issue is important to engineers intending to design the practical solar systems for tomorrow's world. It is also of interest to skeptics who erroneously think "there isn't enough energy or power in solar energy to ever do the job."

Fortunately scientists have studied this problem. Rau [37 , p31] tells us that "the earth itself receives only about 1/2,200,000,000 of the enormous heat radiated from the sun." Other scientific results are in Table 1-2.

TABLE 1-2 The Earth's Annual Energy Income From the Sun

		Approximate Q	Figures 10^{12} KWh
(a)	Gross radiation energy striking earth	5140	1,500,000
(b)	Part of (a) reaching surface of earth	3200	940,000
(c)	Part of (a) converted into wind	90	26,000
(d)	Part of (b) reaching land areas	900	260,000
(e)	Part of (b) spent in evaporating water	1000	290,000
(f)	Part of (e) spent in lifting water vapour	17	5,000
(g)	Part of (f) recovered in form of rivers	0.17	50
(h)	Part of (g) used for power plants	--	0.4
(i)	Part of (d) photo-synthesizing land vegetation	0.15	45
(j)	Part of (b) photo-synthesizing marine vegetation	1.25	375
	Present annual energy consumption of mankind	0.09	2.6

Compiled from data given by Ayres and Scarlott.

SOURCE: Thirring, Energy For Man [Bloomington, Indiana: Indiana University Press, 1976]. Reprinted by permission of the publishers.

One of the world energy units is Q. Global energy units, 1 Q = 10^{18} BTU, are large size--a thousand times the 10^{15} BTU Quad energy unit associated with the U. S. energy system. The corresponding KWH figures are in the right column.

From Table 1-2 we see that over 5,000 Q of solar radiation arrives at the earth above its protective atmosphere. Some of this is reflected back out into space. Some is absorbed coming through the atmosphere because of ozone, dust, scattering, and gases in the atmosphere. About 3,000 Q reaches the surface of the earth where it heats land and water masses and maintains the earth in thermal equilibrium. About 900 Q reaches the land area.[1] This is the absolute maximum man could ever expect to collect. We now skip down to the last line and find that as of the early 1950's, about 0.1 Q was the annual energy consumption of all kinds for all of civilization. Austin and Brewer's [2 , Fig. 3] curve for total world energy use roughly checks the 0.09 figure in Table 1-2 but permits us to find the probable 1977 value. It is about 0.20 Q/year. Using this more recent figure, we find the ratio:

$$\frac{\text{Max. solar energy reaching world's land/yr.}}{\text{Energy consumption of all mankind/yr.}} \sim \frac{900}{0.2} = 4500:1$$

We thus arrive at the conclusion, certain to startle some, that <u>there is about 4,500 times as much solar energy arriving on the earth's land as is used by all mankind for all purposes</u>. Stated another way, 1/4500th or about 0.02 percent, of the land could be used to supply all the earth's energy needs for all purposes <u>if</u> we knew how to convert it at 100 percent efficiency. But now assume we can <u>only</u> capture 1/10 of that usefully. Then only 0.2 percent of the land area would be needed for solar collectors. Finally if we throw in another factor of 10 to, pessimistically, account for the day-night energy storage need, inefficiencies, effects of cloud cover, dust, and other effects, we see that only about 2 percent of the land area, at most, would be needed.

From the above rough order-of-magnitude simple calculations it is evident that, conservatively, solar energy arrives on the land surface several <u>thousands</u> of times greater than our present total world energy needs. We also see that even pessimistically only a small percentage of the land would have to be used to capture solar energy realistically.

Some individuals think that impractical sized land areas would be required to utilize solar energy. The above rough argument reveals the fallacy in such reasoning.

Following somewhat similar reasoning, Rau [37 , p5] concludes that "if only 3 percent of the solar energy incident on one-tenth of the earth's land area were converted into usable power it would be enough to meet the power requirements of a world population of 6 billion--twice the present population."

---

[1] Recall that most of the earth's surface is water.

Hayes [1 , p8] states that "the sunshine that falls each year on U. S. roads alone contains twice as much energy as does the fossil fuel used annually by the entire world." In another paper [42 , p7] he sums the matter up: "Everyday the sun delivers thousands of times more energy than humankind employs from all conventional sources. The trick is to harness that energy to do work at a price people can afford."

More modestly, and less quantitatively, the 1972 Solar Energy Panel [43 , p5] concluded: "Solar energy is received in sufficient quantity to make a major contribution to the future U. S. heat and power requirements."[1]

There are many other ways we could look at the availability of solar energy to perform the useful tasks of the world; but they all lead to the conclusion that there is far more than enough for doing Man's tasks. Best of all, it's clean and inexhaustible, but, on the negative side, it is capital intensive.

<u>Solar Energy Arriving On The United States</u> is of particular interest.

The U. S. is fortunate to have within its borders one of the high areas of the world for receiving solar energy, principally in our Southern and Western states. Some other regions of the world receiving the highest amounts of total solar energy are Saudia Arabia, Northern Africa, South Africa, Northern Chile and Argentina and North Central Australia. The reader will find nice world maps summarized in Palz [30 , pp36-37] based on the most comprehensive analysis presently available, he claims. The results of similar worldwide solar studies also appear from time to time in various publications of the International Solar Energy Society (ISES). We only need to know here that such scientific studies have been made, that the data exists, that such studies are of an ongoing nature because of the many climatic variables, and finally that there are sizeable numbers of people in the scientific community actively working, measuring, and recording the solar energy available throughout the world. This is an important activity, for the amount of solar energy available and its intensity are the design starting points of all useful solar apparatus. The collection of solar energy data throughout the world is primarily done by various meteorological agencies, though there are a few private solar energy radiation measuring stations.

Total solar energy is measured by an instrument called a pyranometer. It is basically a sophisticated thermocouple with one leg heated by solar radiation (black absorber surface) and the other cooled (white radiant surface). It yields an electrical output indicative of the solar intensity at that instant. By integrating the output, say over a day, the amount of solar energy received at that site is effectively measured. The instrument measures total radiation coming into it from all spatial directions and over all useful solar wavelengths; thus it 'sees' a hemisphere of radiant energy arriving. The details of solar

---

[1] The issue of the availability of solar energy was <u>very</u> extensively examined and argued behind the scenes in the Panel's deliberations. The statement here is the epitome of conservatism.

radiation measurement are beyond our scope here, but the reader interested in this area of electro-physics will find clear summary treatments of the many and various solar radiation measuring instruments in Robinson's book [44].

Solar radiation measurements are of three kinds:

- Direct Radiation is that arriving directly from the sun at the measuring instrument. It arrives over the small spatial angle subtended by the sun.

  On clear sky days most of the solar radiation received is of this kind.
- Diffuse Radiation is that arriving at the measuring instrument by other than a direct path. The sun has, in an appropriate way been occluded for this measurement.

  On cloudy days most of the solar energy received is of this kind. This is why we can get a suntan on a cloudy day.
- Total Radiation is direct plus diffuse.

  Total radiation is what is recorded by the majority of solar reporting stations.

When dealing with solar energy availability it is important to know to which of the above three kinds of radiation one is referring. Additionally we must know whether the solar radiation was measured on a:

- Horizontal Plane, i.e. the measuring instrument was level with the ground. Usually such measurements result in less than the maximum solar energy being measured, but the measurement is easy to execute.
- Plane Normal to the sun's direct ray.

  Such measurements indicate the maximum radiant power available at that site, and if the instrument is caused to track the sun's direct path during the day, the result is the maximum energy for that site. Tracking complicates the measurement apparatus, however.

Most of the solar energy data collected in the United States by the National Weather Service is for the horizontal plane case. While such data are of some value, their usefulness is suboptimal. Boes et. al. [45] says that "maps showing only the availability of total horizontal radiation can be seriously misleading." Most well designed solar collectors will be southerly facing and tilted up[1] so that the collectors tend toward being normally oriented.

---

[1] The starting point is to incline it from the horizontal plane at an angle equal to the site latitude. One can then fine tune that angle, depending on whether he wishes to collect more solar energy in summer or winter.

To find a more accurate representation of how solar energy is geographically distributed throughout the U. S., Boes and his group started with the solar data base from the National Climatic Center, Asheville, N. C. where they found hourly solar data for the five years 1958 through 1962. Then, after some editing of the data, hourly values of direct-normal radiation were computed using the recorded measurements of total-horizontal radiation and an empirically verified conversion formula. Mean daily totals of direct-normal solar radiation were calculated by month for each location and averaged over all five years. Finally they used a computer code to automatically plot the solar energy contours.

Selections of their contour maps are shown in Fig. 1-19 and 1-20. Note the contours, called isopleths, are for energy units of KWH/m^2 of absorber area at the surface. They believe the maps showing the availability of direct normal radiation are the first of their kind published. They observed that the direct normal radiation nearly always equals or exceeds the total horizontal. They conclude that, though their results will be refined by better future data, "there is more direct solar energy than has been generally believed; direct-normal radiation availability generally exceeds total-horizontal availability by about 60% in the winter."

Such maps, and their future refined versions, may be of substantial use to those planning solar installations.

Geographical integration of such direct-normal maps should, in principle, result in finding a figure for the total solar energy arriving within the country. Then that total could be examined from month-to-month and its statistics studied. As of 1978 this appears to have not been done. When and if done, it would permit a more accurate assessment of the ratio of total solar energy arriving to the total energy of all kinds used by the nation--something more accurate than the order-of-magnitude analysis we did for the world earlier in this section. The Department of Energy's recent National Solar Data Program appears to be a positive step in this general direction.

Solar energy arrives at the top of the earth's atmosphere at very close to a constant intensity[1] of 1353 W/m^2. How is it then that the solar energy--either total horizontal or direct normal--as seen in Figs. 1-19 and 20 is so variable for the United States? The answer, of course, lies in meteorological conditions. Cloud and weather patterns vary and attenuate the solar radiation in traversing the atmosphere. Thus we'd expect to find a region that has generally cloud-free weather to also receive the greatest solar energy. The Southwest U. S. is well-known for such weather, and the maps in Figs. 1-19 and 20 clearly show this region receives the most solar energy. Additionally, water vapor, haze, dust--any particulates--, pollution gases, natural atmospheric gases, and other atmospheric contaminants tend

---

[1]This is the 'solar constant' measured at the mean earth-sun distance of one astronomical unit of AU. Its symbol is Io. It has been extensively studied by the late Thekaekara, et. al. [46]. Were it substantially less than this we'd freeze. Were it substantially more than this we'd bake.

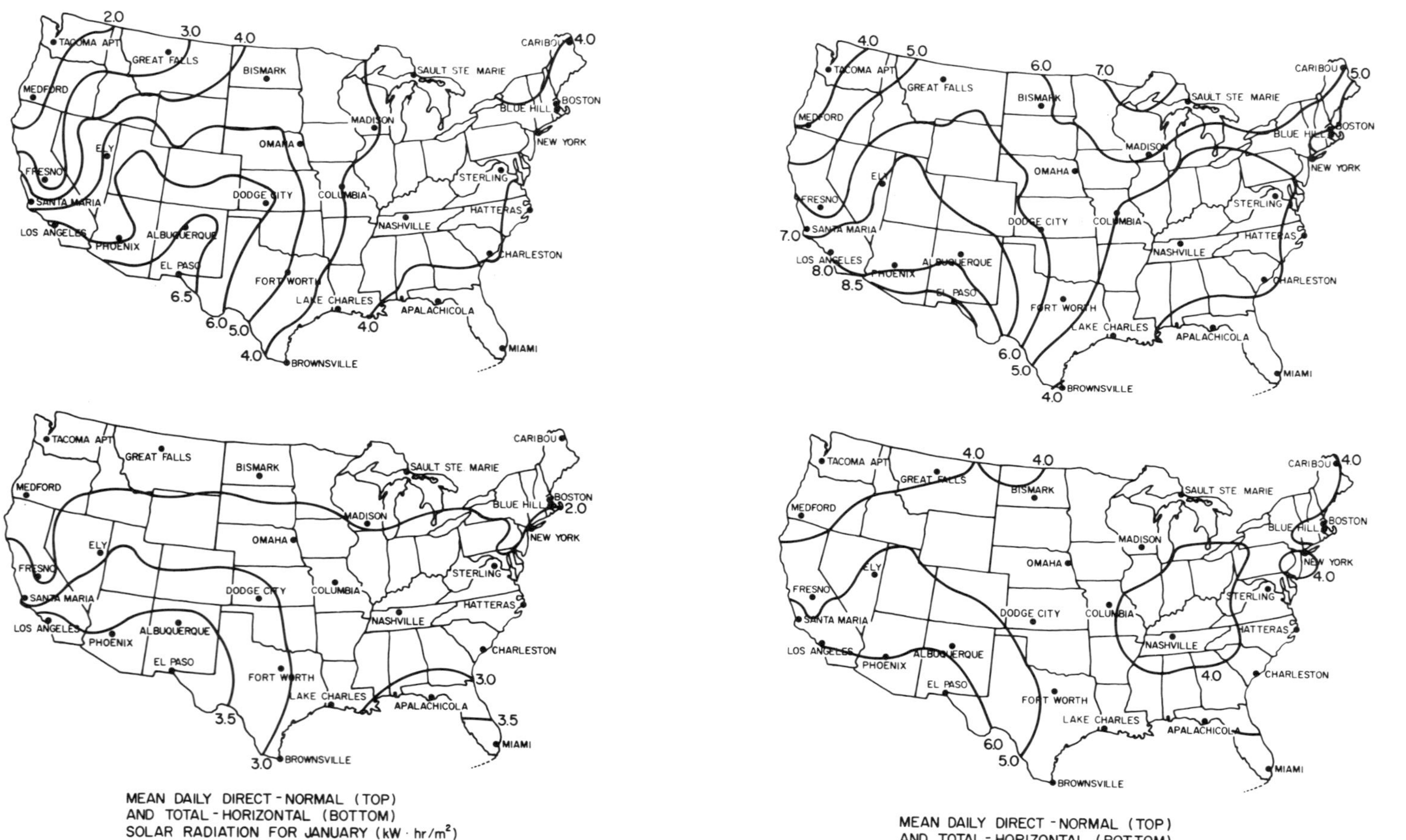

MEAN DAILY DIRECT-NORMAL (TOP)
AND TOTAL-HORIZONTAL (BOTTOM)
SOLAR RADIATION FOR JANUARY ($kW \cdot hr/m^2$)

MEAN DAILY DIRECT-NORMAL (TOP)
AND TOTAL-HORIZONTAL (BOTTOM)
SOLAR RADIATION FOR MARCH ($kW \cdot hr/m^2$)

FIG. 1-19 Mean measured U. S. solar radiation for January and March averaged over the 5 year period 1958 through 1962 for direct-normal case (Top) and horizontal (Bottom). Source: Boes, et. al. [45].

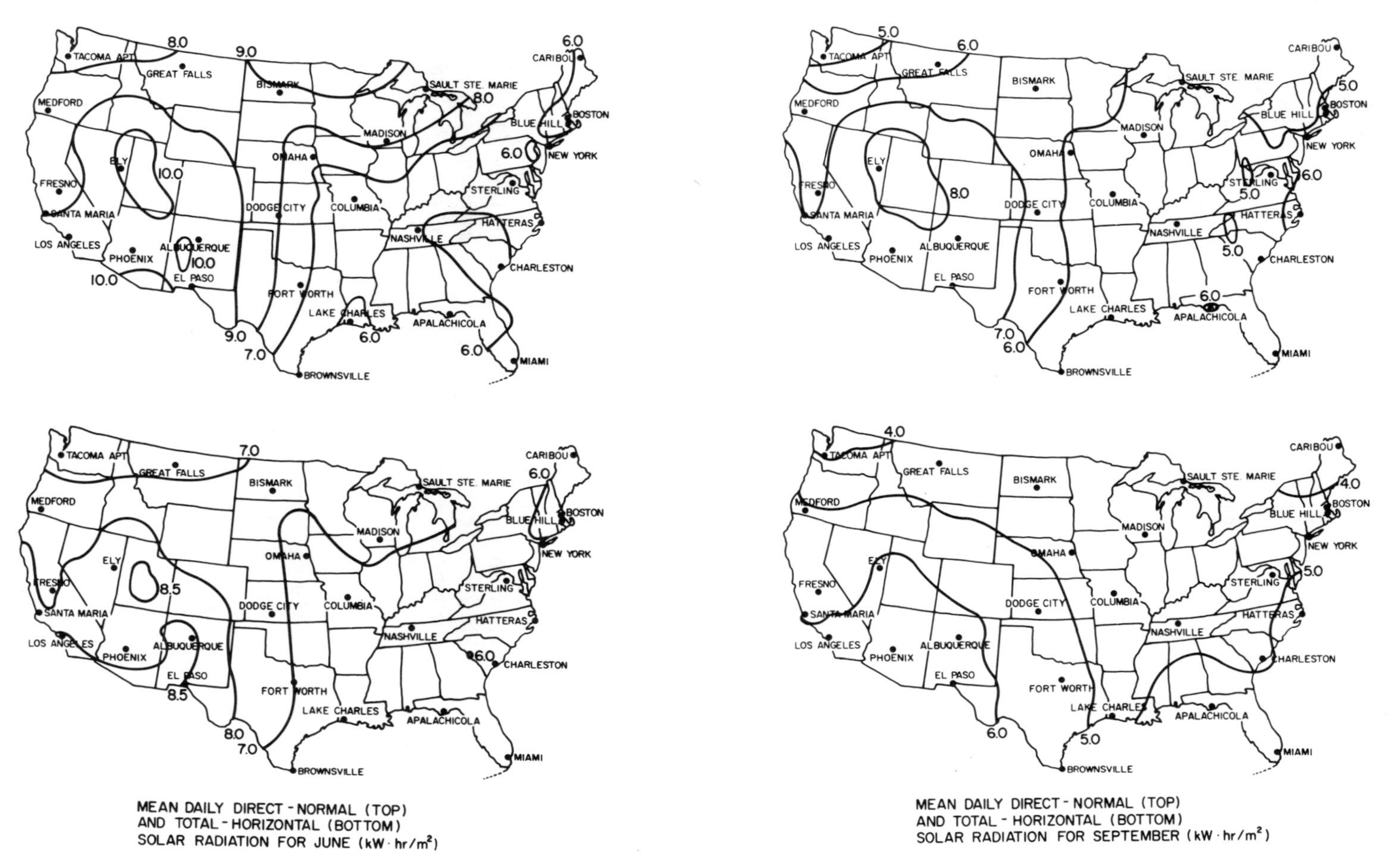

FIG. 1-20 Mean measured U. S. solar radiation for June and September averaged over the 5 year period 1958 through 1962 for direct-normal case (Top) and horizontal case (Bottom). Source: Boes, et. al. [45].

to decrease the solar energy available at the surface. Inasmuch as these quantities vary with altitude, then solar energy also varies with altitude.

Thus solar energy availability varies from site to site primarily because of the variability of weather phenomena. As the weather slowly becomes more predictable as we refine our forecasting techniques, it isn't hard to see that the ability to forecast solar energy availability will also improve. By 2025 solar historians will probably look back on our existing inadequate national system for collecting solar energy data and consider it archaic.

The Sun's Spectral Distribution is important to know about in solar work. Its measurement is intimately tied to measurements of the solar constant, as we shall see. The sun's radiant energy and its spectral distribution have been a source of substantial research by early pioneers in solar science. The measurements are complex and tedious. We present only an overview here.

Solar energy is really nuclear energy--at a distance. The source of all solar energy is the sun wherein the transmutation of helium into hydrogen and vice versa results in electromagnetic radiation occurring from the sun's surface. The sun is "a gigantic thermonuclear bomb in which matter is converted to energy" [37 , p17]. The complex reactions within the sun occur at hundreds of millions of degrees. The physics are extremely complex, and we need not bore into them here. One of the best sources for learning what is known about the sun is Gibson's book [47].

We get an integrated picture of the sun's spectral distribution from Fig. 1-21. Here we're plotting watts/m^2/100 Å[1] vs wavelength. On the abscissa the more modern SI term for light wavelength, λ, is its unit. micrometers, a millionth of a meter, abbreviated μm.

The ordinate is a bit tricky. Think of a 1 m^2 totally absorbant surface oriented normal to the sun's direct ray. Then visualize a fictitious optical filter preceding it with an optical passband of 0.0 1 μm (100 Å = 0.0 1 μm). Thus the fictitious filter allows spectral energy to pass through only in that passband which is centered on a certain wavelength. We measure that radiant power impinging on the 1 m^2 absorber surface at that wavelength. That power[2] gives us a point on the ordinate. The asbcissa is the wavelength at which the passband is centered.

Now visualize our being able to someway vary this 'optical window' over other wavelengths while simultaneously recording the power density impinging on our 1 m^2 absorber. The resulting trace would be the sun's spectral distribution. The ordinate we now give the name solar spectral irradiance[3] and the symbol $E(\lambda)$.

---

[1] Å = Angstrom.

[2] Actually it is more like a power density, as it is spread out over the 1 m^2 area. The SI term for watts/m^2 is radiant exitance.

[3] It is frequently shortened to irradiance in solar literature. The modern unit for it is W/m^2 μm. The units used in Fig. 1-21 are not the modern SI unit. Such details need not bother us at this point, however.

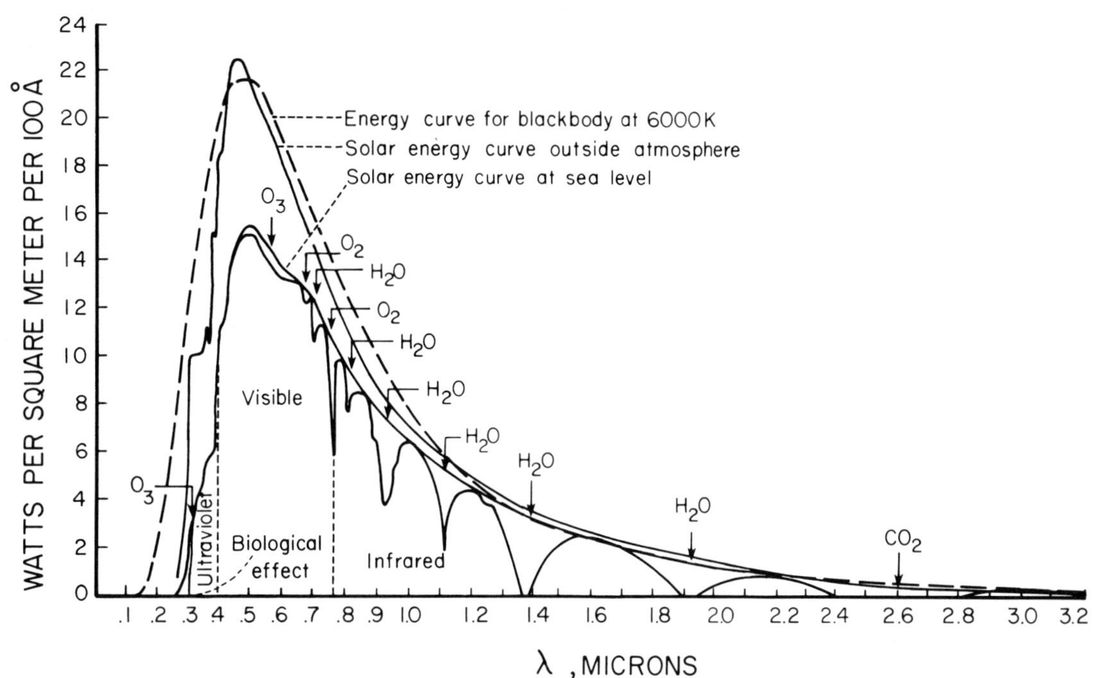

FIG. 1-21 Solar energy spectral distribution above the earth's atmosphere (Air Mass 0) and at the earth's surface (Air Mass 1.0). From Astrophysics, by J.A. Hynek, [48, p272], Copyright © 1951, McGraw-Hill Book Co.

Having settled the units, names, and symbols we now focus on what Fig. 1-21 tells us.

First, we note the solar energy curve outside the atmosphere (Air Mass 0 condition). This is how the sun's energy is spectrally distributed upon arriving at our planet under normal conditions of a quiet sun, i.e. without sunspots or other perturbations. Note that the sun's spectral energy distribution is approximated by that from an equivalent blackbody at 6000K.[1] We also note that the sun's spectral distribution peaks in the visible range. It is much wider also, spanning from ultraviolet through the infrared and on up into radio wavelength ranges.[2] Its useful range is from about 0.2 μm to about 3 μm. About 98 percent of the sun's energy falls within this range.

---

[1] But this doesn't mean the sun's temp. is 6000 K. It's much hotter inside! It means we can model the sun, to a rough first approximation, as if it were a perfect blackbody radiating at 6000 K.

[2] Thekaekara [46 , p31] gives the complete authoritative measured curve. The spectrum extends continuously on up into the radio range, but the total power under the curve beyond 3 μm is negligible for solar utilization purposes.

Next we note the solar energy curve at sea level (Air Mass 1 condition) and observe that it is always below the curve for air mass 0 condition. The reason, of course, lies in atmospheric attenuation; but the attenuation varies with wavelength because of water vapor and various gases in the atmosphere which absorb some or all the solar radiation. Note, for example, the strong attenuation of the ozone layer, O_3, in the 0.3 - 0.4 range. It protects us from the harmful ultraviolet solar rays.

The notches caused by water vapor effects are also strongly evident, particularly in the infrared range. Observe the shielding effect of the carbon dioxide, CO_2, layer at about 2.6 μm. It is this layer which largely due to its opacity restricts the earth's reradiation of the solar energy back into space; It results in livable temperatures on the earth's surface and is the basis of the Greenhouse Effect widely utilized in applied solar technology.

Thus what we have to work with at the earth's surface is a solar spectrum distributed between about 0.2 to 3.0 μm with serrations in the spectrum. It is not a nice smooth curve.

Since we're interested here in utilizing the sun's energy, it is instructive to inquire how the sun's spectral irradiance curve is related to the solar power we desire.

The power density I in a wavelength range $d\lambda$ for an arbitrary spectral irradiance curve $E(\lambda)$ is

$$dI = E(\lambda)d\lambda \quad \text{watts/m}^2. \tag{1-1}$$

Integrating both sides we find:

$$I = \int_{\lambda_1}^{\lambda_2} E(\lambda)d\lambda \tag{1-2}$$

and if we let $\lambda_1 \to 0$ and $\lambda_2 \to \infty$, i.e. we sum the spectral contributions over all wavelengths, then we find

$$\begin{aligned} I = \int_0^\infty E(\lambda)d\lambda &= I_o \quad \text{watts/m}^2 \\ &= 1353 \text{ watts/m}^2 \text{ at Air Mass 0} \\ &= \sim 1000 \text{ watts/m}^2 \text{ at Air Mass 1.} \end{aligned} \tag{1-3}$$

Thus the area under the solar spectral irradiance curve gives us the power density we have to work with; and we note that the power density includes the totality of the various spectral effects for the curve we're working with. The resulting integration gives us the solar constant above the earth's atmosphere and a corresponding 'constant' for the sun's power density at the earth's surface of approximately[1] 1000 watts/m^2. The remarkable implications are explored in the following examples of progressive modeling.

Example 1-1

How many megawatts/km^2 irradiate a 1 km^2 surface normal to the sun? Assume a cloudless sky.

$$P = I_o A = 1{,}000 \text{ watts/m}^2 \times \left(\frac{1000 \text{ m}}{\text{km}}\right)^2 = 1{,}000 \text{ MW/km}^2 \tag{1-4}$$

We instantly see the large solar power showering us. Many large fossil fuel electric generating plants require this area of land and may generate less than 1,000 MW of electric power.

Example 1-2

Roughly how many megawatts of solar energy irradiate the entire continental United States? Assume the entire land could be covered with optimally oriented solar collectors of 1.0 absorptivity. Further assume solar noon, but ignore the 3 hour time shift across the country.

We can't directly use the horizontal land area of the country, for it isn't normally oriented. Instead we consider fictitiously tipping the entire land area up so it is approximately normal. Since the average U. S. latitude is roughly 38°, tipping the area up this amount will be close enough for this approximation. Since the solar power density is a function of the cosine of the angle between the direct sun ray and the normal to the area, we have:

$$P = I_o A \cos\theta \quad \text{MW/km}^2 \text{ of horizontal area}$$

$$= 1{,}000 \cos 38° = 790 \text{ MW for 1 km}^2 \text{ of horizontal area.}$$

Then since the area of the continental U. S. is about 8×10^6 km^2,

$$P_{Nation} \sim 790 \frac{\text{MW}}{\text{km}^2} \times 8 \times 10^6 \text{ km}^2 \sim 6330 \times 10^6 \text{ MW}$$

This is a mind-boggling peak power showering upon us! It is roughly 16,000 times the installed electric power generating capacity of the country.

Example 1-3

Assuming 12 hr. average sunlight per day, and assuming--someway

---

[1] It naturally varies according to atmospheric conditions. The 1000 watts/m2 is the figure most widely used in engineering calculations for solar apparatus. It too is frequently given the symbol I_o. The reader must be careful in theoretical work to distinguish between the two values. It causes little problem to knowledgeable persons.

unspecified--that the U. S. area is orientated normal to the sun at all times throughout the day and further assuming 365 cloud-free days, calculate the annual solar energy income to the continental U. S.

Let w = total solar energy income/yr. to U. S.

$$= 6330 \times 10^{6} \text{ MW} \times \frac{12 \text{ hrs.}}{\text{days}} \times \frac{365 \text{ day}}{\text{yr.}}$$

$$= 2760 \times 10^{10} \frac{\text{MW hr.}}{\text{yr.}} = 2760 \times 10^{13} \frac{\text{KWH}}{\text{yr.}}$$

$$= 2760\ 10^{13} \frac{\text{KWH}}{\text{yr.}} \times \frac{3416 \text{ BTU}}{\text{KWH}} = 94000 \times 10^{15} \frac{\text{BTU}}{\text{yr.}}$$

$$= 94{,}000 \text{ Quads}$$

Example 1-4

Compare the result of Example 1-3 as a ratio to the total present energy used by the United States.

From Fig. 1-12 we found the total 1974 U. S. Energy used due to all sources was about 74 Quads.

Thus:

$$\frac{\text{Max Solar Energy Received/yr. by U. S.}}{\text{Total U. S. energy used (1974)}} = \frac{94{,}000 \text{ Quads}}{74 \text{ Quads}} \tag{1-5}$$

$$= 1{,}270$$

From the above examples, which admittedly are rough orders-of-magnitude only, we see that solar energy showers the highest energy using country in the world with over 1,000 times the amount of total energy we need for all purposes. While this is somewhat lower than the 4,500/1 ratio we earlier calculated for the world, the U. S. uses more average energy/yr. than the rest of the world. One would therefore expect to find a higher denominator for the U. S. than for the world.

Even allowing for seasonal effects, cloud cover, dust, time zone changes, effects of latitude from south to north of the country, and other variables--which would have the collective effect of decreasing the ratio--it is evident, even to doubters, that there is ample solar energy to make a major contribution as was concluded in 1972 by the Solar Energy Panel.

Returning to the theoretical considerations before the above examples, it is evident that when clouds obscure the ray path to the sun they simply result in a lowering of the spectral distribution curve from that shown in Fig. 1-21; from eq. (3) we therefore see that the power density received at the site will also be something less than 1000 watts/m^2. The exact value would be determined by the thickness and type of clouds.

We now have some general theoretical understanding of why the received solar power density at a site may decrease when clouds come by; but let us not forget that Fig. 1-21 was for direct solar radiation and that when clouds come we then receive appreciable solar diffuse radiation and less direct radiation.

We close our treatment of the sun's spectral distribution and its availability by indicating that the reader interested further in solar spectral physics should study Henderson' book [49] which gives an excellent historical outline of these measurements, their problems and difficulties. Robinson [44] looks at other aspects of these complex scientific measurements, while Thekaekara [46] integrates most of the previous investigators work to give us the latest most authoritative irradiance curve for Air Mass 0 conditions.

More recently, Heaney [50], using modern electrical communication engineering spectral analysis procedures and techniques, has shown for the first time that the sun's irradiance curve can be synthesized for analytical purposes as the summation of 18 judiciously chosen plane electromagnetic waves linearly polarized with randomly distributed phases.

Finally, it is clear that our theoretical understanding of how to model and quantify the sun's radiation is in need of further basic scientific research. The sun's radiation, from a theoretical standpoint, is far more complex than has generally been thought or has been presented in this brief overview.

The Sun's Diurnal Variability is another characteristic of solar energy. First visualize an absorber surface normal to the sun receiving the sun's direct radiation. Then the power density is I_o watts/m^2. Now if we tilt the absorber through an angle θ between the sun's ray and the normal to the absorber the power density[1] I incident on the absorber is:

(1-6) $$I = I_o \cos\theta \text{ watts/m}^2$$

Looked at another way, if we held the absorber fixed and normal to the sun at solar noon at our particular site, then as the sun moved across the sky during the day we'd expect to find a cosine type curve traced out for the absorbed power density. Does the theory work out in practice?

Figure 1-22 clearly shows (top curve) that for a nearly clear sky condition the variation is approximately of a cosine form whose peak magnitude is the I_o in eq. (1-6). It isn't exactly of this form near each skirt because of other minor effects which need not bother us here.[2]

---

[1]The term 'insolation' is widely used to describe quantitatively solar power density.

[2]Actually Fig. 1-22 is for a horizontal surface at 40°N latitude in the Washington, D. C. area "under various seasonal and weather conditions." So to be correct $\theta = 40°$ for this case at solar noon. Then there is the additional time variable, which translates to a spatial angle; the mathematical details need not bother us just now.

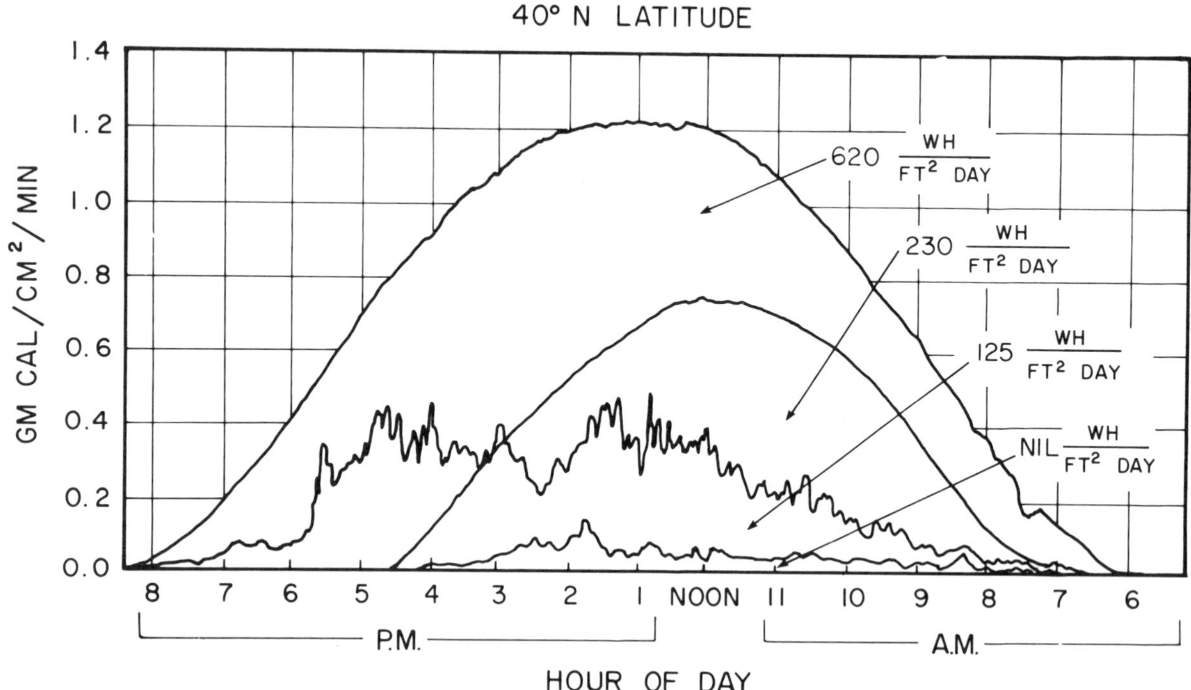

FIG. 1-22 Measured solar isolation at a sea level site under clear sky and various degrees of cloudiness. From Cherry [51]. Used by permission of ASME.

Recall now that from the definition of power as the time rate of using energy, we can write the energy received per m^2 as:

$$W = \int_0^{t = 24 \text{ hrs.}} I dt = \text{Area under the insolation-time curve.} \qquad (1\text{-}7)$$

Thus we see the importance of the insolation in determining the solar energy received at this site.

The received solar energy is simply the area under the curve shown as 620 watt-hours/ft^2/day for that particular site and day.

Naturally as the earth's declination changes with the seasons[1] the amount of solar energy received on the horizontal surface will also vary. The lower cosine curve would have been for a day near wintertime when only 230 watt-hours/ft^2/day were received. Tipping the measuring instrument up to be

[1]Recall this gradual shifting of the earth's axis, called declination, is the principal cause of our seasons.

normal to the sun at solar noon would have substantially raised the curve.

The noisy curve shows the effect of clouds which at any instant decrease the insolation as we've earlier indicated. A single small cloud on an otherwise clear day being blown through the direct sun-site path would simply result in a notch in the cosine curve. General overcastness pulls the entire curve down, and if the thickness varies we obtain a noisy curve. In any case the area under the curve with clouds will be less than for no clouds.

Thus on-site measurements of solar energy, if correctly done, account for the energy in the entire solar spectrum from ultraviolet to infrared as well as account for effects of all time and space angles, the latter including seasonal effects. Finally cloud and meteorological conditions are accounted for in the measurement. It is therefore a complete measurement of the solar energy at that site for that particular day and season.

We hardly need to mention the trivial obvious fact that the insolation curve is zero during the night. Reflected solar energy from the moon and stars is negligible for power utilization purposes.

Note that the absolute amount of solar energy received at the site on a normal clear day is a quantity of energy. The energy quantity will remain the same whether we integrate over the daylight hours or over a total 24 hour period. However if we think of power, or rate at which we use this energy--perhaps from a storage means--then we must be careful to denote the averaging time, i.e. whether it is daylight or a total day (day plus night).

Example 1-5

Compute the daytime average heat power available from a fixed absorber of 1.0 absorptivity oriented normal to the sun at solar noon receiving 1 kw/m^2.

The average value for an insolation of the form of Eq. (1-6) over the positive period is 0.636 I_o. Thus the average power during daylight hours is

$$= 0.636 \times 1000 \text{ watts/m}^2$$
$$= 636 \text{ watts/m}^2$$

To get it at all times, we obviously need energy storage.

Example 1-6

Same as Example 1-5 but average over 24 hours.

Since the time is now doubled but the energy must remain the same as before, it is immediately evident that the new average power is now

$$= 0.636/2 \times 1000 \text{ watts/m}^2$$
$$= 318 \text{ watts/m}^2$$

Thus, if we must average the power over 24 hours, the average value will be slightly less than 1/3 of the peak power available at solar noon; but the energy will be the same regardless of whether we average 12 hrs. or 24 hrs.

The previous basic examples have many profound implications for practical solar hardware where we must be concerned with both capturing a quantity of solar energy and the rate at which we release it, i.e. power.

We now leave the review of solar energy availability to focus on the next topic.

Advantages And Disadvantages of Solar Energy Utilization have not previously been collected systematically and integrated into simple understandable tables. The story is very scattered throughout the literature.

Since, up to now, we've established the copious availabilty, distribution, and basics of collection physics, we naturally would like to see a list made on a single page of the principal advantages of utilizing solar energy, considering the technology as a whole and considering the nation's interest as a whole.

Much of the solar literature was scanned specifically looking for advantages to utilizing solar energy as seen by solar researchers and pioneers. These were assembled and edited, and the result is shown in Table 1-3. These were adapted from an earlier work by Watson [52] who compiled the first known similar list. Table 1-3 is built on that work and extends it considerably.

Numerous words could be used to describe the beautiful virtues of solar energy utilization. We forego these here for brevity, letting Table 1-3 speak for itself. We only point out that we've attempted here to see a composite; the reader new to solar literature is forewarned that such a comprehensive view is not characteristic of most published solar papers which tend to feature only a few of the many substantial advantages of utilizing solar energy.

After a similar fashion, a list of disadvantages of solar energy utilization was accumulated. It is shown in Table 1-4. The disadvantages, somewhat understandably in positive oriented R & D, are frequently hardly mentioned in much solar literature; yet an objective view must acknowledge there are some, however painful such acknowledgment may be to some solar enthusiasts.

The most serious disadvantages in 1978 appear to be in the cost, storage, and social areas. As fossil fuel prices continue to climb solar system costs are slowly becoming competitive[1]. There are vast numbers of competent scientists and engineers working on overcoming some of the difficult energy storage problems, and increasing beneficial changes appear to be occurring in the social area aimed at overcoming the disadvantages listed and hastening the solar age.

On balance, the evidence speaks far more in favor of an all solar energy technology for the nation than against it. What few negatives exist are gentle, tractable, and, given a few years, are amenable to being overcome. The technology is certainly not characterized by some of the more frightening environmental, social, and personal consequences associated with some of the other energy alternatives open to the nation. Fully developing solar energy technology to the point of widespread use in the United States could have far reaching and profound beneficial effects on the nation.

---

[1] The assumption is that solar system costs would increase either not at all or at a lower rate than fossil fuel costs.

TABLE 1-3 PRINCIPAL ADVANTAGES OF SOLAR ENERGY UTILIZATION for solar technology as a whole. Adapted from Watson [52].

Inherent	Technological	Environmental	Social
Cleanliness - inherently so. Availability ● Abundantly available whether we use it or not. ● Daytime availability coincides with man's peak power needs. ● Delivery rate--a gentle 1 KW/M^2 at earth's surface. ● Only known 'energy income'. ● Virtually inexhaustible. A renewable energy source. Conserves fossil fuels. Dependable--consistently so. Intensity--average is predictable for a site. Distributed ● Globally. ● Delivered directly to need site, by-passing conventional energy firms government regulations, and taxing structures. Costs ● No fuel costs. ● Principal cost is apparatus initial capital cost. Globally - maintains earth's heat balance. Applicable for peaceful purposes to help mankind. Safe energy form. No nuclear explosion possibilities.	Multiple Practical Uses possible in many flexible ways. Convertible to heat, cooling, fuel, biomass, food, shaft rotation, electricity and other useful forms. Simplicity characterizes most systems. Systems ● Many beneficial combinations possible e.g. solar cells combined with wind power. ● Highly efficient combined systems. ● Many systems installed could supply significant percent of nation's energy. ● Dispersed (small-scale) or centralized (large-scale) power or mixture of both can be created. ● Minimum of energy transmission problems for dispersed systems. ● Most systems amenable to either customized or mass production implementation. ● Many systems tend toward modularity. Can easily add to for more energy. ● Control of systems--relatively straight-forward. ● No/few moving parts (many systems). ● Lifetime long } For well-engineered ● Maintenance minimal } systems. Materials ● In many cases (thermal systems) can use locally available materials. ● Few exotic materials have to be created. Temperature ● Low, intermediate, or high temperatures achievable. ● Practical systems operate in reasonable temperature ranges. Applications ● Can be used on walls or roofs of new or existing buildings or entirely separate. Land Use ● Large-scale systems could utilize low cost unpopulated land. Invention ● Field opens doors to many beneficial invention possibilities.	Pollution ● No combustion by-products. ● No gaseous effluents to pollute air/water. ● No radioactive residue. ● Essentially no noise. Land ● Minimum permanent damage. Ozone Layer ● Doesn't deteriorate it. Nature ● Compatible with instead of fighting nature. Overall ● Solar technology acceptable to most environmentalists as a 'gentle' energy form with few overall negatives.	Oil Imports ● Wide scale application would promote independence from imported fuels and better trade balances. Applicable to developed or developing nations Life-styles and Social Patterns would be minimally changed in developed countries. Major Residue Problems tend to be solved with some solar systems, e.g. garbage utilization for gas, fuel, oil, fertilizer metals, etc. Time To Create many solar systems is only months/years instead of a decade or more from 'go' decision. Business ● Field amenable to small, large, or mixture of equipment manufacturers. ● New companies/jobs will open as solar industry grows. Dispersed Systems ● Highly favorable, solving land use, energy reliability, transmission, and other problems impacting on society. ● Defensively more impregnable than centralized plants. Terrorists ● Can't build known solar systems into bombs. Philosophically And Aesthetically appealing to large percentage of people. Overall ● Solar energy utilization is socially acceptable with few negatives.

TABLE 1-4 PRINCIPAL DISADVANTAGES OF SOLAR ENERGY UTILIZATION or solar technology as a whole. Adapted from Watson [52].

Inherent	Technological	Environmental	Social
Not Concentrated energy source way we're used to thinking. Geographical Variability of collectable energy. Sensitive to atmospheric conditions. Variable both diurnally and seasonally.	Large Collector Area needed to harvest bulk energy. System Efficiencies are generally lower than desired. Improving them is a major problem. Larger Land Use/useful power required if central stations are created. Transmission Line problems will continue from conventional electric systems if remote solar central stations are created. Waste Heat removal/disposal is a problem for both dispersed and centralized systems. Storage of Energy ● Frequently needed for later use. ● Long-term storage difficult. Costs ● Many 1978 systems more costly than equivalent fossil fuel system. ● Initial cost/energy unit in 1978 tends to be higher than equivalent fossil fuel system. ● Cost/KW depends on plant size. Minimizing it is a problem with most known systems. Tracking ● Essential for some systems. Integration of solar systems into existing buildings and matching them to existing energy systems, e.g. solar-electrics, is a problem. Technology must be fossil fuel subsidized to initially be realized (Fossil fuel to manufacture, etc.). Multidisciplinary Nature of solar technology. Field inherently requires multidisciplinary thinking/people as opposed to specialists in one field.	Environmental Effects of wide ranging nature are implicit in some solar systems, e.g. hydroelectrics and solar satellite power systems. Pollution ● Localized visual nature. ("It's there!") ● Some systems change localized thermal environment, e.g. OTEC. ● Some systems may change global heat balance, e.g. solar satellite power systems tending to melt polar ice caps. ● Some systems (large wind-electric generators) upset direct TV broadcasts. Weather ● Large-scale implementation of some systems, e.g. large windmills, may slightly alter localized weather conditions.	A National Commitment to solar energy utilization is required to have any significant effect. Opposing forces don't want to commit to solar. Some Systems have wide ranging deleterious social effects, e.g. hydroelectric. Centralized solar-electric plants will still have problems of locating remotely from load center, e.g. transmission lines, zoning, land acquisition, costs, etc. Capital Intensive technology, basically, as opposed to our previous minimum first cost criterion. Building Codes require extensive modifications to accommodate solar. Laws ● Need clarifying for a land owner's "sun rights". ● Solar commercialization will require generating many new laws and changing many existing laws. Changeover to widespread solar systems injects pains of implementing any large technological change into society. May impact utilities and energy corporations in intermediate and long run. ● Vested interests view solar energy utilization as a threat. ● Significant time is required to effect a large-scale changeover to solar utilization, i.e. solar is not 'instant solution.' Political Many arguments current over best funds allocation to realize solar technology. Business ● Requires new kinds marketing and sales firms for many systems. ● Monopoly possibilities of solar industry by fiscal 'giants'.

Before solar energy technology becomes widespread, however, we should address the issue of in what form it should appear. We argue this case in the next section.

1-4 Electricity: A Vital Energy Form

The Fundamental Question There are so many varied and useful forms solar energy can be converted into[1] that we should justify converting it to electricity before plunging into the technology details of how to do it.

The fundamental issue is "Why should solar energy be converted into electricity?" The desirableness of energy in electric form is widely assumed a priori, but the issue should be examined with some care.

We should first point out, by way of perspective, that electricity, contrary to what some electrical engineers think, is not the only useful energy form solar energy might take; nor is it even desirable in some instances. For example, it is poor logic and economics to create a solar-electric home system and then use the electric output to heat water or air. Reason dictates that we first ask what we are trying to do, to heat water-or air-, and then ask is it possible to do it directly. Fortunately, these tasks are among the simplest for solar energy to achieve; converting to electricity then to heat makes little sense with today's solar technology.

Other useful energy forms than heat and electricity are light, food, and fuels, all of which can be obtained solarly without direct use of electricity.

So let us acknowledge at the beginning there are many other energy forms man finds useful other than electricity. Each of these has a definite place in the fabric of society where need, reason, economics, and history have dictated that particular energy form. Also let us acknowledge that many of the tasks for which we are presently using electric energy could be done solarly without electricity, e.g. clothes drying, cooking, hot water heating, house heating, house air conditioning, pumping water, and others. The advent of widespread solar energy utilization already is causing technical people to fundamentally rethink each energy-using task and ask what energy form is best for doing that task. The answer, unfortunately, is not always black or white.

Why electricity then? An important perspective is that there are many useful tasks of man which cannot be done except with energy in electric form. Television sets and computers don't work well with coal inputs nor do thousands of other pieces of apparatus we've designed to run off energy in electric form.

Virtually all the advanced nations of the world have come to recognize energy in electric form as desirable and necessary to the quality of life as we know it today. They have therefore invested fantastic sums in electric energy generation, transmission, and distribution systems across their nation. Nations, industries, and commerces presently literally run on electricity. A close relationship exists between electric energy availability and economic health.

---

[1] We make no attempt here to give perspectives on the entire solar energy field and its many possibilities. Daniels' book [38] does so. See also Farber [53] and reference [43].

The Principal Advantages Of Energy In Electric Form[1] include:

- Inherently Clean And Non-Polluting form of energy at the use site. No smoke, fumes, air pollution, gases, etc. are emitted at the use site. These turn up in concentrated form, however at the generation site where they can be more easily dealt with all at once.

- Versatile and can be used for many distinctly different tasks from baking cakes to instantly communicating. No other single form of energy--except possibly solar--which, relatively easily, permits such a vast array of varied tasks to be accomplished.

- Safe systems, when properly installed and maintained, result. There are no gases to leak out and explode while one sleeps. Electric systems, well maintained, don't normally kill people or cause fires.

- Controllable systems result so that desired power and energy delivered can be controlled according to our needs and desires. If we need light, because of the controllability advantage of electric energy, we get it by turning on a switch.

- Predictable systems, within limits, can be created. All the principal governing laws--technical and economic--are known.

- Manageable systems, within prescribed limits, exist; our knowledge of how to better manage them grows yearly.

- High Conversion Efficiency from mechanical energy to electrical energy and vice versa. Generator efficiencies of 95 percent are common. The principal inefficiency in existing systems is going from fuel to shaft motion to drive the generator.

- Easy To Get Rotating Fields which are of incalculable usefullness in the world's electric motors.

- Easy To Transmit the energy via high voltage AC or DC lines spanning hundreds of miles if necessary. This feature alone makes it possible to locate the dirty generation plant remotely from the clean use site.

- Instant Availability as soon as we turn on a switch.[2]

- Low or High Power can be delivered according to the customer's needs.

---

[1] We imply here the electric industry as it exists today throughout the world. Some of these would be modified slightly were we to create extensive solar-electrics at some future time; but in the main the conventional electric energy system furnishes us a valuable reference point from which we can start such an examination.

[2] It is interesting to observe that electric energy is the only commodity which is consumed the instant it is generated; supply and demand exactly match instant-by-instant

- Many Choices of phases and load voltages, depending on the needs of the customer being served. The 3 phase AC system is widespread from which all the useful combinations needed can be obtained.

- Smooth Power Flow can be achieved with 3 phase systems. The result is quieter motors and less stress on the materials therein.

The net result of these features, unique to energy in electric form, is a taking of the load off man's back and a comfort and convenience unparalleled and unknown to the world before the present era, thanks to electric energy.

Energy in electric form contributes immeasurably to our quality of life. It is so woven throughout the fabric of contemporary society, as recent blackouts forcefully remind us, that no thoughtful person would seriously consider eliminating electric energy systems[1] energy.

In the next several sub-sections we assemble a few of the key facts describing the magnitude and characteristics of the present electric industry in the United States. It is important to both solar-electric researcher and newcomer to the field to have some perspectives on this important gigantic industry; ultimately all solar-electrics will have to compete against it for delivering useful KWH's to the busbar or, perhaps, in the future directly to the customer at the need site.

As in the past, we again remind you we shall only cover enough highlights here for a preliminary feel of this important industry. Details of the utility industry will be found in the references and elsewhere.[2]

Electricity And Nation's Total Energy for the past, present, and future would give us a quick perspective of this important energy form.

Fig. 1-23 plots the U. S. energy consumption as a function of time. The figure indicates that the electric power industry's share of the nation's entire basic energy use has grown steadily from approximately 11 percent in 1920 to 20 percent in 1960. The forecast is that it will reach 31 percent by 1980. Whether an increasing fraction of America's energy turns up in electric form depends on many complex economic[3] and technical factors[4] difficult to forecast, but society appears to be saying it wants more and more of its energy in electric form. What maximum percentage of the nation's energy should be in electric form will undoubtedly be the subject of much future public debate. Nagel [56] sums the matter up: " We must accept the fact that a rapid increase in electrification is consistent with the necessary transition away from excessive dependence on oil and gas, the scarcest of our endemic fossil fuels."

---

[1]Readers interested in studying modern electric energy systems in more detail are referred to Elgerd [54] and other authors in this area.

[2]Reference [55] contains many helpful facts on electric utilities. The pamphlet is updated annually.

[3]The electric energy market is price sensitive. It also competes with other energy forms, notably gas and oil.

[4]We should realize there are powerful pro-electrical forces working to increase the fraction, e.g. hydrogen energy proponents.

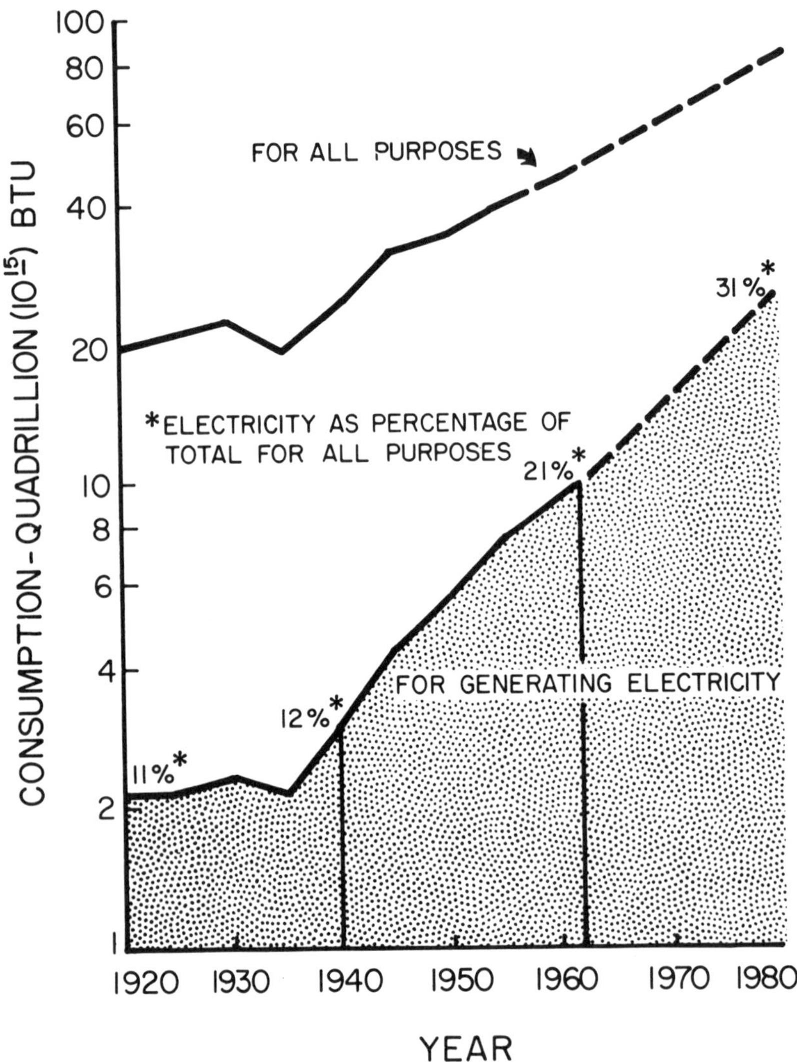

FIG. 1-23 Electricity as a fraction of total U. S. energy. Data source: 1920-1960 based on U. S. Bureau of Census data and Edison Electric Institute heat rates. From [57].

It is clear that energy in electrical form constitutes a major portion of the present U. S. energy consumption. The electric energy industry is therefore a major user of energy quads, as Fig. 1-23 shows. One notes that the number of quads now used for electric energy alone is about equal to the total U. S. Energy used in the 1920's.

Electricity And Nation's Productivity. Is there any relationship between the energy input we commit to the electric industry and the nation's productivity as measured by the GNP? The answer lies in Fig. 1-24. We see that if one considers electricity as the independent variable the GNP historically has increased with rises of energy input for electricity. The forecast is for this trend to continue.

It is evident that America's productivity of useful goods and services is tied in complex ways to our use of energy in electric form. Should we choose to throttle the energy to be used for electric generation, then one would expect limits to the GNP to occur. The nation's energy planners obviously don't have such limiting phenomena in mind, though, judging from Fig. 1-23 where the forecast is for a nearly constant growth rate until at least 2000. Whether such growth will, in fact, occur may ultimately be tied to fuel availability and whether we as a nation want to commit it to electricity generation.

Financial Aspects of the Electrical Industry are staggering. "In terms of the dollar value of plant and equipment the electric power industry is by far the nation's largest" [57]. The total value of plant and equipment of only the investor-owned companies was $173.9 billion[1] in 1975 [55]. Newcomers to the energy area may not realize that the electric power industry in the United States as it presently exists far exceeds the nation's investment in the petroleum industry.

Some feel for the tremendous amounts of capital required to accommodate the annual growth of the electric industry can be had from Fig. 1-25. Here we see plotted the annual capital investment dollars for the total electrical industry (in billions) as a function of time.

First we note the tremendous annual investment being currently made by the nation in the electric industry--of the order of 20 billions of dollars[2]--and all this is just to accommodate the needed system growth to meet anticipated customer needs; it doesn't consider the total invested over previous years of growth.

Second, we note the large capital investment made annually in electric generation. It is the largest investment category of the industry, as expected, with distribution systems being second.

Third, we note the clear effects the 1973-74 oil embargo had on the electric industry as indicated by the notch in the curve.

Fourth, we note the near exponential growth of capital required to finance the electric industry. It is clear this is a heavily capital intensive industry; there is grave concern within the industry over where the staggering amounts of capital required to finance the industry's growth will come from.

---

[1] This figure is 2 1/2 times the 1965 amount! Total value of plant and equipment includes value of plant devoted to gas, steam, water, traction, etc., besides that devoted to electrical plant and equipment in the so-called 'combination companies'.

[2] For added perspective, compare this with the approximate 0.5 billions of dollars the government is budgeting for all solar energy R & D in 1978 of which solar-electr is but a part.

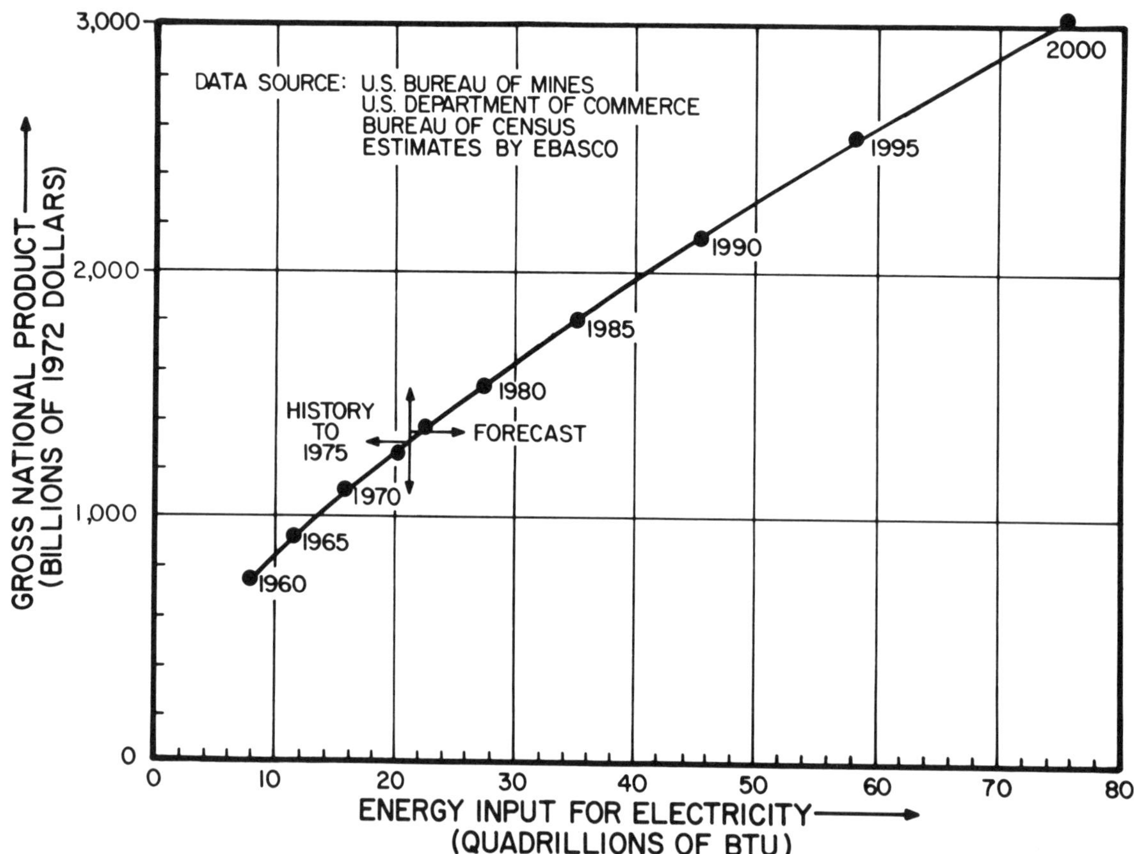

FIG. 1-24 Relation between GNP and total energy input for conventional electricity generation in the United States.

Fifth, we observe the growing decay caused by inflation which clearly impacts the electric industry in a big way.

The electrical industry updates its capital expenditure curves annually, and curves of the type shown in Fig. 1-25 periodically change.

Another way of acquiring a financial perspective on this enormous industry is to compare the capital investment per employee in the electric utilities with similar investments in several other industries. This is done in Table 1-5.

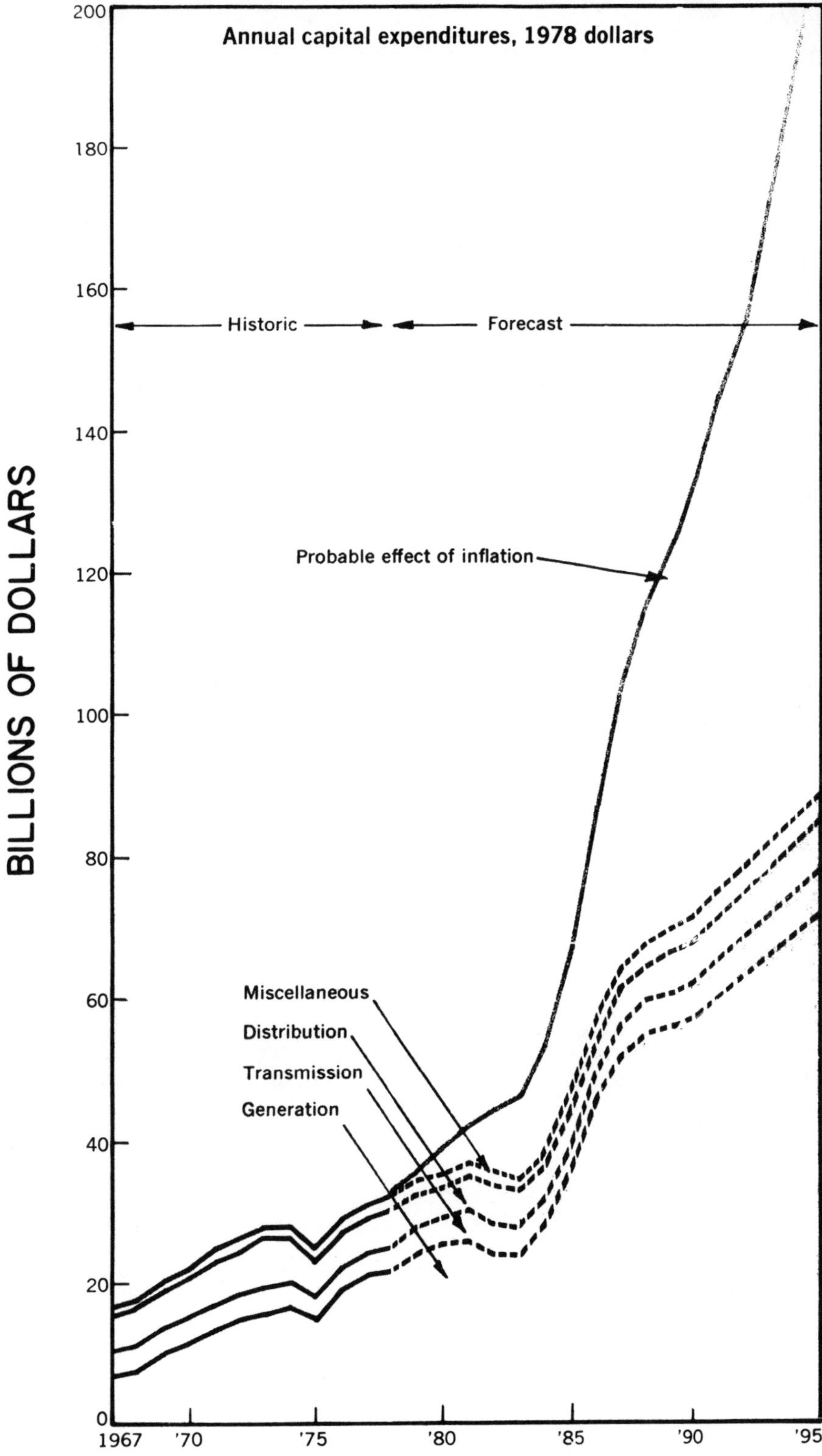

FIG. 1-25 Annual capital expenditures for America's electric power industry [58]. Reprinted from September 15, 1978 issue of Electrical World © Copyright, 1978, McGraw-Hill, Inc. All rights reserved.

TABLE 1-5 Investment Per Employee In Various Industries*

Electric utilities	$250,700
Petroleum**	203,517
Tobacco	88,928
Motor vehicles	60,703
Chemicals	51,321
Primary Metals	40,744
Other transportation equipment	31,625
Nonelectrical machinery	30,685
Paper and allied products	29,739
Electrical machinery	28,453
Food and beverages	28,178
Lumber and wood products	26,484
Stone, clay and glass	23,931
Instruments	23,252
Rubber and miscellaneous plastics	19,814
Fabricated metals	19,095
Printing and publishing	18,386
All manufacturing	31,580

*Capital invested is total assets less investments in governmental obligations and securities of other corporations. It is stated after deducting all reserves such as for depreciation, etc. Data are for 1972, the latest available for all industries.

**Consists of petroleum extraction, refining, and pipeline transportation.

SOURCES: Electric utility industry data estimated by Edison Electric Institute on the basis of electric utility company annual reports. Data for other categories from the Conference Board.

We immediately notice that the investment per employee in the electric utilities is about 8 times that for all manufacturing; the absolute amount even exceeds that for the petroleum industry.

The highly capital intensive nature of the present electric utility industry is evident. Naturally, there are many more fiscal details to this large, complex, and powerful industry; but the above fiscal perspectives are adequate for now.

Production and Sales of Electricity are of interest in our search for perspectives on the electric energy industry.

The key facts are in Fig. 1-26 where we see both the energy requirements and the peak load power plotted as a function of time over enough years for us to gain perspective. In 1976 the electric industry supplied about 2,000 million KWH of energy and a peak power of about 400 million KW. Since the scales in Fig. 1-26 are logarithmic, a straight line for either energy or power indicates

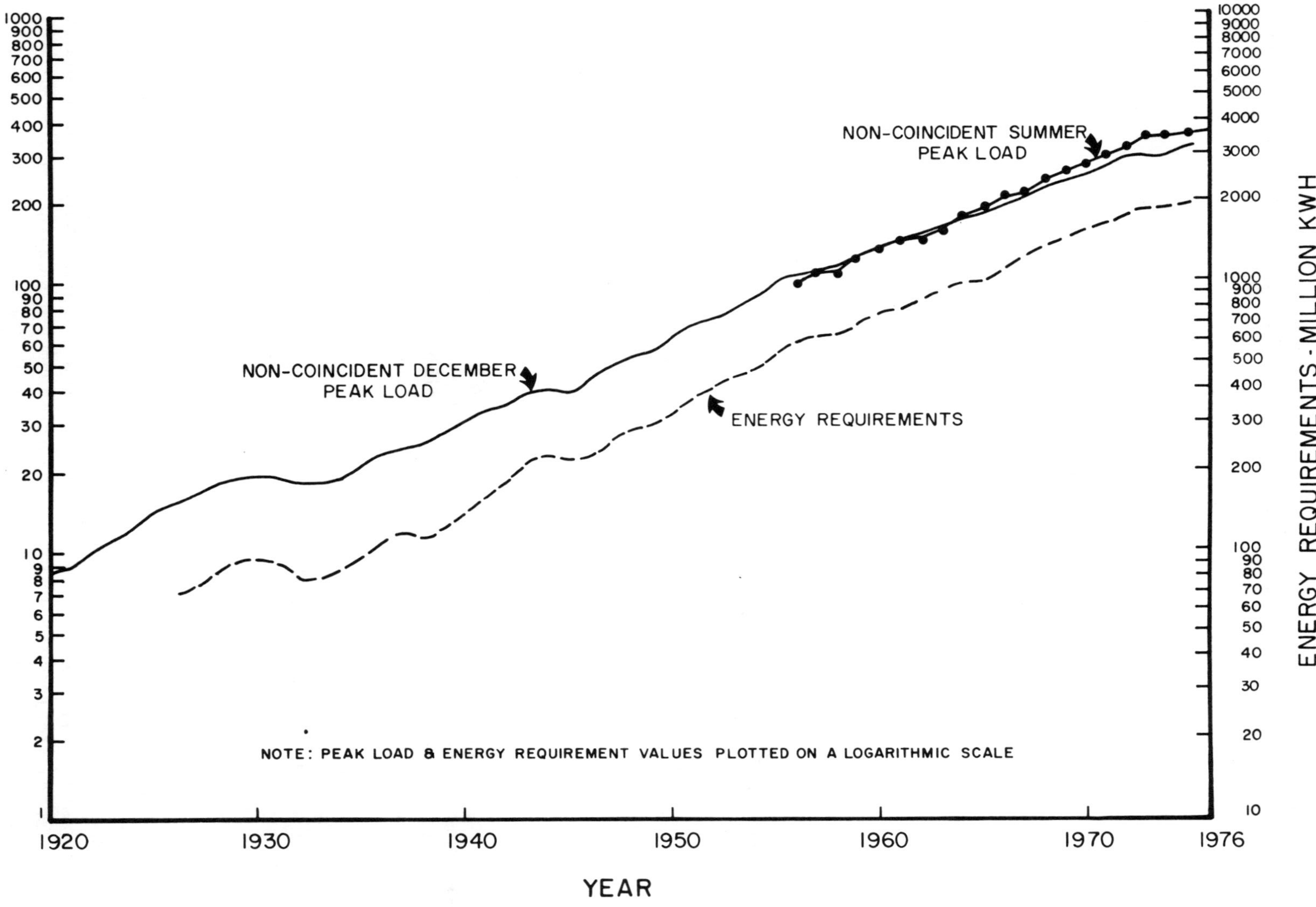

FIG. 1-26 Non-coincident summer and winter peak loads and annual energy requirements for the total electric utility industry (excluding Alaska and Hawaii). From [59].

exponential growth. The peak power load growth averages about 7 percent per year. This corresponds to a doubling time of about 10 years. Thus the electric industry grew by a factor of approximately 32 from 1920 to 1970, doubling every 10 years. The economic, business, technical, political, and other implications of such a growth industry are truly staggering; but, as we indicated earlier, it is the largest single industry in America.

"The future rate of growth of electric power demand is currently the subject of great controversy and uncertainty" [59]. Underlying the growth rate issue is the more fundamental issue of what percentage of our nation's total energy needs should be in electric form, an issue we earlier touched on.

We naturally, in our search for perspectives on the electric industry, inquire about the relative standing of various countries producing the major part of the world's electric energy. These facts are in Table 1-6 where we see the United States, by far, produced more electric energy in 1975 than any other country, about a third of the total produced in the world. Russia is the second leading producer.

Thus we see that not only is the electric industry the largest single industry in the United States, but it is the largest in the entire world. Thus when we think of the electric energy business we are dealing with a gargantian industry. The mere size and productive capacity of it in the United States gives some indication of the importance society attaches to energy in electric form at this point in time. The magnitude of the fuel supplies required to sustain such an enormous industry in the United States stretch our imaginations to visualize.

How the electric utility customers are distributed between investor-owned companies, cooperatives, and governmental agencies is shown in Table 1-7. Over three-fourths of the electric industry's customers are supplied by the investor-owned utilities. Thus we immediately acquire another important perspective on the industry; it is dominantly owned by investors and not the government. However, we note that an appreciable portion of the power generated is by various governmental agencies, the Tennessee Valley Authority (TVA) and the various municipally owned utilities being the chief such power producers. Finally, note that the customers are not necessarily distributed in proportion to the installed generating capacity or total energy generated. A few customers may use appreciable energy and vice versa.

In passing, we should be aware that the electric utility industry has voluminous data on its customers, sorted by region, type utility, amount of power capacity installed, energy used per year, and many other detailed facts. Such data are beyond our need here for perspectives, however; but do be aware that the utilities know far more about the customers they are serving than is commonly thought.

A perspective on the sales of electric energy in various categories is revealed in Table 1-8. The U. S. industrial sector is the largest user of electric energy with residential use closely following. The latter should come as no surprise, as we appear to increasingly acquire more and more devices and systems in our homes which require electric energy.

TABLE 1-6 World Production of Electricity - 1973

Country	Production Kilowatthours in Millions	Percent of World Production
United States*	1,958,745	32.4
Russia	914,653	15.1
Japan	470,082	7.8
Germany (Western)	298,995	4.9
United Kingdom	282,128	4.7
Canada	262,272	4.3
France	174,080	2.9
Italy	145,518	2.4
Poland	84,302	1.4
Sweden	78,080	1.3
Germany (Eastern)	76,908	1.3
Spain	75,765	1.2
Estimated other countries	1,232,138	20.3
	6,053,666	100.0

* Includes Alaska and Hawaii

SOURCES: Statistical Abstract of the United States 1975. United Nations Statistical yearbook 1974. EEI Statistical yearbook.

A sampling of useful key facts about electric energy-using customers is in Fig. 1-27. Here the electric power capacity per capita is plotted as a function of time. The electric energy per capita is also plotted in the same figure. We note that in 1973, for example, the average person had about 2.1 KW generating capacity installed[1] and he used about 9,030 KWH/yr. of energy in electric form.

[1] The average power he actually used would be somewhat less than this.

TABLE 1-7 Percentage Comparison of Production, Sales, Customers - 1975*

	Installed Generating Capacity	Total KW hr Generated	KW hr Sale to Ultimate Customers	Ultimate Customers
Investor-Owned Companies	78.7%	77.6%	77.7%	77.8%
Cooperatives	1.5	1.8	6.0	8.8
Governmental Agencies				
Federal (TVA, Bonneville, etc.)	9.9	11.5	3.9	**
Municipal governments	5.7	4.3	10.2	11.7
Power districts, state projects	4.2	4.8	2.2	1.7
	100.0	100.0	100.0	100.0

* Includes Alaska and Hawaii

** Less than 1/10 of 1 percent

SOURCES: Federal Power Commission and Rural Electrification Administration reports. Edison-Electric Institute.

TABLE 1-8 Annual Sales of Electricity by Consumer Groupings*

Classification	Kilowatthours in Millions	Percent
Industrial	691,578	38.8
Residential	595,025	33.3
Commercial	428,304	24.0
Others	69,625	3.9
Total	1,784,532	100.0

* Includes Alaska and Hawaii. Data for 12 months Ended June 30, 1976.

SOURCE: EEI Statistical Yearbook

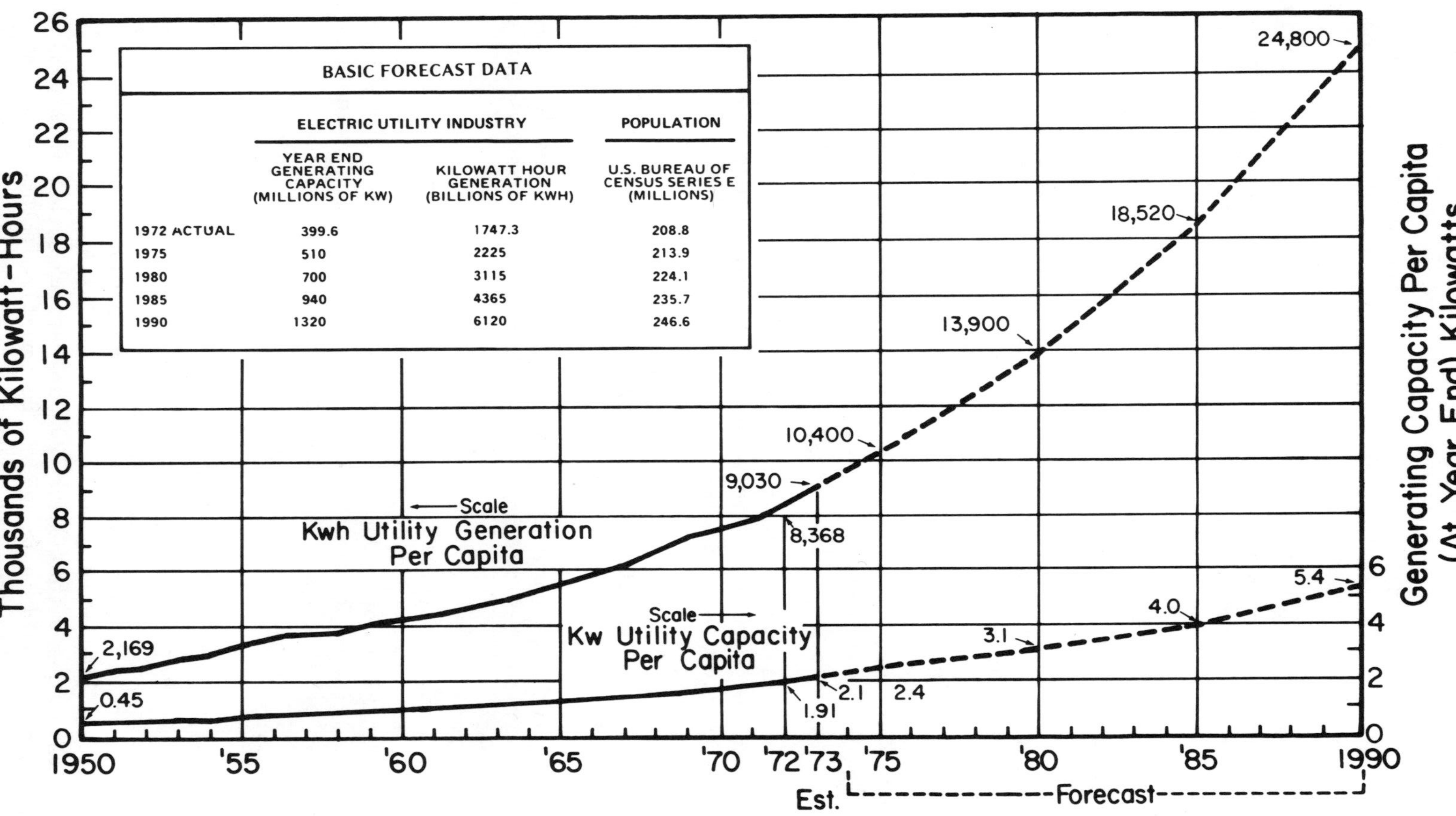

FIG. 1-27 Per capita electric generation and capacity for the total United States electric utility industry. Source: Based on U. S. Department of Commerce, Bureau of the Census, E. E. I. data, capacity and generation estimates by Ebasco. From [60].

We note the positive slope on both these curves; they remind us again of the oasis analogy of Sec. 1-1. We observe, for example, that the electric energy 'drink' the average person took in 1973 was over 4 times that taken in only 1950. Whether we truly need this much energy and power/capita in electric form is currently being debated; it undoubtedly will continue. There are numerous moral and ethical questions involved, as is the entire subject of conservation at the personal level.

More recent curves have a slight perturbation because of the 1973-74 OPEC oil embargo. But the effect is of a minor nature. Americans went right on using electric energy without changing the positive slopes of the curves significantly over several years.

The Costs of Electric Energy to the customer are of much importance, both to the present electric energy industry and to future competing alternate energy sources.

Our search for perspectives through the previous sub-sections has now lead us to the endpoint of the electric energy system, the customer. The customer is the one the industry is there to serve. It tries hard to meet his electric power and energy demands. Since the customer ultimately pays for both the electric energy used and the capital invested in the industry, we should here summarize what the present costs are in terms of a typical American home and in terms of the cost per KWH of electric energy; the latter fiscally 'fuels' the entire electrical industry.

We first focus on a typical home. Naturally the amount of electric power required and the energy used will depend on the number, kinds, and frequency of use of electrical devices and systems--which may depend on the local weather. Some electric apparatus, e.g. a freezer may require relatively little power but use it a high fraction of the time with the result that a large amount of energy is used. Others, e.g. an electric oven, may require large power but it is used infrequently so that the total energy used may be modest.

Some quantitative feel for the energy used per year by various household appliances and electrical devices is in Table 1-9. It also shows the estimated yearly costs for each.

The table is highly instructive to study, especially for newcomers to the energy field. One can quickly deduce from it relative priorities for problem areas worthy of attacking for reducing energy use or substituting via an alternate energy source. Then one can quickly estimate whether the energy saved will justify the expense of an alternative energy source.

Example 1-7

What household function should we first focus on to achieve solarly to save the customer the most money per year?

The answer, from Table 1-9 is obviously water heating. It uses about 5800 KWH/yr. and costs about $174/yr. for electric energy. It is easy to see that 1000-1200 dollars expended in a soalr water heater would in 6-7 years be amortized; thereafter no yearly energy cost would be incurred, for the sun's energy is free once the initial apparatus has been paid for.

TABLE 1-9 ANNUAL OPERATING COSTS FOR TYPICAL HOME APPLIANCE AND ELECTRICAL DEVICES. From Jurgen [61]. Copyright © 1974 by The Institute of Electrical and Electronics Engineers, Inc. Reprinted, by permission, from IEEE SPECTRUM, Vol. 11, No. 6, June 1974, pp62-63.

Appliance	Frequency/Type of Use	Energy Consumption per year (kWh)	Yearly Cost*
Water heating			
Storage heater		5800	$174
Quick-recovery heater		4811	$144.33
Water pump		231	$ 6.93
Space heating/cooling			
15 000-Btu air conditioner	500 hours per year	1150	$ 34.50
Burner and pump+		810	$ 24.30
Burner and fan+		810	$ 24.30
Burner only+		410	$ 12.30
5000-Btu air conditioner	500 hours per year	410	$ 12.30
Dehumidifer		377	$ 11.31
Portable heater		176	$ 5.28
Attic window fan		170	$ 5.10
Humidifer		163	$ 4.89
Kitchen appliances			
14-foot^3 no-frost refrigerator-freezer		1548	$ 46.44
21-foot^3 manual defrost upright freezer		1392	$ 41.76
30-inch electric range, self-cleaning oven		1068	$ 32.04
Dishwasher	46 cycles	330	$ 9.90
Microwave oven		190	$ 5.70
Electric skillet	10 hours per month	182	$ 5.16
Coffeemaker	12 cups, 8 times per week, kept warm for one hour	92	$ 2.76
Toaster oven	Toast 4 slices of bread per day, bake 2 hours per week, top brown twice per week	92	$ 2.76
Hot plate		90	$ 2.70
Deep fryer		83	$ 2.49
Electric kettle	Boil one quart of water, 10 times per week	58	$ 1.74
Electric griddle	3 times per week, 17 minutes per use	56	$ 1.68
Two-slice toaster	8 slices of toast per day	40	$ 1.20
Food mixer/blender		15	45¢
Food waste disposer	108 cycles per month	4.8	14.4¢
Trash compactor	108 cycles per month	2.4	7.2¢
Portable mixer	208 times per year, 3 minutes per use	1	3¢
Electric carving knife	150 times per year, 3 minutes per use	0.7	2¢
Can opener/ice crusher	Open 10 cans per week, crush 14 ice cubes twice per week	0.3	1¢

Appliance	Frequency/Type of Use	Energy Consumption per year (kWh)	Yearly Cost*
Laundry appliances			
Electric clothes dryer (14-pound capacity)	7-pound load, 33 cycles per month	1212	$ 36.36
Self-cleaning steam/spray iron	2 hours per week on normal heat setting	88	$ 2.64
Automatic washer (14-pound capacity)	7-pound load, 33 cycles per month	76	$ 2.28
Entertainment products			
19-inch color television, hybrid chassis	5.7 hours per day	444	$ 13.32
25-inch color console, solid-state, instant color	6.6 hours per day	360	$ 10.80
(with instant color off)	6.6 hours per day	334	$ 10.02
19-inch color television, solid-state instant color	5.7 hours per day	258	$ 7.74
12-inch monochrome television, hybrid chassis		187	$ 5.61
10-inch color television, hybrid chassis	3 hours per day	169	$ 5.07
19-inch monochrome television, solid-state	5.1 hours per day	129	$ 3.87
Automatic stereo phonograph	2 hours per day	68	$ 2.04
Stereo component system with FM/AM and FM stereo, tape player, four channel	3 hours per day	54	$ 1.62
FM/AM digital clock radio	24 hours per day for clock, 2 hours per day for radio	27	81¢
AM clock radio	24 hours per day for clock, 2 hours per day for radio	18	54¢
Bathroom appliances			
Salon-style hair dryer	2 times per week	44	$ 1.32
Hand hair dryer	4 times per week, 10 minutes per use	24	72¢
Electric toothbrush (continuously charging)	2 times per day	12	36¢
Lighted makeup mirror	10 times per week, 15 minutes per use	2.3	7¢
Heated shaving-cream dispenser	1 time per day	0.3	1¢
Miscellaneous electrical devices			
Swimming pool pump, 1 hp		2160	$ 64.80
1/2 hp		1080	$ 32.40
1/4 hp		540	$ 16.20
Vacuum cleaner		46	$ 1.38
Electric clock		17	51¢
Floor polisher		15	45¢
Sewing machine		11	33¢
Heating pad		10	30¢

*Based on 3¢ per kilowatthour. Average rate for 1973, according to Edison Electric Institute was 2.3¢ per kilowatthour.

+Electricity costs for operating oil and gas burners. Does not include cost of fossil fuels.

Conversely one can quickly deduce which areas are not worthy of our focus, e.g. finding an alternate electric toothbrush.

Thus such a table--based on facts and assumptions[1]--can be a useful tool to us in seeking perspectives on the customer and his needs in a typical American home. Developers of solar-electrics may find Table 1-9 helpful for reference purposes when developing practical solar electric apparatus for small-scale home use, including estimating initially energy storage required and, indirectly power required.

A fundamental parameter in electric energy systems of interest to both customer and electric systems people is the delivered cost /KWH of electric energy. This is sometimes referred to as the 'busbar cost', though the latter term may ignore transmission and distribution costs.

A perspective on electric energy cost is presented in Fig. 1-28. Here we see the cost vs time over several decades. The time of 'cheap energy' was roughly up to 1973. Note that even in a time of cheap energy electricity costs in 1930 were several times the 1977 3.62 cents per KWH. There has been a gradual decline in the cost principally because engineers were busy lowering the heat rate[2] of their generating plants. That they were successful in picking away at this difficult technical problem is attested to by the gradual decrease in electricity cost over the 1930-1970 years. This is a good example of making gradual solid improvements over a long time period as opposed to trying to make spectacular progress in a great hurry and fail. The power engineers were succeeding; then came 1973.

When the oil embargo was placed on the U. S . in 1973-74, electricity prices started to rise, as the curve clearly shows. The reason was twofold: (1) a majority of U. S. electric generation was oil fired--a concerted national environmental clean-up away from coal and toward oil fired plants had been made, and (2) with radical increases in oil prices the utilities could not possibly absorb the rising fuel prices; they therefore were financially forced to pass the fuel increases on to customers as higher electric rates shown.[3] Many people believe oil prices will continue to increase in the future. When and if this occurs, electric rates will almost certainly continue climbing; our era of 'cheap energy' is no longer with us!

---

[1] Note that in Table 1-9 Jurgen was careful to state his assumptions for no. of hours or no. of cycles. Naturally if these assumptions are changed, the energy used and hence the cost will change. You begin to see why an electrical systems engineer would be concerned with the duty cycle for any appliance and for them in aggregation for the house.

[2] Heat rate = Btu of fuel/KWHe generated. It is a measure of the efficiency of a generating plant, although it is more like the inverse of efficiency. Current heat rates are about 9,000 BTU/KWH for a well designed plant. It is desirable that the heat rate be low.

[3] U. S. utilities are regulated by Public Service Commissions and Federal Agencies which by law set electric rates and limit utility companies profits to a modest rate. The utility therefore cannot absorb large fuel cost increases in their profits.

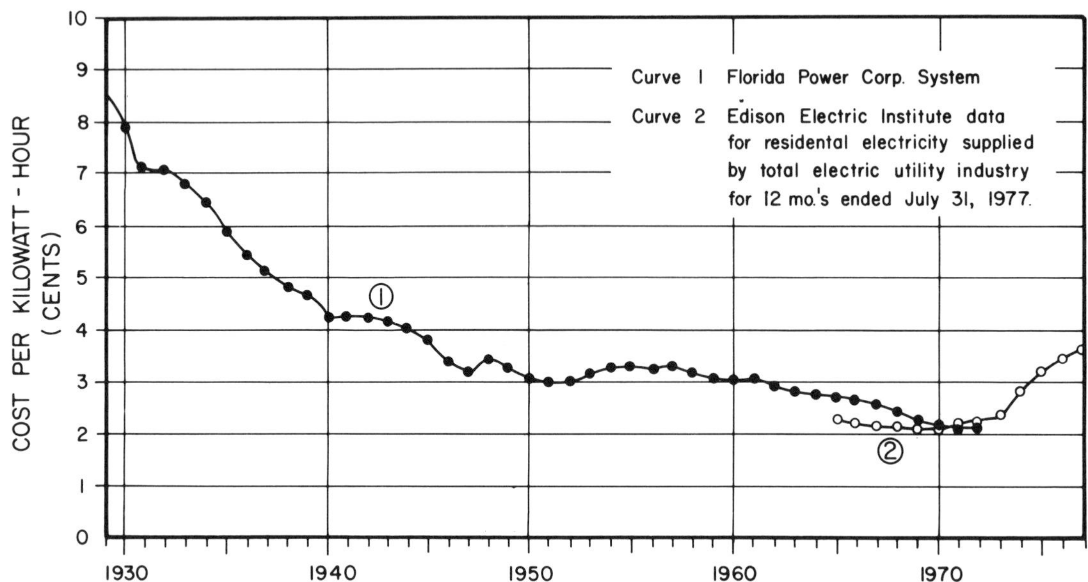

FIG. 1-28 Yearly average revenue per KWH of electricity sold for a selected utility and, more recently, for the utility industry. Curves like this for a specific area are what solar-electrics must ultimately compete with as far as the customer is concerned.

Fig. 1-28 is potentially helpful to developers of solar-electrics, for the current price/KWH of electric energy is the competition for this new technology. The importance of minimizing the cost/KWH of energy delivered to the customer cannot be underestimated for either existing electric generation systems or future solar-electrics. Cost has a way of being a tough competitor, especially when one competes with an industry which has already undergone almost 100 years of refinements and improvements as has the electric industry.

We should not leave the issue of costs of electric energy without recording the fact that in the mid 1970's there was substantial public discussion over the issue of declining rates with increasing electric energy use by a customer (the present system) vs increasing rates for more energy used. The latter is intended to motivate energy conservation by customers. The fate of this issue is unclear presently.

Fuel For Electric Generation and its tightening nature has been generally discussed in Sec. 1-2. It is abundantly evident to observers outside the utility industry that the major future problem with obtaining an adequate supply of electric power and energy for the nation is with the front end of the system, namely fuel; quantity, type, quality, availability, price, and environmental side effects are just a few of the many complex variables with which a utility has to cope.

Unfortunately, the perspective of the central importance of the fuel is sometimes lost sight of by some utilities which appear more concerned with the details of 'expanding the system'[1]. Focus on the fuel will undoubtedly grow in these transition years we are now in.

We previously indicated the rising fuel costs were the principal reason for electric rates increasing in recent years, as shown in Fig. 1-28. The dramatic fuel cost increases for the utilities is revealed in Fig. 1-29. The relative constancy of fuel costs prior to the early 1970's is evident as is the sharp rise about the time of the 1973 oil embargo. The forecast is for all three major fossil fuels to continue increasing in price. Note that the curves are not saying the prices of fuels will be anything like they were in the era of cheap energy prior to about 1970.

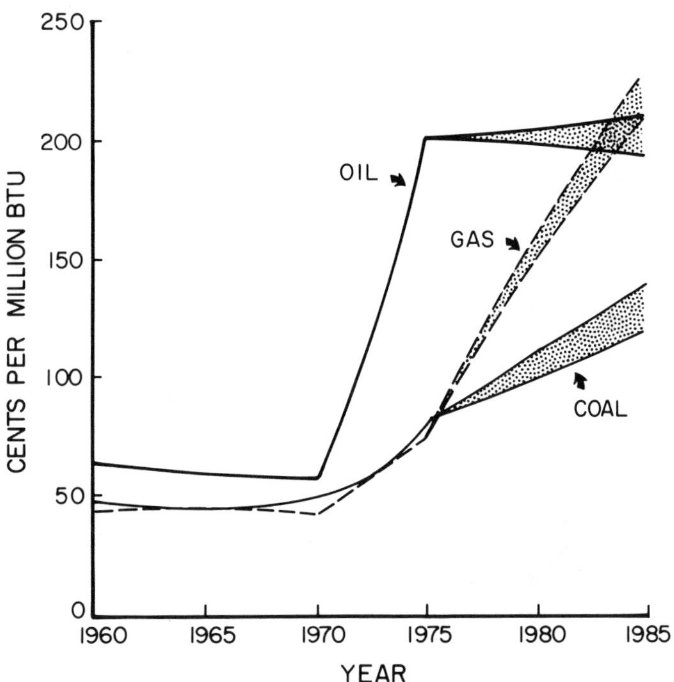

FIG. 1-29 Cost of fossil fuels to steam electric plants in constant 1975 dollars. From [59].

If the forecasts are reasonably accurate, it is obvious electric rates of Fig. 1-28 will continue rising. Thus increases in fuel prices increase electric rates which ripple through the nation's economy and business system and adversely affects the U. S. trade balance, but the basic forcing function is fuel price.

[1]The counter argument can be made and defended, however, that the utilities are in the middle trying to establish a balance between fuel supply and service to their customers.

It is interesting to speculate what effects the future arrival of a large scale solar-electric technology might have on the costs of fossil fuels remaining to be used for non-electric generation purposes. The current problem of fuel costs would clearly be non-existent were we into the era of solar-electrics implemented on a wide scale.

The amounts of fossil fuels used by all electric utilities is summarized in Fig. 1-30. With the exception of gas, the trends are upward; gas is increasingly in short supply.

Thus it is evident from Fig. 1-30 that we see here on the front end of the nation's electric energy industry the effects of customers demanding more and more electric energy; again we are reminded of the oasis analogy of Sec. 1-1. The ever rising need for more fossil fuels for electric utilities cannot continue indefinitely with a finite energy capital.

Having now completed our brief overview of the gigantic conventional electric utility industry in this section, we now turn to joining the solar Sec. 1-3 with it and discuss perspectives on the new unconventional solar-electric field.

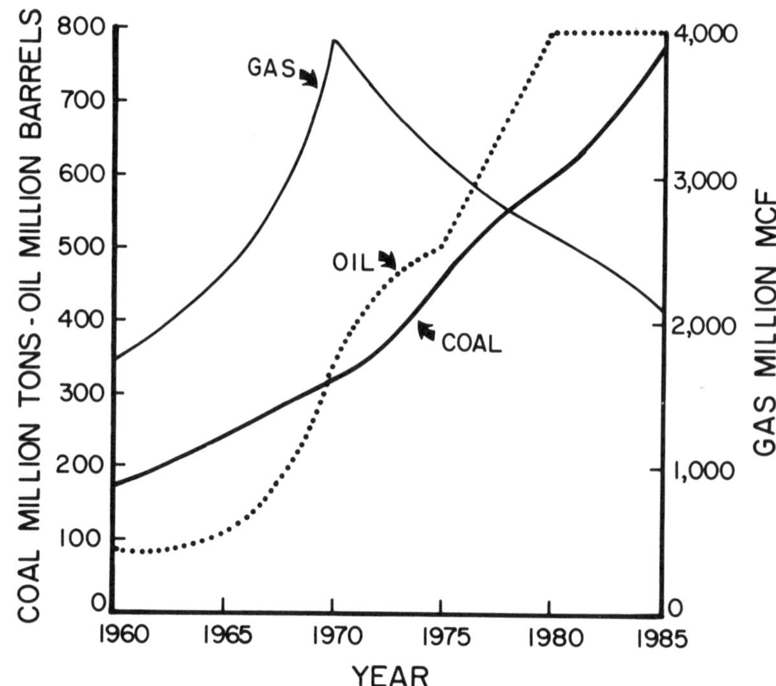

FIG. 1-30 Fossil fuel consumption by electric utilities. Fuel demand projections are based on the general assumption that electric energy demand will grow at an average annual growth rate of 8% per year[1] during the period to 1985. Source [59].

[1]This percentage estimate is subjected to continual readjustment by the utility industry as the total energy situation changes.

1-5 Solar-Electrics -- A Part of Larger Solar And Electric Energy Technologies

Place And Organization of Solar-Electrics. We earlier said that solar energy utilization is a part of the larger field of alternate energy sources. Solar-electrics is the sub-set of advanced technologies for obtaining electricity solarly.

Thus the newly emerging field of solar-electrics has roots in the broader field of solar technology as well as in conventional electric generation technology as shown in Fig. 1-31.

The broad field of solar technology includes solar water heating, solar residential heating, solar air conditioning, solar distillation and desalination, solar cooking, solar furnaces, solar ponds, solar biological area--including fuels, solar pumping, solar engines, solar collectors of various kinds, solar energy storage, solar-electrics, and many other highly useful and beneficial areas. Solar technology is already a large interdisciplinary field; it is rapidly growing and is expected to continue growing.

Conventional electric generation technology, as reviewed in Sec. 1-4, encompasses the totality of means now widely used to supply bulk electricity. It principally uses fossil and nuclear fuels. As an industry, it is the largest in the world. Its principal reason for being is to supply the customer with the electric energy and power he needs when he needs it at a price he is willing to pay. To do this it is broadly concerned with the details of how to convert fossil fuels into electricity. It is an available, proven, cost effective technology.

Solar-electrics is concerned with converting solar radiation into useable electricity. It may achieve this on either a small scale, i.e. dispersed solar-electric systems, or on a concentrated basis, i. e. centralized systems. The field involves know-how from both solar technology and conventional electric technology.

Solar-electrics, as a field of technical endeavor is relatively new compared to the other two and is still largely in the R & D phase. Because of the large body of solar-electric knowledge already existing and the many specialities thereof we are justified in calling it a separate and distinct field. It is also rapidly growing as of the mid 1970's; it appears likely to grow substantially more and to become increasingly important to the future electric energy supply for the United States and the world.

How is the newly-emerging solar-electric field organized? The answer lies in Fig. 1-32. We immediately see that each specialty is concerned with converting solar radiation into electricity. The first six methods might be thought of in a 'devices' context; aggregations of these, perhaps in various mixes and combinations, are encompassed in the last 'systems' block.

The prominence of energy storage and generator technologies is evident from Fig. 1-32. The electrolysis-storage-fuel cell loop is an interesting part of solar-electrics; clearly this loop could be applied to any of the other areas were it expedient to do so. It offers the possibility of highly efficient energy storage during times when there is little or no insolation.

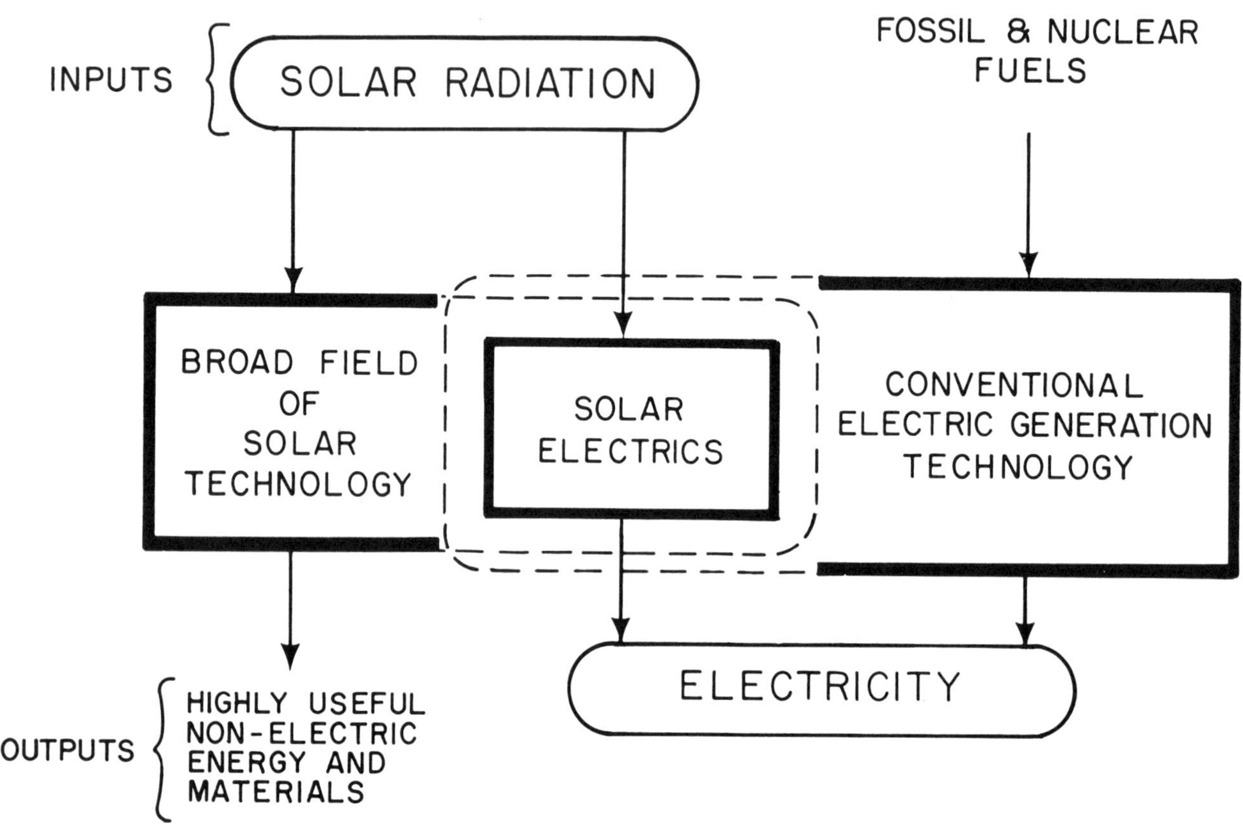

FIG. 1-31 Perspective on the technical place of solar-electrics. It is a part of two larger technologies. Both have legitimate interests in solar-electrics.

We shall explore in more detail several solar-electric areas of Fig. 1-32 in later chapters. Don't lose sight of the perspectives afforded by Fig. 1-32 where clearly each sub-specialty is competing with all the others for a viable niche in the future production of electric energy. Some may lose out.

Thus the mid 1970's perspective is that there are seven major solar-electric options from which to choose for a given application. This multiplicity of options is another unpublicized advantage to solar-electric utilization; some freedom potentially exists in making trade-offs between options to arrive at the best one for a given application. One must be familiar with all the specializations, though, and the local site characteristics before making a final choice.

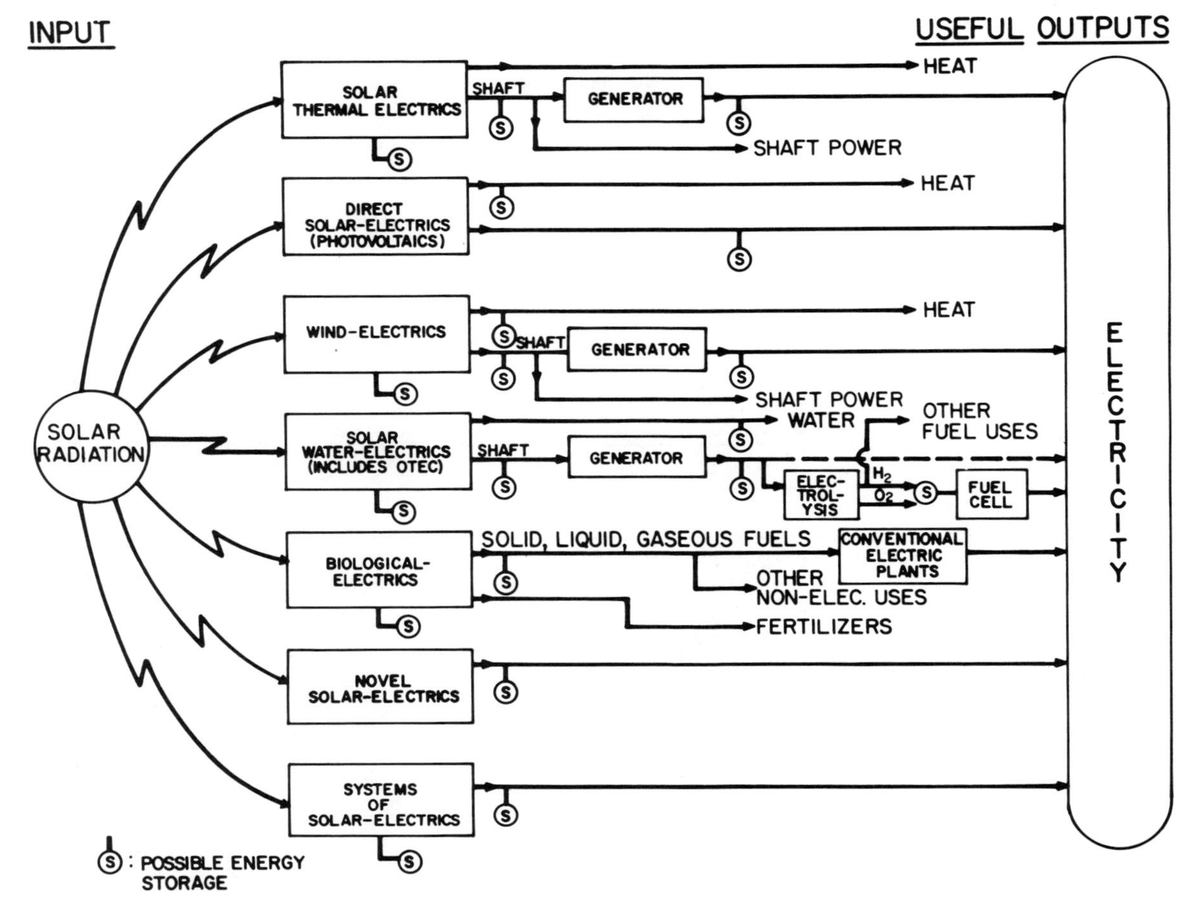

FIG. 1-32 One way of integrating solar-electrics. Other structures can also be created. Details of each will be revealed in later chapters. Note that while our chief interest here is in electricity, other useful nonelectric outputs also occur. Also observe the important roles of energy storage and generator technologies.

Land Area For Solar-Electrics is an issue of importance, particularly as land prices have generally tended to rise in recent decades. We preliminarily dealt with the land issue in Sec. 1-3. We now present an easier more direct method for estimating the land area required to produce the nation's future electric energy via solar-electrics. The land area is only for that of the solar collectors.

Fig. 1-33 is a nomograph allowing us to find the total collector land area required, depending on the conversion efficiency.

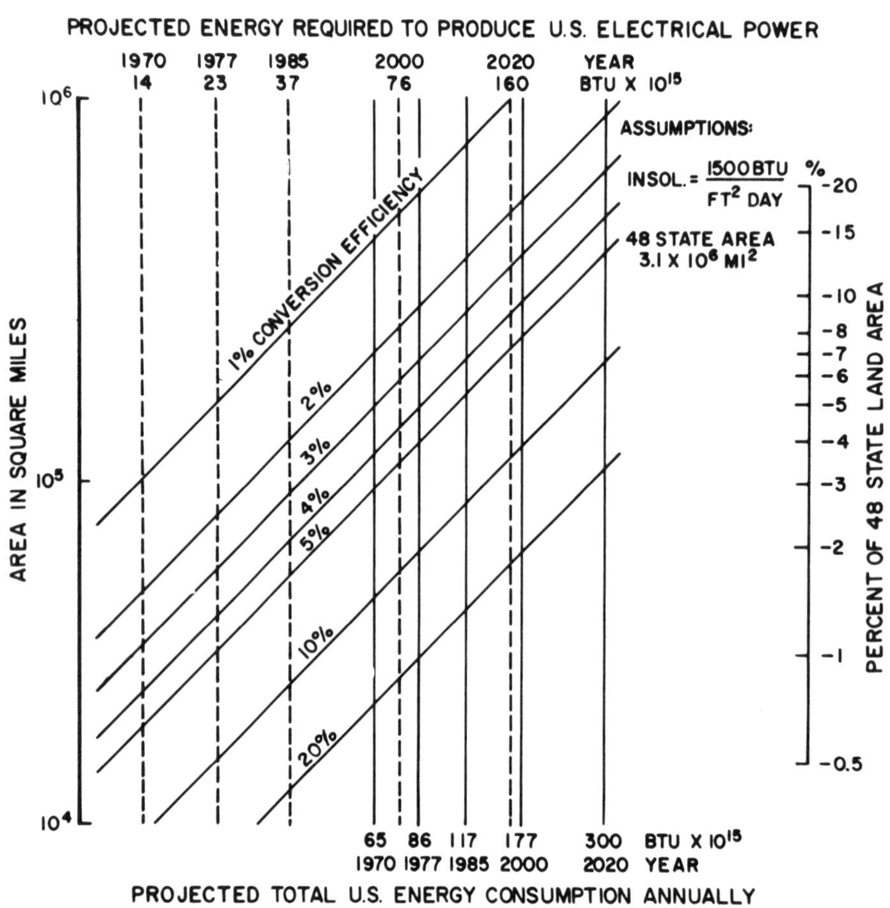

FIG. 1-33 Land area to meet (1) total U. S. energy needs or (2) U. S. total electric energy needs. Source [43, p24].

Example 1-8

Assuming 10 percent solar-electric conversion efficiency, what percent of the U. S. land area would be required to meet the total 1977 energy needs of the country?

Using the scale at the bottom of Fig. 1-33, we quickly see the answer is about 2 percent.

We notice in passing for the 1977 abscissa that if the conversion efficiency could be improved to 20 percent the land area would only be about 1 percent. Thus improvements in overall solar electric conversion efficiency turn up in the form of less land area required and may result in less total system cost.

Example 1-9

Same as Example 1-8 except for the total 1977 electrical energy needs of the country.

Using the top scale of Fig. 1-33 and the 10% line, we find about 0.5 percent of the U. S. land area would be required.

These examples reinforce the notions of Sec. 1-3 showing that there is in fact an adequate supply of solar energy to either run the entire country or the electric needs of the nation. Technological improvements in the conversion efficiency are obviously desirable, tending to reduce the required collector area and hence the precious land area.

Solar-Electric Cost is an important parameter in determining the future commercial debut of solar-electrics. There are many different ways of comparing electricity costs from these new technologies. From a systems viewpoint it is desirable to compare alternate advanced technologies.

One viewpoint, representative of utility thinking, on costs for some solar-electric options has been provided by Starr [62] in Fig. 1-34. Starr plotted capital cost vs capital charges for the various technologies (dotted areas) using capacity factor[1] as a parameter. Thus for a fixed capital cost, increasing the capacity factor markedly decreases the capital charges, a desirable direction to move. The capacity factor is of considerable importance to the electric utilities, as it influences the electric energy capital charge. The importance of the capacity factor appears not widely recognized throughout the present-art solar-electric R&D literature where many researchers are implicitly assuming station capacity factors of 1.0.

Starr's 'solar' refers to solar-thermal central stations, the so-called 'power towers'. Photovoltaics were not considered because their capital cost presently far exceeds the options shown.

---

[1]Capacity factor = annual avg. elec. power output/gross station capacity.
~ 80 percent or higher for a present-art base load coal plant.

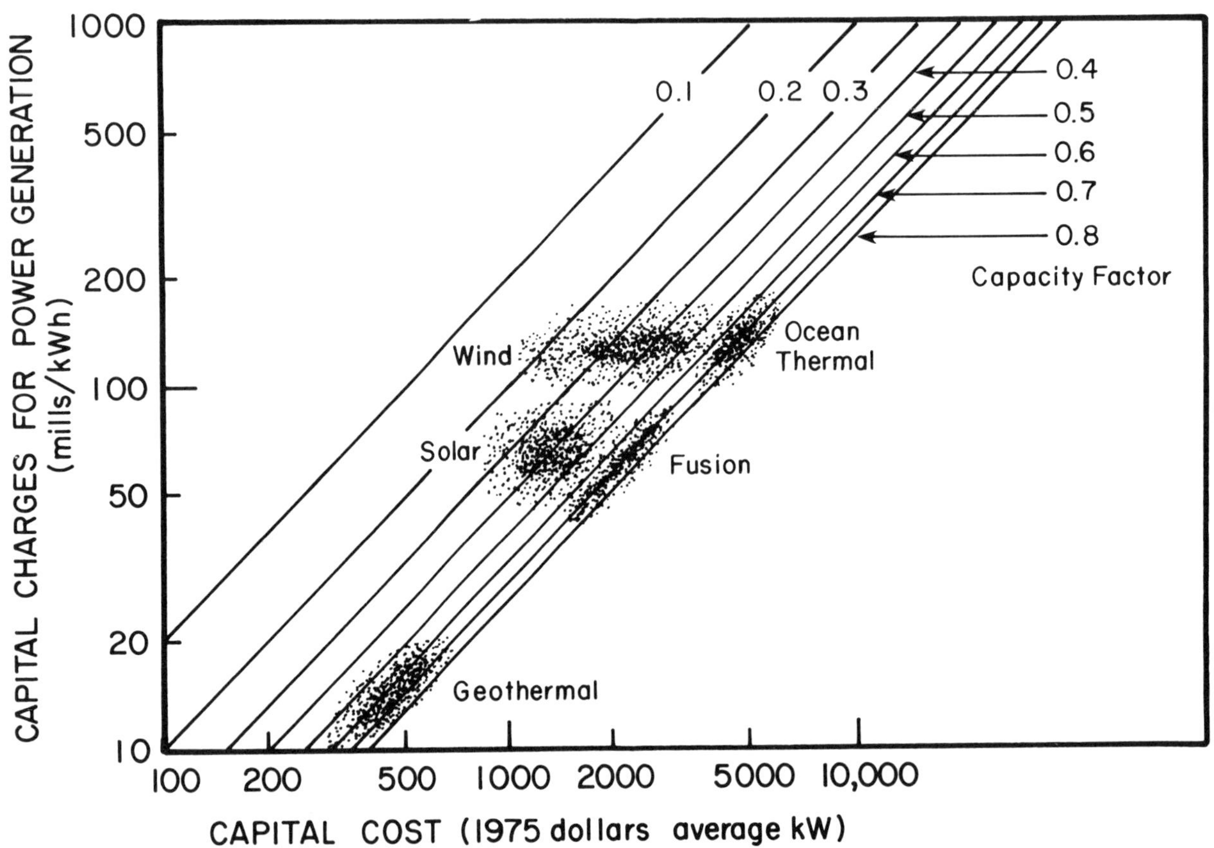

FIG. 1-34 Capital costs and capital charges for new electric power generation options. From Starr [62].

The capital costs for a typical fossil fueled conventional power plant are roughly in the $300/KW range, i.e. just below that labeled geothermal in Fig. 1-34.

The general need for substantial cost reductions in all solar-electrics is evident from Fig. 1-34 before they can compete effectively against existing electric energy systems. This result is not surprising, for solar-electrics is so new and the field has not yet had time for substantial cost reductions to be made as in the conventional electric energy system.

An additional feel for conventional and one solar-electric cost can be had from Eldridge's [63] Table 1-10. We immediately note that in computing the cost of electricity there are many other cost factors involved than the initial capital, the fuel cost being one major one for the fossil fuel plants. The reader is referred to Eldridge's original report for the assumptions implied for the Table. One of these is an 80 percent capacity factor for each option. Another assumes that solar cells at $0.50/ft^2 could be made available[1]; if they existed, then clearly, according to Eldridge, the solar cell generated electric energy cost would be below that of all the conventional fossil fuel plants. It remains to be seen whether solar cells can be made this cheaply; nevertheless his figures reveal the potentialities in solar cells as one solar-electric option.

Those interested in further exploring the costs of conventional fossil and nuclear electric generation and the many detailed factors involved will find an able treatment in Hill's book [64]. The detailed treatment of costs is beyond our scope here; but researchers developing solar-electrics may need to probe the cost area in some depth for conventional plants in order to compare alternative solar-electric devices, systems, and final busbar electric energy costs for each.

Cost of electrical energy via solar-electrics is an important factor in determining the future growth of any single solar-electric specialty area. The numbers continually change in each area as new assumptions are made and as technical progress is made. Do note, however, the importance of computing the cost over the life cycle, 20 years for Table 1-10, for the expected plant. The initial capital cost does not tell the whole story; such comparisons alone will probably always make solar electrics appear poorly, but let us not forget the 'solar fuel' over the life of the solar plant is free. The latter factor alone could make a huge difference in determining the competitive cost effectiveness of any solar-electric method.

The Introduction And Integration of Solar-Electrics into the nation's electric energy system is of great concern for the nation and for this growing new technology. There are, as expected, three viewpoints:

- The Optimists who, erroneously, think that solar electrics can be introduced as an overnight 'step function' into the nation's energy scene. According to these proponents, many of whom are in the sciences instead of engineeering, our greatest problem is to decide to commit the nation to a future path of solar-electrics. They imply that the path from laboratory to full-scale utilization is simple and mostly one of 'doing'. They fail, largely, to understand the technology transfer process and the time it takes; therefore their timing estimates are always far too short.
- The Pessimists who see solar electrics as so laden with problems that we can't possibly ever develop the technology to a practical reality--or if we can it will take so long to do it either we might as well not try or the fossil fuels will have run out by that time. These proponents place the commercial arrival of solar-electric as well into the 21st century, if at all.
- The Realists who factor the truth from both the above views and who set about as quickly as possible taking steps toward making solar-

[1] Solar cell manufacturers, naturally, disagree over what this value could realistically be. This is a natural result of an incompletely developed present market of photovoltaics.

TABLE 1-10 Comparison of 20 year cost of various types of electrical energy systems. From Eldridge [63]

TYPE OF SYSTEM	INITIAL INVESTMENT	FUEL COST	OTHER O&M COSTS	COST OF CAPITAL	TAXES & INSURANCE	TOTAL COST	COST OF ELECTRICITY
Hydroelectric	\$257-\$806/KW	--	\$34-\$206/KW	\$228-\$710/KW	\$288-\$903/KW	\$807-\$2631/KW	5.8-18.8 mills/KWH
Gas Fired	\$257/KW	\$483/KW	\$55/KW	\$228/KW	\$288/KW	\$1311/KW	9.4 mills/KWH
Oil Fired With Environmental Protection	\$300/KW	\$610/KW	\$98/KW	\$266/KW	\$336/KW	\$1610/KW	11.5 mills/KWH
Coal Fired With Environmental Protection	\$297/KW	\$664/KW	\$80/KW	\$264/KW	\$333/KW	\$1638/KW	11.7 mills/KWH
Nuclear Fueled With Environmental Protection	\$321/KW	\$516/KW	\$261/KW	\$285/KW	\$360/KW	\$1743/KW	12.5 mills/KWH
Solar Cells \$0.50/ft^2	\$220/KW	--	\$33/KW	\$195/KW	\$246/KW	\$649/KW	5.0 mills/KWH

electrics a reality. These realize that the progress we make is largely a matter of will and dollars committed to the task; fewer dollars and competent researchers, developers, and engineers result in taking longer than if more dollars and competent people were applied.

Starr [62], representing the utility industry through EPRI, has considered the general problem of integrating the growing solar-electric field into the electric utility systems of the nation. His considerations are principally limited to the solar-thermal power tower central station concept. His view is succinctly embodied in Fig. 1-35. He sees the transition as consisting of the discrete steps shown with some overlap between steps and all having higher and higher peak costs as the systems being created become larger and more complex. He argues the early R&D phases must be federally supported, the engineering and demonstration phases should receive support from both government and utilities and the final commercial phases should be the responsibility of manufacturers and utilities.

According to Starr's schedule, it would be well past 2000 A.D. before we could expect to see utility integration of solar-electrics occur on a growing basis. He appears to have given little thought to what could be done to substantially speed up this necessary but ponderous process; Starr's schedule would place the nation dangerously close to the exhaustion of the world's oil supplies, as we have argued in Fig. 1-6, and without an in situ solar-electric system delivering appreciable fraction portions of the nation's electrical energy.

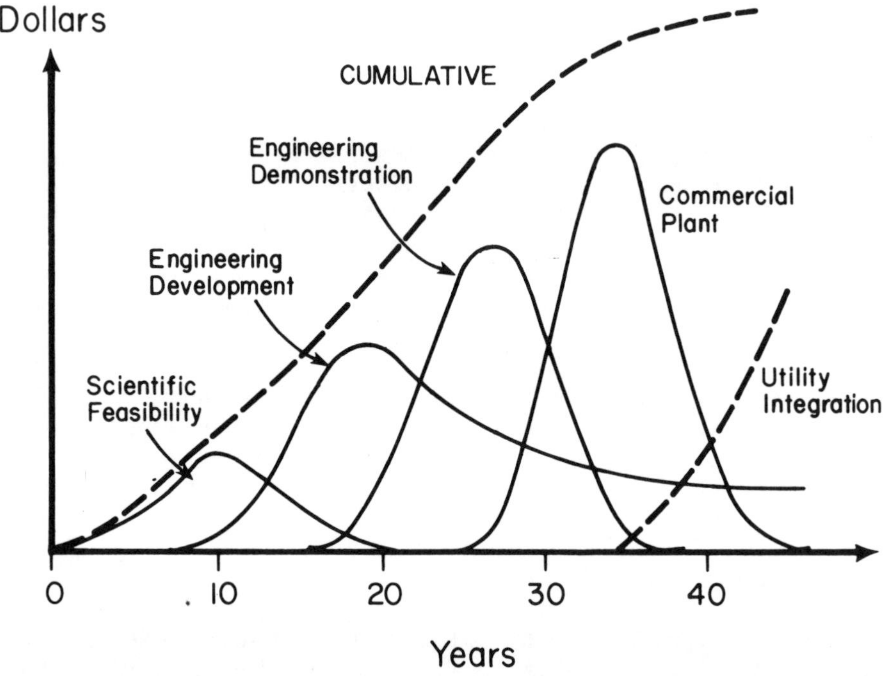

FIG. 1-35 R&D phases toward eventual utility integration of solar-thermal power towers, according to Starr [62].

It is obvious major action has to be initiated to speed up the total process; we just don't have 40 years of "business as usual" to inject solar-electrics into the nation's commercial energy enterprises! The stakes are too high. Undoubtedly this important planning issue will be argued more extensively in the future by the nation's energy leaders.

We conclude this section by again saying that solar-electrics is rooted in both solar and electric technologies; it will take bold imaginative engineers now and in the future to integrate both fields and to do it expeditiously before the nation is caught with most of its oil gone. The transition period will undoubtedly be a dynamic one as we create the gentle energy systems of the future, solar-electrics.

1-6 Dispersed Vs Centralized Utilization

Having acquired some feel for the overall organization of solar-electrics, in this section we come to grips with the critical issue of whether it is best to create nationally n small-scale geographically dispersed solar-electric generators, where n is large--in the millions, or whether we should create m large-scale centralized solar-electric systems, where m is small--a few hundreds.

The dispersed vs centralized issue is unique to solar-electrics, primarily because of the ease with which electric energy can be transmitted over great distances. The issue does not arise in strictly solar-thermal utilization because of the technical and practical difficulties in transmitting heat energy beyond a few kilometers. The difficulty makes large-scale geographically dispersed centralized solar-thermal stations whose sole output is heat an academic question. It is therefore immediately obvious to those in solar heating that solar systems should be small-scale and dispersed, i.e. a solar thermal collector at each need site. It is not so obvious for solar-electrics.

The dispersed vs centralized issue is of great future systems importance. While the roots of the issue lie in philosophic and moral areas, the future decision for solar-electrics to go all dispersed or all centralized--or more probably a mixture--is of critical importance in integrating solar-electrics into the nation's energy system. The issue may well determine the future technical and economic growth of solar-electrics. The issue is currently being examined, for example, in the government's Domestic Policy Review (DPR) [65].

A more basic issue common to both decentralized and centralized solar-electrics, as Schumacher [66] has pointed out, is the question of an appropriate scale. "For every activity there is a certain appropriate scale--" [66, p62]. Finding the right scale for solar-electrics is the crux of the problem.

As the arguments on this issue are scattered in the solar literature, we attempt to present in this section the first integration of principal arguments for decentralization of solar-electrics and those for centralization. We attempt to keep the arguments applicable to solar-electrics; but it should be realized that in the literature those few who even argue the issue seldom are so crisp. Rather the issue tends to get argued in a generalized way. Considerable filtering and judgment is necessary to determine which issues, if any, may apply only to solar-electrics. Even then the arguments may or may not be applicable to every type of solar-electric generation means.

The collection of the merits of small vs large solar-electrics is all the more important to bring into focus while the industry is young, small and growing; perhaps by thoroughly airing the case at this early stage serious major errors--with all their later social consequences--can be avoided. Thus we present the following collection of arguments not so much as a final disposition of the issue as to stimulate its discussion, for, after all, the non-resolution of this important issue is a part of the state-of-the-art of solar electrics in the mid seventies.

You are also forewarned that because centralized conventional electric systems are so well-known whereas dispersed electric systems are practically nonexistent, we therefore strongly make the case here for the less well-known dispersed systems which appear so appropriate for solar-electrics. We necessarily must bridge the technical-social interface; we shall find, as Hayes [1, p73] has pointed out, that "the advantages of decentralization are more social than technical."

Case For Decentralized Solar-Electrics Decentralization potentially applies to direct-solar-electrics (photovoltaics), wind-electrics, solar-water-electrics, biological electrics, and systems of solar-electrics. It does not apply to the Solar Satellite Power System (SSPS), as it is difficult to conceive of such systems being dispersed to the individual small-scale customer level.

Starting with the need, in a generalized sense Schumacher [66, p31] sees that "we need methods and equipment which are:

- Cheap enough so that they are accessible to virtually everyone.
- Suitable for small-scale application; and
- Compatible with man's need for creativity."

He argues that "Today, we suffer from an almost universal idolatry of giantism[1]. It is therefore necessary to insist on the virtues of smallness--where this applies." [66, p62]. He sees that "There is wisdom in smallness." [66, p33] and that "In small-scale enterprise, private ownership is natural, fruitful, and just." [66, p250].

Schumacher further argues that "Any third-rate engineer or researcher can increase complexity; but it takes a certain flair of real insight to make things simple again." [66, p146] and that "To go for giantism is to go for self-destruction." [66, p150].

With that introduction, we now present the detailed arguments collected from multiple sources favoring the case for decentralizing future solar-electrics.

- Decentralization Matches Distributed Nature of solar energy to the distributed pattern of energy consumption [43, p55].

---

[1]The reason, of course, is because of the economics of scale which favor large plants over small ones.

- Decentralization Is Simple and Elegant "If gigantic power plants were displaced by thousands of smaller units dispersed near the points of end-use--technology would again concern itself with simplicity and elegance--." Hayes [1, p63].
- Decentralization: Ultimate in Power Directly to People, bypassing most existing complex, confusing (to customer), and costly centralized electric energy "facilities designed to serve large urban conglomerations," says Hayes [42, p32].
- Decentralization Promotes Customer Energy Independence from electric utility systems provided the total small-scale system has been well engineered including adequate on-site energy storage. Marvin [67] succinctly states, "Solar appeals to those who dream of individually controlled energy supply facilities, e.g. windmills, photovoltaic arrays, solar heating and cooling." For those who think it can't be achieved, Hayes [42, p30] gives us a data point: "Recent reports indicate that 17 million people use biogas for cooking and lighting in Szechwan province (China) alone."[1]
- Decentralization Matches Modular Nature of some solar-electrics, e.g. photovoltaics, inherent in decentralized on-site generation. It is easy to add more modules to get more power--without having to wait 10-12 years for a central plant to be constructed. Hayes [1, p63] sees that "small units could be added incrementally if rising demands require them." He also sees [42, p2] that "The technology is most sensibly applied in decentralized fashion--perhaps incorporated into the roofs of buildings."
- Decentralization and Cost of Electric Energy. Meadows [68] finds "The increased backup capacity required with large plants makes smaller units more valuable than large ones. One estimate indicated that one kilowatt of dispersed generating capacity is worth two and a half times more than a kilowatt of centralized generating capacity, in terms of backup capacity needed (Lovins [69])."

Hayes [42, p30] underscores low cost potentialities by saying, "Small-scale wind turbines to produce electricity cost as little as $500 per kilowatt for 15 kilowatt generators."[2]

The Department of Energy's national R&D goal for photovoltaics is $500 per kilowatt with some discussion to possibly lower it. Photovoltaic arrays in 1978 cost approximately $8,000-$12,000 per kilowatt.[3] The entire photovoltaic industry is working to meet--or lower--the $500 per kilowatt figure. When reached, photovoltaics will be competitive with conventional electric energy generation and a natural for dispersed solar-electric systems.

Bossong [70] sums up the viewpoint of some utilities: "Using the euphemism 'capacity displacement', the utilities argue that solar

---

[1] One may validly question, however, at what standard of living.

[2] It is interesting to compare his wind figure with the capital costs of conventional electric generation plants, as in Table 1-10. Unfortunately, his assumptions implied in the cost figure were not stated.

[3] The figure is even higher if power conditioning and energy storage costs are included.

houses will fall back upon the services of their local utility only during the peak periods. Because peak power is generally more expensive for a utility to provide, the argument runs that solar houses should pay a higher rate for the electricity they use than houses relying fully on the utility." He says this conclusion is debatable however.[1] There is also the unanswered issue regarding the time of day pricing for the needed power, especially in relation to the peaks and valleys of the demand curve.

We shall definitely see much more effort devoted to reducing the costs of solar-electrics to a level affordable by the small-scale customer. Researchers and engineers are aware of the need to reduce costs of all species of solar-electrics. They are collectively diligently working on this problem which most believe is tractable.

- Decentralization Eliminates Transmission Lines, their capital costs, their energy losses, and their maintenance costs. Palz [30, p229] reports that "The cheapest low power/low voltage overland connection line costs about $4,000/km." Meadows [68] finds that "Nearly 70 percent of the cost of electricity delivered to residential and small light and power customers is associated with transmission and distribution costs." He also finds "Transmission of electricity on 345 kilovolt power lines for a distance of 1000 miles would result in losses of 46 percent while a 100 mile trip would yield losses of only 4.6 percent."

The extensive land costs or leasing costs along with ecological costs associated with conventional centralized electric energy systems would, naturally, be nonexistent with fully decentralized solar-electrics.

- Decentralization Serves Remote or Urban Areas No costly transmission lines would have to run for remote sites operating exclusively solar-electric. This factor alone would probably--with government help--permit many of the nation's poor to have some limited electrical energy where they do not now, e.g. in Appalachia. Hayes [42, p32] sums it up: "If the greatest good for the greatest number is to be the aim, highest priority among investments in new energy sources should be given to decentralized facilities designed to provide on-site power to remote rural villages where most people live."

Decentralization is also a natural for urban areas, e.g. photovoltaics, provided the air pollution does not attenuate the insolation. The advent of such decentralized solar-electrics to urban areas would give added impetus to the desirable movement to clean up the nation's air.

- Decentralization Eliminates Fuel Transport Problems, as no fuel is involved (Palz [30, p229]) as in a conventional centralized electric plant.

An extension of this argument leads to elimination of the large oil and gas storage tanks and their safety hazards to those living

---

[1] This points up one of the many facets of integrating solar-electrics into conventional systems. More in-depth research into the issue raised is needed.

[2] A representative U.S. figure is about $10,000/mile (Florida Power Corp.)

nearby. Such hazards simply would not exist in an all solar-electric decentralized national energy system.

- Decentralization Tends To Eliminate Distribution Network at the local level, including its capital cost, losses, and maintenance expenses as during storms and natural phenomena. About 50 percent of the utility industry's total capital investment is in distribution networks. On-site solar-electric generation with adequate energy storage which is entirely independent of the electric grid needs no distribution network, as power would be locally generated. If storage is not to be installed, then the distribution network of the local power grid would have to be called upon at night.
- Decentralization Increases Defense Hardness A nation composed of 70 million energy independent solar-electrical sources would be practically impossible for any would-be enemy to knock out. In contrast, a few hundred highly centralized electric power plants are prime military targets and much easier to nullify in this ICBM age. "To decentralize power sources is in a sense to act upon the principle of 'safety in number'," Hayes [1, p63] says.
- Decentralization Minimizes Land Required, as solar collectors e.g. photovoltaic arrays, can be placed on rooftops, thus using the land for dual purposes. Central conventional electric plants require extensive safety buffer zones of land--in addition to land for transmission lines. All this land would not be needed for fully decentralized solar-electric systems with energy storage.
- Decentralization Increases System Reliability A national electric grid system fed by millions of independent small-scale solar-electric generation sources would appear to be far less likely to have massive wide area blackouts and brownouts than the present electric grid with a few thousands of large-scale power plants, any one of which can and does fail; when it does the system stability is seriously upset. Meadows [68] reinforces this point.
- Easy Integration With Utilities "--they would be much easier than large new facilities to integrate smoothly into an energy system." Hayes [1, p63].

  There are, however, unique technical problems involved with integrating solar-electrics with the conventional power grid. We shall expose these in later chapters.
- Easy Integration Into Buildings so as to be aesthetically pleasing, e.g. photovoltaic panels on rooftops.[1] The ease of integration feature for photovoltaics may not hold for all solar-electric systems, however; physically integrating a windmill into a typical suburban home so it isn't an eyesore poses a nontrivial problem.
- Dispersed Systems Would Be More Easily Mass Produced--millions of on-site solar-electrics--than a few giant power plants. Hayes [42, p29] thinks that "smaller windmills would lend themselves more easily than large ones to mass production."

---

[1] Palz [30, p230, 234, 235] graphically shows that this idea is at least as old as 1929.

- Shorter Lead Times--by an order of magnitude--for adding on more capacity than the present 8-12 years it takes to get a centralized conventional electric plant built. Hayes [1, p63] believes that "small simple sources could be installed in a matter of weeks or months; large complex facilities often require years and even decades to erect." Meadows [68] thinks that shorter lead times associated with dispersed systems would also reduce "interest costs, inflation effects and misestimation of future energy demands." He concludes: "Thus large facilities are much more subject to inflation."
- Decentralizations Tends To Use Local Materials and Labor greatly benefitting thousands of local communities and saving energy for transportation of materials that otherwise would have to be hauled in. An example is small hydroelectrics which Hayes [42, p28] claims number 50,000 in China in units of 100 KW or less and account for one-third of all hydro-electricity in China.

  Schumacher [66, p55] philosophizes that "--production from local resources for local needs is the most rational way of economic life."
- Dispersed Systems Tend To Produce Many Jobs Jobs would be widely dispersed throughout the land rather than concentrated in a few urbanized industrialized cities. Costle [71] points out that "satisfying some of our energy needs through the use of sunlight would generate thousands of new jobs for sheet metal workers, installers, carpenters, plumbers, pipefitters, engineers, architects--the list goes on. It has been calculated that $2 billion invested in solar energy--would provide four times as many jobs as the same amount invested in nuclear plants--."
- Stimulus To Small Businesses by the thousands. Small businesses are in 1978 receiving only 0.6 percent of government prime contracts, awards and grants in photovoltaics while laboratories received 61.6 percent [72, p100].
- More Compatible "than centralized technologies with social equity, freedom, and cultural pluralism," says Hayes [1, p13]. He further sees [p63] that "If small-scale, decentralized renewable-energy technologies were embraced, few aspects of modern life would go unaffected. Farms would begin to rely on wind power, solar heaters, and waste conversion technologies to supply a large fraction of their energy needs---."
- Safety Enough solar-electric options exist that for a given application some optimization for safety can be made.

  The centralized option, Solar Satellite Power Systems (SSPS), however raises serious microwave radiation safety hazards. "Microwaves, even at low levels, have been shown to cause central nervous system disorders, cataracts, and genetic changes," Ottinger [73] claims.

- Less Disruption By Pressure Groups Hayes [1, p65] opines that "Where energy production is centralized, those seeking to coerce or simply to disrupt the community can acquire considerable leverage." The coal mine strike of the winter of 1977-78 which caused serious disruption to mid U. S. electric energy systems is an example. Such tactics would not be possible with dispersed solar-electrics.

- Decentralized Solar-Electrics Can't Easily Be Monopolized "But the solar resource can turn into a monopoly if the government subsidizes high-technology, sophisticated energy harvesting," says Smith [74]. Smith suggested decentralization. Bossong [70] sees the possibility of utilities swiftly monopolizing the solar energy field and driving small industries out of business.

  Meadows [68] points out "that appropriate technologies do not lend themselves easily to monopolistic control and its attendant high profit margins."
- Sale of Excess Power generated by the localized decentralized solar-electric converter could, on occasion, be sold back to the utility grid, thus offsetting some utility-provided energy costs.[1]

  Naturally utilities are less than enthusiastic about this sale-back possibility. Recent cases in Florida, New York and other places have already indicated utility positions on this issue which had to be settled in the courts in favor of the individual solar-electric 'customer'. These cases, however, suggest that, at least on a small-scale basis, such dispersed solar-electric generators can be successfully integrated with the electric utility power grid if necessary.
- Decentralized Solar-Electrics Tends To Return People To The Countryside instead of further crowding the urban areas with their many problems. Hayes [1, p47] envisions that "Decentralized co-generation using wood would also fit in well with current worldwide efforts to move major industries away from urban areas."
- Decentralization Would Help Set Export Patterns for small-scale technology to the Third World peoples much in need of electric power. Dispersed solar-electrics could help "create the kinds of decentralized societies that must become the worldwide norm if civilization is to be sustained in humane form," says Hayes [42, p16].
- Permits Use of Rejected Energy for other useful purposes. For example, for photovoltaics in "decentralized use, the four-fifths or more of the sunlight hitting them that such cells currently cannot convert into electricity can be harnessed for space heating and cooling, water heating, and refrigeration," Hayes [42, p27]. How to do it was first shown in [43, Fig. 13].
- Decentralization Stimulates Inventors and Creators Beattie [75] says that "Appropriate technology ideas do not have to be new ideas, but can involve applications of existing technology. The technology is usually small-scale and decentralized, and should be simple to install, operate and maintain." Marvin [67] says that solar "appeals to the tinkerer in us--to the do-it-yourselfer who can conceivably match his own energy production to his energy use." Decentralized solar-electrics could be a great stimulator to man's individual creativity.

---

[1] This approach raises the whole question of 'back-up power' needed to meet utility reliability requirements. This issue is treated in Chapter 5.

We draw the case for decentralized solar-electrics to a close by citing a few generalized comments[1] on the issue:

> "To the extent that energy needs can be met with lower quality sources or decentralized equipment, the centralized options should be avoided." Hayes [1, p23].
>
> "Instead, national priorities should be directed to the development and deployment of diverse, small-scale systems carefully matched in size and thermodynamic quality to end-use demands." Meadows [68].
>
> "Based on information in OTA's study released this summer (1977) that found small-scale on-site electric devices too important to be ignored--" [76].
>
> Meadows' forceful paper [68] points out many misconceptions that reduce interest in small-scale technologies.
>
> Decentralization suffers from a lack of commitment to solar by the utilities. They "maintain the posture that solar technologies will not play a major role in the nation's energy picture for decades to come," says Bossong [70].
>
> "What keeps utilities from plunging into providing individual systems for customers--as opposed to looking into central station solar power--is the difficulty in finding common grounds in solar systems so that a rate structure can be devised," says Freeman [77].

We conclude, from the collection of arguments--which cannot be exhaustive here--that solar energy utilization is inherently amenable and matched to the idea of dispersed solar-electric systems. Dispersed solar-electrics appear, on balance, to have enormous potential beneficial effects for the nation.

Fortunately, the Department of Energy has recently begun to pursue some solar-electric R&D projects aimed at eventually giving us the option of having decentralized solar-electrics also feeding the electric power grid.

<u>The Case For Centralized Solar-Electrics</u> is made only weakly here because of the widespread existence of centralized power embodied in the classical conventional electric power system.

Probable appropriate solar-electrics most affected by the centralized issue involve power towers, large-scale hydropower, ocean thermal energy conversion (OTEC), satellite power systems, biological-electrics (energy plantations), and perhaps large-scale photovoltaics. As before we try to restrict the arguments to those potentially applying to solar-electrics as much as possible.

- <u>Centralization</u>: <u>The Historic Pattern</u> evolved to date for the conventional electric energy industry has been one of centralized generation. It is thus easy and natural for utilities to continue thinking 'centralized' when considering integrating solar-electrics.

---

[1] Some of these tend to be highly controversial, the reader is warned.

The centralized concept involving a few hundreds of high power capacity plants is largely unquestioned by the utility sector.

- Centralization: Permits Plant Location In Maximum Insolation Area, thus tending to capture a larger percentage of solar energy than otherwise (Palz [30, p231]) and maximizing the return on investment. An example is the Power Tower plants to be located in the Southwest U. S.[1]
- Centralization Realizes Economies of Scale which result in electric energy cost decreasing as plant size increases--or at least there is an optimal size range for minimum cost. This is the principal reason utilities have evolved toward progressively larger conventional generating plants.
- Centralization Promotes Reliability for large interconnected systems. If one part of the system fails temporarily, the remainder can be briefly overloaded to give the customer uninterrupted service. Reliability of most modern conventional power systems well exceeds 99 percent, depending on the locale and system. Backup may be required, though.
- Large Power And Energy can be produced in centralized plants to meet the nation's future need for large bulk power and energy. Recall from Fig. 1-23 that large Quads of electric energy will be required to run the country. It is therefore natural to think in terms of a few large solar-electric facilities to meet the power need.
- Centralization Could Supply Back-Up Capability for millions of small-scale solar-electric supplied dwellings. Using the utility power grid in times of low on-site insolation would obviate the need in many cases for large numbers of individualized small-scale costly electric energy storages.

  Doing this, however, on a large-scale could seriously worsen the utility systems load factor problem and increase the busbar power costs. Using the utility grid as a backup in this way points up one of the many technical difficulties a utility can be expected to have with integrating large numbers of solar-electric generators into the existing power grid. This facet of the integration problem requires further R&D study and actual on-line experience before its seriousness can be assessed.
- Centralization Uses Cheap Land, generally in unused desert regions. [Palz 30, p231].
- Any Adverse Environmental Effects associated with a centralized plant can, in general, be far removed from the principal population.
- Safety For Population is achieved--to some degree--by distance when explosions or other industrial type accidents occur. Safety may be a factor, for example, in locating hydrogen and methane storage tanks where the gas is to be used for electricity generation.
- Financing R&D "Central governments can much more easily spend their revenues on a handful of gigantic projects than on millions of smaller ones." Hayes [42, p35].
- Centralization Benefits Large Corporations, especially aerospace

---

[1] See Figs. 1-19 and 1-20 for why. Location of centralized power plants in remote areas relatively removed from the urban concentrations of people where the power demand is poses obvious major transmission problems.

firms which can acquire large multi-million dollar R&D contracts instead of the substantially smaller multiple contracts which would have to be associated with developing small-scale decentralized technology. Centralization of solar-electrics, if the industry is to move in that direction, means large business for a relatively few firms, usually located in urban areas. Small businesses may only be involved as suppliers of components, goods, and services--all small in comparison.[1]

- Ease of Transporting Electric Energy via high voltage lines is one important factor making centralized solar-electric plants attractive.

  We should note in passing, however, that contrary to what some think, the transportation of energy in electric form is among the most expensive ways known to transmit energy. Table 1-11 shows the cost to transport energy 100 miles via various routes. Clearly the cost per million BTU transmitted for a 200 KV AC line is 6.6 to 9.0 cents--an order of magnitude higher than the same energy shipped via oil in a tanker ship. So while it is easy and fast to ship energy electrically, an important perspective is that this is a very expensive means of transporting a unit of energy.

We close the issue of decentralized vs centralized solar-electrics by observing that, in all likelihood future decentralized solar-electric installations may coexist with centralized solar-electric plants and either or both may feed the electric power grid of the future.[2] Alternatively if we first develop fully decentralized small-scale solar-electrics, then we may well see these appear commercially in the increasing scales of:

- Individual solar-electrics for houses.
- Neighborhood solar-electrics where an appropriate modest sized solar-electric facility powers a localized neighborhood.
- Centralized solar-electric plants supplying whole cities or parts thereof but located relatively close to the system's load center.

Some such complex mix appears inevitable.

Whatever the scale resulting from the current debate, the electrically tying together of so many diverse generation and distribution means is certain to pose numerous system integration problems for these new solar-electrics. The body of systems knowledge for how to achieve this complex integration successfully appears to be in a very elementary state in 1978, and much more research appears needed in systems integration.

Hayes remarks [1, p13] provide a fitting conclusion to this section:

"But energy sources are not neutral and interchangeable. Some energy sources are necessarily centralized; others are

---

[1]Some undoubtedly, will argue that the same relationship between large and small business may also occur if decentralized solar-electrics are developed and that this relationship is implicit in the U.S. business fabric.

[2]A recent review of the Department of Energy's solar program by a Solar Working Group [78, p14] concluded, correctly, that "The Working Group has not been persuaded that centralized and decentralized systems cannot exist side by side--to great advantage."

TABLE 1-11 Cost of Transporting Energy, by Form of Energy and Means of Transportation

Form of Energy	Means of Transportation	Transportation Cost per 100 Miles	
		¢ Million Btu	Mills/kwh
Nuclear fuel	Railroad	Under 0.03	Under 0.003
Oil	Tanker Ship	0.1 to 0.5	0.01 to 0.05
	Pipeline	0.4 to 1.6	0.04 to 0.16
	Barge (Average)	0.5	0.05
	Railroad Tank Car (Average)	4.3	0.43
	Truck (Average)	7.4	0.74
Natural Gas (Gas)	Pipeline	1.1 to 2.4	0.11 to 0.24
Natural Gas (Liquefied)	Tanker	0.5 to 0.9	0.05 to 0.09
	Barge	0.6	0.06
	Railroad	2.7	0.27
Coal	Slurry Pipeline	1.0 to 2.0	0.1 to 0.2
		1.5 to 5.0	0.15 to 0.5
	Railroad: Integral Train	0.7 to 2.0	0.07 to 0.2
	Shuttle Train	1.5 to 3.5	0.15 to 0.35
		1.8 to 3.6	0.18 to 0.36
	At Filed Tariffs	3.0 to 7.6	0.3 to 0.76
		3.6 to 8.3	0.36 to 0.83
	Western States (1964 Average)	3.6	0.36
Electricity At 10000 BTU/KWH	High Voltage Trans. Line		
	700 KV AC	3.0 to 3.6	0.3 to 0.36
	500 KV AC	3.6 to 4.8	0.36 to 0.48
	345 KV AC	4.8 to 6.6	0.48 to 0.66
	200 KV AC	6.6 to 9.0	0.66 to 0.90
	Unspecified Voltage	7.8 to 30.0	0.8 to 3.0

Source: Originally from Federal Power Commission, The 1970 National Power Survey, Part III. Above table appears in reference [21] which contains data sources for each transportation system, assumptions, and other notes vital to those studying energy transportation in-depth.

necessarily dispersed. Some are exceedingly vulnerable; others are nearly impossible to disrupt. Some will produce many new jobs; others will reduce the number of people employed. Some will tend to diminish the gap between rich and poor; others will accentuate it. Some inherently dangerous sources can be permitted widespread growth only under authoritarian regimes; others can lead to nothing more dangerous than a leaky roof. Some sources can be comprehended only by the world's most elite technicians; others can be assembled in remote villages using local labor and indigenous materials. Over time, such considerations may prove weightier than the technical criteria that dominate and limit current energy thinking."

The issue of decentralized vs centralized solar-electric will undoubtedly be argued further in the future; the collection here of the relative merits of each--as an issue--is probably only the beginning. Hopefully the reader can add constructively and objectively to clarifying this important issue. Its outcome determines the kind of solar-electric hardware which should eventually be built, and hence the issue's outcome will markedly shape the future of solar-electrics.

1-7 Multiple Conversions--A Recurring Principle

We bring this introduction to a close by briefly showing a principle which occurs repeatedly throughout solar-electrics. The principle has no formal name; here we refer to it as the principle of cascading efficiencies. Firmly grasping this simple principle gives us considerable perspective on all solar-electric systems.

In Fig. 1-36 we see multiple stages of energy conversions with the output of one stage being the input to the next. The collection of stages might

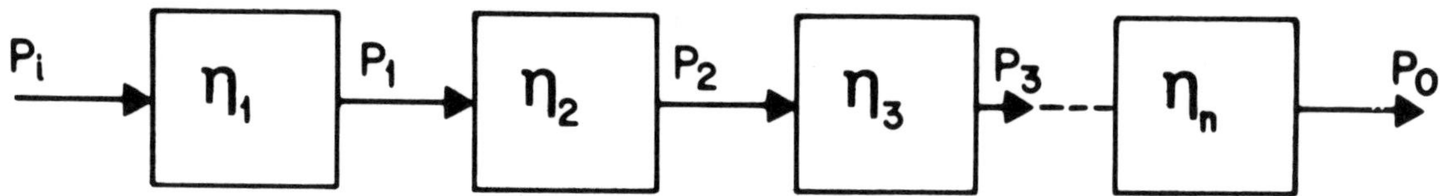

FIG. 1-36 Cascaded Energy Conversions

be a complete system whose overall conversion efficiency, η_o, we would like to know. The efficiency of each individual stage is assumed known--or could be found.

From elementary considerations, the definition of efficiency of an arbitrary apparatus is

(1-8) $$\eta = \frac{\text{Useful Power Output}}{\text{Power Input}}.$$

Applying this to Fig. 1-36 we can write

(1-9) $$\eta_o = \frac{P_o}{P_i} = \frac{P_1}{P_i} \times \frac{P_2}{P_1} \times \frac{P_3}{P_2} \times \cdot \cdot \cdot \frac{P_o}{P_{n-1}}.$$

We immediately recognize this as simply

(1-10) $$\eta_o = \eta_1 \, \eta_2 \, \eta_3 \cdot \cdot \cdot \eta_n.$$

Eq. (1-10) is the principle of cascading efficiencies. It says that if we have an energy conversion system composed of cascaded stages the overall system efficiency will simply be the product of efficiencies of all the stages. Note that the overall efficiency will always be less than the <u>lowest</u> efficient stage in the chain.

<u>Example 1-10</u>

What is the overall system efficiency of a photovoltaic solar-electric system composed of 15 percent efficient solar cells, 85 percent efficient power conditioner, 80 percent efficient battery energy storage, followed by a DC-AC inverter of 90 percent efficiency?

Applying the principle of cascading efficiencies, we can immediately write:

$$\eta_o = 0.15 \times 0.85 \times 0.80 \times 0.90 = 0.092$$

$$\eta_o = 9.2 \text{ percent.}$$

From Example 1-10 we also note the overwhelming effect the <u>low</u> efficiency term has, the photovoltaic array in this case, in reducing the overall system conversion efficiency.

The above considerations lead us to two useful corollaries.

- <u>Corollary 1</u> In an energy conversion system of n cadcaded stages, each having a known efficiency, but a single stage has an efficiency markedly lower than the others, then R&D should first focus on increasing the lowest if the goal is to improve the overall system efficiency.

- Corollary 2 The overall system efficiency of n cascaded stages, where n is large, possibly may be markedly increased by, in an appropriate way, eliminating as many stages as possible.

  This principle of stage minimization leads to the notion that the most direct energy conversion system would be preferred, i.e. the one with the fewest n number of stages, a single stage being the lower most desirable limit.

  Application of this corollary may lead to substantial system simplification in some cases; it also can lead to posing new worthwhile invention problems. An example would be a one-step solar-to-60-cycle-AC converter. It has not yet been invented.

The above notions are beautiful in their simplicity; nevertheless they are of great usefulness in solar-electric engineering. Occasionally these ideas are forgotten when large, complex (and usually very costly) solar-electric systems are proposed involving far too many energy conversions for the overall system to be very efficient.

This completes our introduction to solar-electrics. We have dealt with many important general matters and issues which apply to solar-electric engineering as a whole.

In the following Chaps. we examine each subfield in more detail; but the technologies following rest on the fundamental ideas and issues dealt with in this chapter.

REFERENCES

1. Hayes, Denis, "Energy: The Solar Prospect," Worldwatch Paper 11, March 1977. Available from Worldwatch Institute, 1776 Massachusetts Ave., N. W., Washington, D. C. 20036. This paper adapted from the author's book, Ref. [5].

2. Austin, Arthur L., and John W. Brewer, "World Population Growth and Related Technical Problems," IEEE Spectrum, Dec. 1970, pp. 43-54.

3. von Foerster, H., P. M. Mora, and L. W. Amiot, "Doomsday: Friday, 13 November, A.D. 2026," Science, Vol. 132, Nov. 1960.

4. Wilson, Carroll L., Energy: Global Prospects 1985-2000, [New York, N. Y.: McGraw Hill Book Co., Inc., 1977], 291 pages. A report on the Workshop on Alternative Energy strategies, a project sponsored by the Massachusetts Institute of Technology.

5. Hayes, Denis, Rays of Hope: The Transition To A Post Petroleum World, [New York, N. Y.: W. W. Norton & Co., Inc., 1977], 240 pages.

6. Usmani, I. H., "The Challenge of Energy," Key-Note Address, International Conference on Alternative Sources of Energy, Miami Beach, Florida, December 5-8, 1977. The author is Energy Advisor to the United Nations.

7. Hill, James H., and F. Clark Huffman, Energy In Perspective, An address before The Society of Sigma Xi; State University College, Fredonia, N. Y., May 15, 1975. NTIS No. TID 26900. (His source for the graph is not cited. It is believed to have originated in a government report.)

8. Felix, Fremont, "Energy Options for the United States," IEEE Spectrum, Vol. 10, No. 9, Sept. 1973, pp. 63-68.

9. The Alcan Pipeline, Bank of America advertisement, The Wall Street Journal, Thursday, December 15, 1977, p. 13.

10. Hubbert, M. King, "The Energy Resources of the Earth," Scientific American, Vol. 224, No. 3, Sept. 1971, pp. 61-70.

11. Gaucher, L. P., "Energy Sources of the Future for the United States," Solar Energy, Vol. 9, No. 3, 1965, pp. 119-126.

12. Mangone, Gerald J., Editor, Energy Policies of the World, [New York, N. Y.: American Elsevier Pub. Co., Inc., 1976], 387 pages.

13. Thring, Hans, Energy For Man: Windmills to Nuclear Power, [Westport, Connecticut: Greenwood Press, Publishers, 1958].

14. Bueche, Arthur M., "The Hard Truth About our Energy Future," General Electric Co. publication, n.d.

15. Sailor, Vance L., "Present and Near-Future Demand for Energy in the United States," BNL-21236, April 8 1976. NTIS NO. CONT - 760479-1.

16. Energy: The Ultimate Resource, Study submitted to the Task Force on Energy, Subcommittee on Science, Research and Development of the Committee on Science and Astronautics. U. S. House of Representatives, Ninety-second Congress, First session, October 1971. Government Printing Office.

17. Felix, Fremont, "U.S. Using Electricity More Efficiently," Electrical World, Dec. 1 1974, p. 59.

18. Ramsdell, A. J., "Statistical Papers of the United Nations World Energy Supplies," Series J, No. 17.

19. Dupree, Walter G., "United States Energy Through the Year 2000," Bureau of Mines - USDI, December 1975. NTIS PB-250600.

20. Hill, James H. and F. Clark Huffman, Energy In Perspective, An address before The Society of Sigma Xi, State University College, Fredonia, N. Y., May 15 1975. NTIS No. TID 26900.

21. Data for Use in the Assessment of Energy Technolgies, AET-8, April 1972. A report submitted to the Office of Science and Technology, Executive Office of the President, under Contract OST-30, Associated Universities, Inc., Upton, New York.

22. Gaucher, Leon P., Energy Requirements of the Future, presented at Int'l. Solar Energy Society Conference, May 10-14 1971, NASA Goddard Space Flight Center, Greenbelt Md. He originally published the graph in Gaucher, L. P., "Energy In Perspective," Chemical Technology, March 1971.

23. A National Plan For Energy Research, Development & Demonstration: Creating Energy Choices for the Future, ERDA-48, Vol. 1 of 2, June 28 1975.

24. Energy Alternatives: A Comparitive Analysis. A study prepared for the Federal Energy Administration and National Science Foundation by University of Oklahoma, May 1975. NTIS No. PB 246 365.

25. Weaver, Kenneth F., and Emory Kristof, "The Search For Tomorrow's Power," National Geographic, Vol. 142, No. 5, November 1972, pp. 650-681.

26. National Energy Outlook. 1976 Executive Summary, Federal Energy Administration, FEA/N-76/100, Available from Government Printing Office.

27. Bronheim, Harold, Robert Nathans and Philip F. Palmedo, Users Guide for Regional Reference Energy Systems, November 1975. NTIS No. BNL 20426.

28. Rose, David J., "Energy Policy In the U.S.," Scientific American, Vol. 230, No. 1, January 1974, p. 20-29.

29. Dance, W. D., "Energy: The Facts versus The Fiction," An executive speech reprint, General Electric Co., May 19 1976.

30. Palz, Wolfgang, Solar Electricity: An Economic Approach To Solar Energy, [Woburn, Mass.: Butterworth (Publishers), Inc., 1978], 292 Pages.

31. Friedlander, Gordon D., "Energy's Hazy Future," IEEE Spectrum, May 1975, pp. 32-40.

32. Felix, Fremont, "Energy Independence: Goal for the 80's," Electrical World, March 1 1974, pp. 52-56.

33. Tanner, James, North Slope Oil Output Allows U. S. to Cut Oil Imports Estimated 9% From Year Ago, The Wall Street Journal, Thursday April 20 1978, p. 2.

34. Nixon, Richard M., Address to the Congress of the United States, June 4 1971.

35. The National Energy Plan: Summary of Public Participation, Executive Office of the President, Energy Policy and Planning. Government Printing Office, Undated (CIRCA 1977).

36. Farrelly, P. J., "The Impact of Oil Price Increases on the Competitiveness of Non-conventional Energy Technologies," Paper presented at Int'l. Conference on Alternative Energy Sources, Miami Beach, Fla., December 5-7 1977.

37. Rau, Hans, Solar Energy, [New York, N. Y: The Macmillan Co., 1964], 171 Pages.

38. Daniels, Farrington, Direct Use of the Sun's Energy, [New York, N. Y.: Ballantine Books, Inc. 1974]. First published 1964 by Yale University Press.

39. Kemper, J. P., "Pictorial History of Solar Energy Use," Sunworld, No. 5, August 1977, pp. 17-32. An excellent summary, with extensive references, of who created what and when.

40. Delyannis, Anthony, "Greek Solar Energy Program," in F. deWinter and J. W. deWinter, Description of the Solar Energy R&D Programs in Many Nations, February 1976 NTIS report SAN/1122-76/1.

41. Pytlinski, J. T., "Solar Energy: Past and Present Achievements," Presented at Miami Int'l Conference on Alternative Energy Sources, Miami Beach, Fla., Dec. 5-7, 1977. Sakkos' achievement was first reported in Time, November 26 1973, P60, according to Pytlinski.

42. Hayes, Denis, Energy for Development; Third World Options, Worldwatch paper 15, December 1977. Worldwatch Institute, 1776 Massachusetts Ave., N. W., Washington, D. C. 20036.

43. An Assessment of Solar Energy As a National Energy Resource. Prepared by the NSF/NASA Solar Energy Panel, December 1972, NTIS No. PB 221-659.

44. Robinson, N., Solar Radiation, [New York, N. Y.: American Elsevier Publishing Co., Inc., 1966], 347 Pages.

45. Boes, Eldon C., Distribution of Direct and Total Solar Radiation Availabilities For the USA, August 1976. NTIS No. SAND-76-0411.

46. Thekaekara, M. P., et al, Solar Electromagnetic Radiation, May 1971. NASA SP-8005, Goddard Space Flight Center, Greenbelt, Md.

47. Gibson, Edward G., The Quiet Sun, NASA 1973. Available from Government Printing Office as Stock No. 3300-0454.

48. Hynek, J. A., Astrophysics, [New York, N. Y.: McGraw-Hill Book Co., 1951]. Cited in Gibson [47, p. 14].

49. Henderson, S. T., Daylight And Its Spectrum, [New York, N. Y.: American Elsevier Publishing Co., Inc., 1970], 277 Pages.

50. Heaney, James A., A Theoretical Analysis of Electromagnetic Radiation From the Sun as Applied to Electromagnetic Wave Energy Converter, Masters Thesis, University of Florida, 1977.

51. Cherry, W. R., "The Generation of Pollution - Free Electrical Power from Solar Energy," Transactions of the ASME, Journal of Engineering for Power, Vol. 94, Series A, No. 2, April 1972, pp. 78-82.

52. Watson, W. K., Potential Impact of Solar Energy Utilization on Florida Utilities, Master's Thesis, University of Florida, 1975.

53. Farber, Erich A., "Solar Energy: Conversion and Utilization," Building Systems Design, June 1972.

54. Elgerd, O. I., Basic Electric Power Engineering, [Reading, Mass.: Addison-Wesley Publishing Co., 1977].

55. 31 Answers to 32 Questions about the Electric Utility Industry, (pamphlet), 1976/77 Edition, Edison Electric Institute, 90 Park Ave., New York, N. Y. 10016.

56. Nagel, T. J., "Electric Power's Role in the U. S. Energy Crisis," IEEE Spectrum, Vol. 11, No. 7, July 1974, pp. 69-72.

57. National Power Survey, A report by the Federal Power Commission, October, 1964, Government Printing Office.

58. "29th Annual Electrical Industry Forecase," Electrical World, September 15, 1978, p76.

59. Factors Affecting the Electric Power Supply, 1980-85. Executive Summary and Recommendations, Federal Power Commission, Bureau of Power, Washington, D. C., 1 December 1976. NTIS No. PB 264 760.

60. 1973 Business and Economic Charts, Ebasco Services, Inc. Research Dept., Two Rector St., New York, N. Y. 10006.

61. Jurgen, Ronald K., "What to Tell Your Neighbors," IEEE Spectrum, Vol. 11, No. 6, June 1974, pp. 61-65.

62. Starr, Chauncey, Role of Solar Energy In Electric Power Generation, paper presented at AAAS Annual Meeting, Denver, Colorado, February 20-25 1977. Available through EPRI.

63. Eldridge, Frank R., Topics for Research on Solar Energy Systems, Mitre Report M 73-9, The MITRE Corporation, January 1973.

64. Hill, Philip G., Power Generation, [Cambridge, Mass.: The MIT Press, 1977].

65. Reported in Solar Energy Intelligence Report, Vol. 4, No. 12, March 20 1978, page 73.

66. Schumacher, E. F., Small Is Beautiful: Economics As If People Mattered, [New York, N. Y.: Harper & Row, Publishers, 1973].

67. Marvin, Henry H., Quoted in: Solar Energy Research & Development Report, (A Dept. of Energy publication), May 1 1978.

68. Meadows, Dennis, Fallacies that Block the Search for an Alternative Energy Plan, Banquet address before the Miami International Conference on Alternative Energy Sources, Miami Beach, Fla., December 5-7 1977.

69. Lovins, Amory B., "Scale, Centralization, and Electrification in Energy Systems," First draft of paper prepared for the 20-1 October 1976 Symposium Future Strategies of Energy Development, Oak Ridge Associated Universities, Oak Ridge, Tennessee. Cited in Meadows [68].

70. Bossong, Ken, "The Case Against Private Utility Involvement in Solar/Insulation Programs," Solar Age, January 1978, pp. 23-27.

71. Costle, Douglas, Quoted in: Costle Calls for More Solar and Conservation, Decentralization of Energy Supplies, Solar Energy Intelligence Report, Vol. 3, No. 40, November 14 1977, p. 258.

72. Small Businesses Seen Needing Help to Get or to Stay in Photovoltics, Solar Energy Intelligence Report, Vol. 4, No. 15, April 10 1978, p. 100.

73. Ottinger, Richard, Quoted in: Solar Energy Intelligence Report, Vol. 4, No. 15, April 10, 1978, p. 100.

74. Smith, Otto, Quoted in: Solar Energy Intelligence Report, April 17 1978, p. 111. His remarks were from a paper presented at the Second Annual Helioscience Institute Conference on Alternate Energy, Palo Alto, Calif., April 1978.

75. Beattie, Don, Quoted in: Doe Receives 1,120 Proposals for Small-Scale, Renewable Energy Projects, Solar Energy Intelligence Report, Vol. 3, No. 46, December 26 1977, p. 299.

76. Utility Ownership of On-Site Solar Devices Called Attractive, Solar Energy Intelligence Report, Vol. 4, No. 3, January 16 1978, p. 15.

77. Freeman, John K., Quoted in: Utility Role In Solar Uncertain but National Needs May Make it Necessary, Solar Energy Intelligence Report, Vol. 3, No. 40, November 14 1977, p. 257.

78. Balcomb, Douglas, "The Department of Energy Solar Program: A Question of Balance," Solar Age, Vol. 3, No. 5, May 1978, pp. 12ff.

79. Krenz, Jerrold H., Energy: Conversion and Utilization, [Boston, Mass.: Allyn and Bacon, Inc., 1976]

80. Modern Energy Technology, Vol. I, 1975, Research and Education Association, 342 Madison Ave., New York, N. Y. 10017.

81. Lincicome, Robert A., "Energy Crisis: Surviving the Critical Years 1975-1985," Electric Light and Power, October 1974, pp. 7-12.

CHAPTER 2

DIRECT SOLAR-ELECTRICS

"It is important to think of photovoltaic solar energy converters as a system rather than as 'solar cells'. What is required of a photovoltaic converter is the delivery of electric power to satisfy the demands of a given load. Such loads may range from a small, single purpose device, such as a navigation light, to the peak power demand of a community or an entire utility network".

Martin Wolf [1, p46]

- Introduction
- Prior Photovoltaic Methods
- Present Art and Problems Being Addressed
- Site Selection Considerations
- System Integration of Photovoltaics
- Economic Aspects
- Barriers to Photovoltaic Integration
- Future R & D Needed for Integration

2-1 Introduction

In Chap. 1 we obtained perspectives on the energy problem, saw the potential in solar energy for meeting world and U. S. energy needs, and saw how the important emerging field of solar-electrics is structured.

In this Chap. we focus on one of the sub-fields which many believe to be both the most important and most promising, direct solar-electrics. We will define it shortly.

Our Purpose And Scope in this Chap. is to identify and summarize those topics believed to be most relevant to a systems engineer seeking to integrate direct solar-electrics into the existing electric networks. The viewpoint here is chiefly that of the systems engineer and not that of the device-orientated solid state physicist. Therefore substantial filtering of the existing technical literature on direct solar-electrics had to be made for this chapter.

As direct solar-electrics encompasses the entire field of photovoltaics it is necessary for us to first understand the operation, characteristics, and equivalent circuit of the basic module, the solar cell. We shall treat these in this section, and in Secs. 2-2 and 2-3 we briefly summarize the prior and present art of solar cells relevant to the systems engineer. Space and time do not permit us to make a detailed review of the extensive prior art of photovoltaic research and development nor would such a detailed study be appropriate for the present work. References will be cited, however, where those desiring such information can find an entrée to it.

The latter part of the Chap. summarizes the state-of-the-art of 'systems integration' of photovoltaics, our principal interest here; references are cited where those needing more systems-orientated details can quickly find them without having to survey the entire photovoltaic field, as was done for this research.

The Literature On Photovoltaics is extensive. In 1978 it, in toto, far exceeds in number and volume the literature of any other sub-field of solar-electrics. It is so extensive and previous research publications have spanned so many information media that the systems engineer may have great difficulty making an initial entrée into the literature. We therefore summarize a few of the most helpful sources of information for those desiring to enter the photovoltaic research literature for the first time at a deeper level than is contained in this Chapter.

- The U. S. Photovoltaic Program A high level review of DOE's program has recently been made by the Solar Working Group, now disbanded. They sought significant imbalances and recommended program modifications. The results are summarized by Balcomb [2, p33], the group's technical advisor.

  The 1978 document defining the national photovoltaic plan is in [3]. The 1977 program is in [4], and the 1976 program is in [5]. Herwig described the 1975 program in [6].

  The U. S. photovoltaic program had its genesis with the Solar Energy Panel [7] in 1972, growing out of the National Science Foundation's (NSF) Research Applied to the National Needs (RANN) program. Though there was active photovoltaic work prior to 1972 by a few Universities, industrial concerns, and government groups, as NASA, [8] there was no substantial nationally coordinated terrestrially-oriented photovoltaic program before 1972.

- Bibliographies and Abstracts of the most current government sponsored research reports on photovoltaic work can be quickly gleaned from a weekly subscription to reference [9]. A similar monthly abstract of photovoltaic R & D is in [10] where photovoltaics is specifically categorized. The Department of Defense (DOD) has accumulated a 1958-1976 bibliography on solar cells and solar panels [11]. The DOE's general bibliography on solar energy is in [12] which has photovoltaic citations and abstracts. The massive index to it, categorized by corporation, author name, and subject, is in [13].

  An extensive photovoltaic bibliography arranged chronologically has been accumulated by Backus [14, Appen. B].

- Brief Survey Articles on photovoltaics occasionally appear which attempt to integrate this whole complex field. One excellent such survey by Rosenblatt [15] appeared in 1974. Wolf [1, Ch. 7] also surveyed the field in 1975 in an AIAA publication.

- Books And Textbooks by various authors discussing photovoltaics have appeared since about 1963. The earlier books included photovoltaics as an integral part of direct energy conversion which grew out of the space program. Helpful books dealing with various aspects of photovoltaics are: Chang (1963) [16, Ch. 6], Daniels (1964) [17, Ch. 16],

Sutton (1966) [18, Ch. 1], Walsh (1967) [19, Ch. 6], Altman (1967) [20, Ch. 4], Kettani (1970) [21, Ch. 8], Merrigan (1975) [22], Angrist (1976) [23, Ch. 5], Krenz (1976) [24, Ch. 7], Backus (1976) [14], and Palz (1978) [25, Ch. 3].

As mentioned before, the photovoltaic literature is very extensive; most of it deals with the detailed solid-state physics of some particular type converter and is of lesser interest to the systems engineer seeking to apply photovoltaics. We shall be citing in this Chap. additional specific information sources of most potential value to the systems engineer.

Definition of Direct Solar-Electrics The connotation of the term 'direct' is that the process of conversion to electricity is direct in that it avoids lossy thermal cycles as well as conversion based on the rotating shaft. As Fig. 1-32 showed, there are other sub-fields of solar-electrics based on these non-direct concepts.

Thus the term 'direct solar-electrics' encompasses the set of methods for converting solar radiation directly to electric power without benefit of heat cycles or rotating shafts, i.e. the process is commonly referred to as 'direct' because the solar radiation directly causes electric charges to flow. In a practical sense direct solar-electrics, as indicated in the previous sub-sections, translates to photovoltaics[1] which usually implies a solid-state solar-electric converter of some kind.[2]

In light of the above, 'direct solar-electrics' presently implies photovoltaics or solar cells; direct solar-electrics is a much broader field, however.

Classification of Photovoltaics in a broad generalized way is embodied in FIG. 2-1. Most of the R & D presently underway is in the upper block focussing on the photovoltaic device, and substantially less--though by no means zero--R & D has and is being devoted to the lower systems area. Considerable previous systems type R & D in the latter area has been done in connection with various past space programs, and some--but by no means all--of that two decades of art and knowledge could be applied to the terrestrial problem; the system requirements between terrestrial and space use are substantially different, however.

---

[1] Palz [25, p179], quite correctly, points out the term 'photovoltaic' is misleading, as it was originally adopted to differentiate between the photovoltaic effect and the photoconductive effect. The photovoltaic effect is a potential power producer and not just a producer of a 'photo voltage' whereas the photoconductive effect is a power dissipator. The latter is commonly used in light sensing devices.

With 1978 technology the term 'photovoltaics' translates to solar cells--a term frequently used in the literature as a synonym. The term 'solar cell' may, perhaps later, also evolve into a misnomer if continuous sheets of large area photovoltaics now being worked on become a reality.

[2] It is recognized that there are liquid photovoltaic converters. Such devices are presently laboratory curiosities from the systems engineer's viewpoint and are of a minor practical importance in 1978.

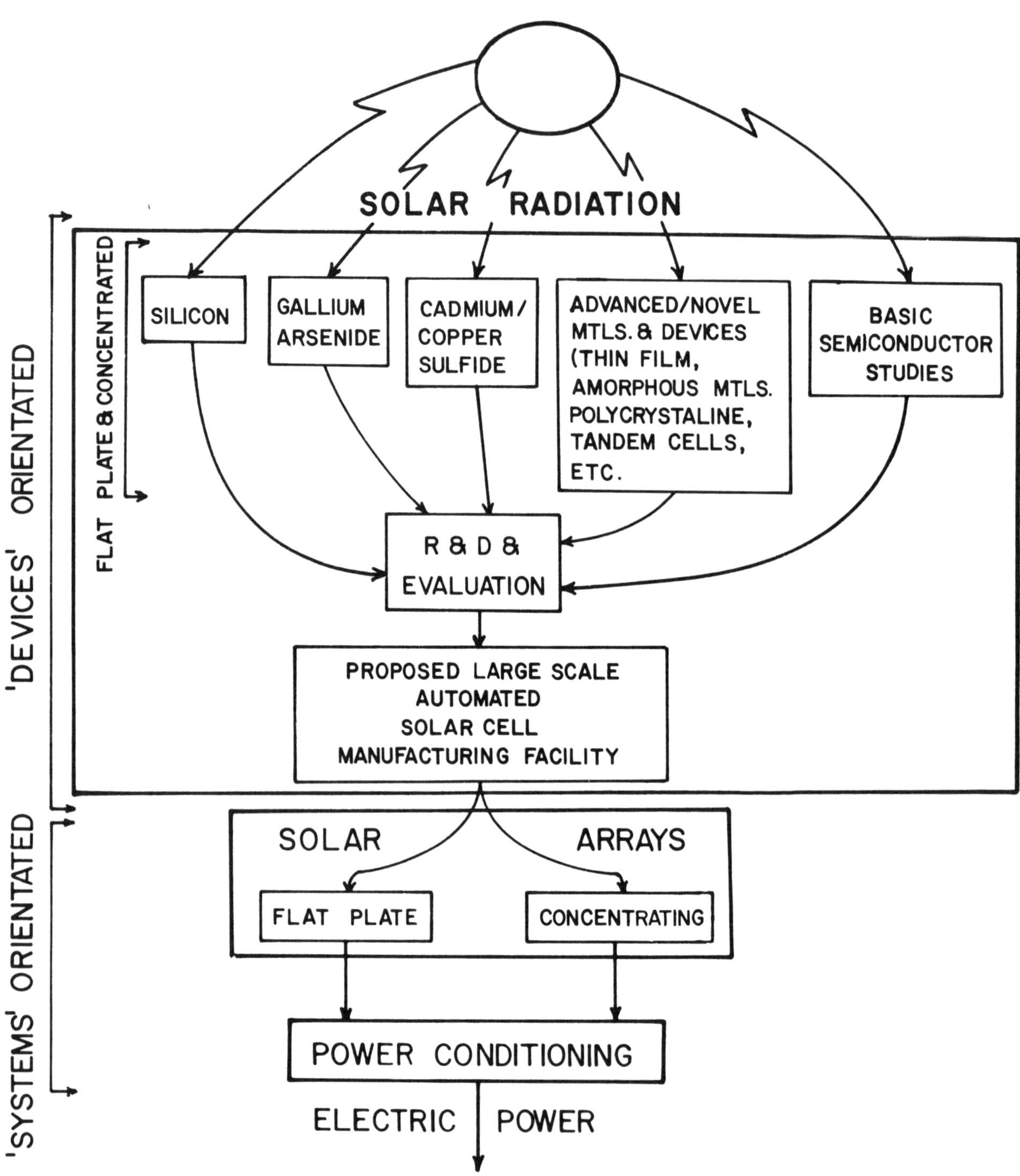

FIG. 2-1 One generalized way of structuring the terrestrial photovoltaic field in 1978. Other more detailed structures could be created. Silicon technology is currently being emphasized.

In the devices area there are advocates of each major technology shown in Fig. 2-1 diligently working to bring that technology to a viable state. Each of those shown has a substantial history which we need not go into here. In spite of this extended history it is still too early, many feel, to make a hardened choice for which single material or combination is 'best'. Such issues are currently being vigorously argued within the photovoltaic R & D community; there is a general expectation that at some time in the near future some one of these technologies--perhaps silicon--can move into the large scale automated factory stage. Presently commercial solar cells are made on a relatively small scale by several manufacturers using considerably less than fully automated manufacturing techniques. Cell cost is therefore higher than it might be under such mass production techniques.

In the systems area, the flat plate arrays (the simplest type) are capable of receiving both direct and diffuse solar insolation while the more complex concentrating arrays must generally have some type of tracking means because they absorb only the sun's direct radiation; these are more complex mechanically.

The power conditioning involves the interface between the DC array output and the utility grid or local power system[1]. Its function is to convert the DC to AC of proper waveform, frequency, voltage and power.

We shall have more to say about these systems aspects of photovoltaics in Sec. 2-5 where we deal with the integration problem.

The Advantages of Photovoltaics are important for the systems engineer to have clearly in mind. These are widely recognized throughout the photovoltaic literature, and we collect and summarize here the chief advantages seen in 1978. We do not attempt to differentiate between the different cell types, preferring to consider them as a set, nor do we try to rank the advantages here. They are a special case of the more general list for all of solar energy we accumulated in Table 1-3.

Photovoltaics:

- Have No Moving Parts in the classical mechanical sense which can wear out or give trouble. Positive and negative electrical charges within the material itself are the only 'moving parts', and these do not cause wear out insofar as is known.
- Contain No Fluids or Gases which can leak out as in some solar-thermal systems.
- Consume No Fuel to operate, as the sun's energy is free.
- Have Rapid Response In Output to input radiation changes; no long time constant is involved, as on thermal systems, before steady state is reached.

[1]Naturally some decentralized applications may not require power conditioning, or a minimal amount, if a separate DC wiring system is run throughout the building and only loads operable from DC are used thereon. This system, for example, has been used on the University of Delaware's solar house. Some DC systems use batteries across the output; they sometime require power conditioning to prevent excessively charging the battery.

- Operate At Environmental Temperatures (flat plate ones) with no inordinately high temperatures involved [26]--even for concentrating photovoltaics. This advantage implies a general tractability of materials, as materials do not have to be developed or sought out to work at elevated temperatures as is necessary in the thermal solar 'power tower' concept for the absorber-boiler.

- Basic Process Is Not Carnot Efficiency Limited as the solar cell is not a heat engine; in fact its performance generally deteriorates at elevated temperature.

  The Carnot efficiency constraint is the chief reason why we have classical fossil fuel electric power plants (heat engines) today with only about 35-40 percent efficiency at the most.

- Produce No Pollution in the classical sense of effluents or noise. A photovoltaic converter produces absolutely no noise. True, there is heat rejected, but ways exist now for utilizing this.

- Have Long Lifetime [27, p. 25] when carefully built, encapsulated (or otherwise protected) and installed correctly. "They have unlimited life", Rappaport [26] claims. Silicon, for example, of which many cells are made is one of the most enduring materials known to man. While photovoltaic array lifetime data is by no means complete yet from the systems engineer's viewpoint, it is already clear from available spacecraft data and some terrestrial array tests that solar cells can be engineered to have long lives. This feature of photovoltaics is particularly appealing to utility type application where 20-30 year life is desired; no one knows however whether solar cells in practical arrays will, in fact, last this long.

- Require Little Maintenance [27, p. 25] once properly manufactured and installed. Occasional washing off the collected dirt or dust or removing snow and ice accumulations are about the only maintenance required, assuming no vandalism.

- Can Be Created From Silicon, the second most abundant element in the Earth's crust [27, p. 25].

- Are Modular in nature, permitting a wide range of solar-electric applications as:

  - ▲ Small Scale for remote applications and residential use.

  - ▲ Intermediate Scale for businesses and neighborhood supplementary power.

  - ▲ Large Scale for centralized 'energy farms' of kilometer size arrays for supplementing utility power in a fuel saving mode.

Hayes [27, p25] points out that because of their modular nature "little is to be gained by grouping large masses of cells at a single collection site", however. The arguments advanced earlier in Sec. 1-6 regarding decentralization particularly apply to photovoltaics.

- Relatively High Conversion Efficiency Photovoltaic cells have given the highest overall conversion efficiency from sunlight to electricity yet measured [26, p36]. Fifteen percent efficient cells are common in 1978, and experimental cells in the range of 20 percent have been reported in the literature, though such cells would be costly. Some tandem cells are being researched whose efficiency is expected to be in the 27-35 percent range [28] [29] [30] [31] [32] [33] [34] [35].

- Have Wide Power Handling Capabilities from microwatts to kilowatts or even megawatts when modules are combined into large area arrays. Solar cells can be used in combination with power conditioning circuitry to feed power into a utility grid.

- Easy to Fabricate, "being one of the simplest of semiconductor devices", claims Rappaport [26].

- Have High Power-to-Weight Ratio [26]. This characteristic, more important for space applications than terrestrial, may be favorable for some terrestrial applications. The roof loading on a house top covered with solar cells, for example, would be significantly lower than the comparable loading for a conventional liquid solar water heater. Therefore ordinarily no extra structure--hence cost--would be needed for photovoltaics because of the high power-to-weight ratio.

- Amenable To On-Site Installation, i.e. decentralized or dispersed power; thus as Rappaport points out, "the problems of power distribution by wires could be eliminated by the use of solar cells at the site where the power is required" [26].

- Can Be Used In Combination With Thermal Systems providing the advantages:

  - ▲ Up to 60 percent of the available solar energy can be utilized;
  - ▲ The thermal collection uses the same land area as occupied by the buildings, and
  - ▲ The components fulfill several functions, yielding a more cost effective system [7, p55].

- Can Be Used With or Without Sun Tracking, making possible a wide range of application possibilities.

- Reliable Power Source when properly engineered and installed so that operation is within the cell performance envelope.

- Permits Electrical Energy Storage from whence the energy can be released months later if need be[1]; this is in sharp contrast to solar-thermal systems where thermal energy storage beyond a few days is impractical.

It is clearly evident that photovoltaics have a particularly appealing unique set of advantages; these, many believe, could and should be exploited on a much larger scale to eventually help relieve some of the nation's electrical energy problems dealt with in Chapter 1.

Their principal disadvantages in 1978 are their high cost and the fact that, in many applications, energy storage is required because of no insolation at night.

Physical Aspects of A Solar Cell are shown in Fig. 2-2.

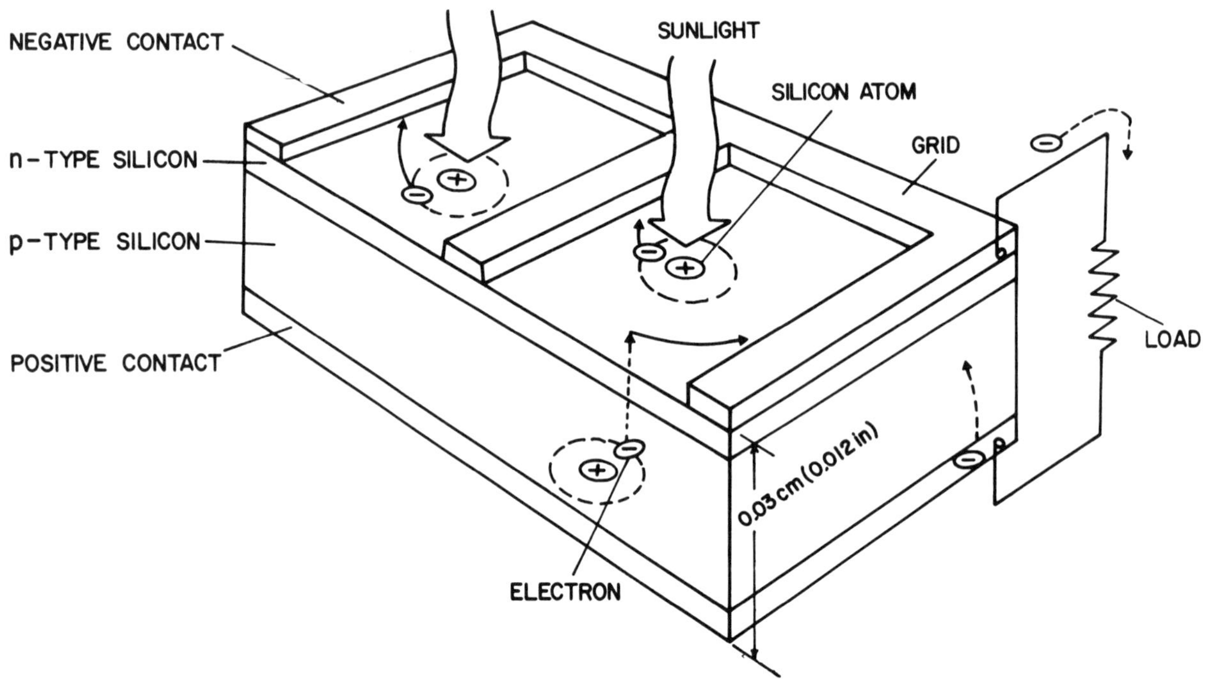

FIG. 2-2 Some physical aspects of a typical rectangular type solar cell. From [36].

First we note it is rectangular in shape. The rectangular module of a few centimeters by a few centimeters grew out of the space programs where a

[1]At the price of a small energy loss in the battery--getting energy in, leakage loss, getting it back out.

rectangular shape permitted maximum utilization of the exposed area and hence minimum array size and weight. The rectangular geometry has been carried over into terrestrial photovoltaics; but here the area packing factor is not so critical, and other lower cost geometries, e. g. round, semicircular, and other shapes, have been created. Many of these latter shapes can now be found commercially available. The point is that photovoltaic cells can be fabricated in a wide range of physical shapes. The maximum size for any one cell is presently limited, however, to that which can be cut from the man-made grown silicon ingot which can be up to several centimeters in diameter. There is the further limitation of size caused by possible cell cracking; large cells tend to get cracked more easily than smaller cells, basically because of the thin fragile nature of the cell.

Solar cells are physically thin devices. The one shown in Fig. 2-2 is only 0.03 cm (0.012 in.) thick--about the thickness of a few sheets of paper. Current research is aimed at making them even thinner to minimize the semiconductor material required, hence cost.

The cell is composed of a thin layer of N type silicon[1] only a few micrometers thick and a much thicker substrate of P type silicon.[2] Semiconductor materials are the only ones suited for photovoltaics. Electrical contact to the thin top layer is effected through the comb-like upper metal contact grid shown. There are many other types of contact grids possible other than that shown. The grid must make effective contact to the thin upper N layer while simultaneously not blocking out a large fraction of the sunlight.

The back side of the cell is usually covered entirely by a suitable metallic contact for the other electrode.

Many physical variations of the cell shown in Fig. 2-2 are possible and desirable for some applications. The upper surface might be texturized, i.e. chemically etched so its surface microscopically appears like continuous pyramids; the process minimizes light reflection from the upper surface with a resultant efficiency improvement. The upper surface might be completely coated with transparent encapsulant to minimize moisture and atmospheric corrosive effects; alternatively the entire cell might be under a glass or plastic pane for the same reason. Integral cover glasses, originally developed for space quality solar cells to protect the cell from high energy particle bombardment and irreversible damage, are not needed on terrestrial cells because the earth's atmosphere largely absorbs what high energy particles reach the vicinity of the earth; but, as indicated for the terrestrial application, some type of corrosion, moisture and mechanical protection is needed for the solar cell if it is to have long life and constant efficiency throughout life.

<u>Operation Of A Solar Cell</u>, at an elementary level, is a simple and straightforward process [26] [37, sec. II]. The key idea is to make the solar radiation cause charge separation, i. e. to cause the presence within the semiconductor junction of separate negative and positive charges. If separation is achieved, presumably these electrical charges could be made to flow through an external load, thereby delivering useful electrical power.

---

1. N type semiconductor material has electrons as the principal or majority electrical charge carrier.

2. P type semiconductor material has positive charges or holes as the majority carrier.

Consider the thin junction region in Fig. 2-2 between the upper N layer and the P type substrate. Assuming the incident solar radiation is in the right wavelength range[1] and that it arrives at the junction through the thin upper layer, then it may dislodge an electron from the crystal lattice. The instant the electron is separated an equal positive charge or hole is left behind. Thus we have the situation where the solar radiation has caused a positive and a negative charge called a hole-electron pair to appear in the junction region. They are therefore free within the crystal lattice to be acted on by whatever localized electric field exists. Fortunately, because of the nature of the P-N junction a potential barrier exists across it which is of the right polarity to cause an internal electric field which sweeps the mobile electrons to the N type upper electrode and the mobile holes to the P type substrate. The net effect of many such photon-atomic interactions is the external flow of electrons through the load as shown in Fig. 2-2. Thus useful electric power appears, having been generated directly from the sunlight. Some thermal heat is generated within the substrate also, and it is generally desirable that the heat be conducted out of the cell thermally. Otherwise the cell temperature rises, impairing the charge separation process.

The detailed operation is much more complex and involves an intimate knowledge of quantum physics and modern solid state theory. It is a very interesting and technically sophisticated area with its own peculiar language, symbols, etc. While it is essential for the device-orientated researcher in photovoltaics to be knowledgeable of such details, they are of lesser importance and interest to the systems-orientated engineer. Accordingly we leave the treatment of such matters to the textbooks and references.

<u>Typical Solar Cell Characteristics</u> are of major importance to the systems engineer to know about for the specific cells to be used.

From the explanation in the previous paragraphs it is evident the solar cell consists of a large area P-N junction. We would therefore expect it to behave like a semiconductor diode when unilluminated, i.e. under dark conditions. By measuring the I-V curve for an unilluminated cell we indeed find the general shape of the classical diode shown in the upper curve of Fig. 2-3. In the first quadrant it has the normal forward characteristic of all semiconductor diodes. Note that it falls in a power dissipating quadrant. The junction also has a small reverse leakage current, Io, in the third quadrant, also a power dissipating quadrant. Thus the unilluminated PN junction is a power dissipator having I-V forward and reverse characteristics in the first and third quadrants respectively.

When the junction is illuminated it has the effect of shifting the I-V characteristic downward as shown in Fig. 2-3. Thus a portion of the characteristic now is in the fourth power-producing quadrant. The cell is now a <u>deliverer</u>

---

[1] Only a portion of the sun's irradiance curve shown in Fig. 1-21 is operative in this charge separation process. The longer wavelengths don't have enough energy to cause the electron to leave the valence band, jump the energy gap, and arrive in the conduction band.

of power instead of an absorber of power. Note that in this region the voltage is in the forward direction but the current flows oppositely to what it would in the forward biased mode. The shape and location of the I-V characteristic in the fourth quadrant is of major importance to users of solar cells; since its illuminated characteristic in the first and third quadrants produces no power and is of little interest, usually only the fourth quadrant is shown.

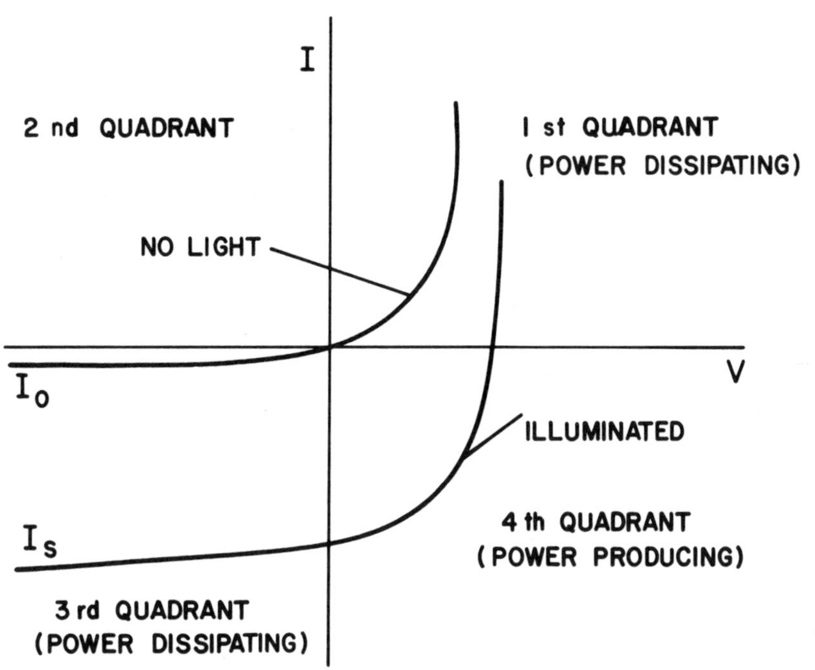

FIG. 2-3 Effect of light on P-N junction characteristic. After Rappaport [26].

Since it is awkward to work with fourth quadrant characteristics if large amounts of data are involved, it is the custom in photovoltaics to 'flip over' the fourth quadrant as shown in Fig. 2-4. Everybody then 'understands' that in the strict mathematical sense the curve really is in the fourth quadrant, but it is more convenient to present and use the curve in the first quadrant.

Fig. 2-4 shows the I-V characteristics for several space quality 2 x 2 cm solar cells. Note that solar cells have an open circuit voltage in the range of 0.6 volt. The short circuit current, naturally, is a function of the cell area, larger areas having higher short circuit currents. These particular curves compare three different kinds of cells under space conditions, i.e. air mass 0[1], and using a sunlight simulator. Their performance would be slightly modified under terrestrial conditions of air mass 1 or higher, the normal condition of most interest to the systems engineer; these curves however do illustrate the approximate best of the current state-of-the-art of non-laboratory space quality solar cells.

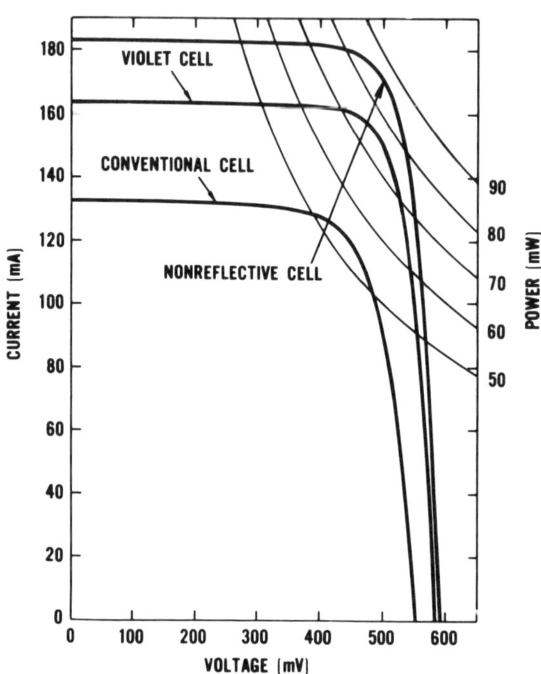

FIG. 2-4 I-V characteristics of conventional, violet, and non-reflective cells at 25°C under simulated AMO illumination. From Allison, et al. [38 , p1039]. Used by permission of The Institute of Electrical and Electronics Engineers, Inc.

---

[1] Air Mass 0, (AMO) signifies operation in a vacuum, i.e. no atmosphere around the cell. Air Mass 1 (AMI) signifies operation at the earth's surface with the sun directly overhead, i.e. the sun ray traverses the earth's atmosphere via the shortest possible path. This mode naturally accounts for atmospheric effects in the solar spectrum as shown in Fig. 1-21. Solar cell efficiencies measured under AMI conditions are usually a few percent higher than under AMO conditions.

The cell efficiences and power outputs at their maximum power points[1] are:

	P_o (mW)	η (Percent)
Non reflective cell	85	15.6
Violet cell	74	13.6
Conventional cell	56	10.3

It is evident that for the same insolation the improved cells yield both higher power outputs and higher conversion efficiency. It is interesting that the nonreflective cell surface is covered with pyramids two to four micrometers high formed by etching the surface, i.e. the surface is optically rough and not smooth.

The effect of changing insolation on a typical solar cell is shown in Fig. 2-5. Note that the cell characteristic moves in the direction of increasing power output as insolation increases. We also note that the short circuit current increases linearly with insolation whereas the open circuit voltage changes little.

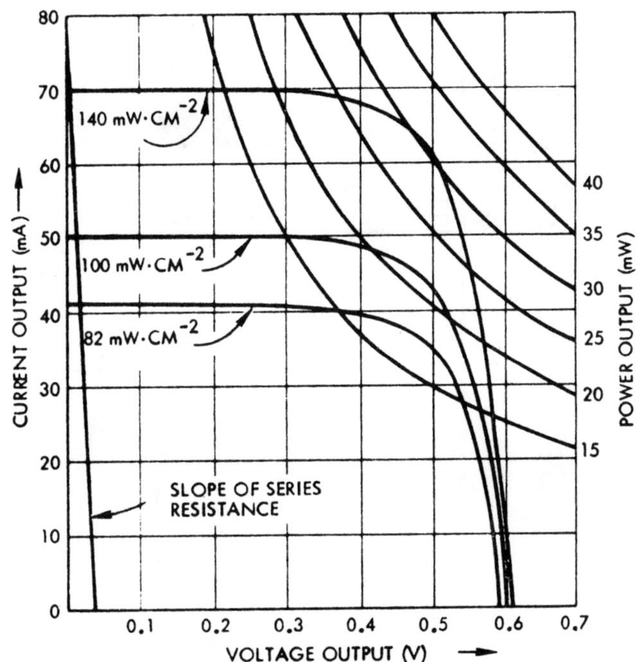

FIG. 2-5 Typical I-V curves of a 1 x 2 cm solar cell at three different illumination levels, assuming normal incidence, constant spectral distribution and constant cell temperature. From [39, 3.5-1].

[1] The maximum power point is where the highest hyperbola of constant power shown barely 'kisses' (osculates is the term a mathematician would use) the cell characteristic curve. This is always in the approximate vicinity of the knee of the curve.

Recall from Section 1-3 of Chap. 1 that the sun's insolation at the earth's surface is normally taken to be 1 kW/m^2. This translates to 100 mW/cm^2, a more convenient unit to work with in small area photovoltaic cells. Thus in Fig. 2-5 the 100 mW/cm^2 middle curve would be the nominal for terrestrial cells. Naturally the shape, location, and numbers for an arbitrary cell illuminated at 100 mW/cm^2 will vary, depending on cell type, manufacturer, and other variables.

The systems engineer should be aware of the fact that even though solar cell characteristics are generally presented in the form shown in Fig. 2-5--always under a fixed and known set of conditions--in an actual application conditions usually are quite different. For example, for a fixed array of cells only at solar noon would the cells be as shown in Fig. 2-5. At other times the radiant power density falls off in a cosine form, as shown in Fig. 1-22. This implies that at different times of the day the cells in the array have different I-V characteristic curves; but a complete family of cell characteristics would permit us to deal with this problem, as we shall soon see.

<u>The effect of cell temperature</u> on a solar cell is illustrated in Fig. 2-6.

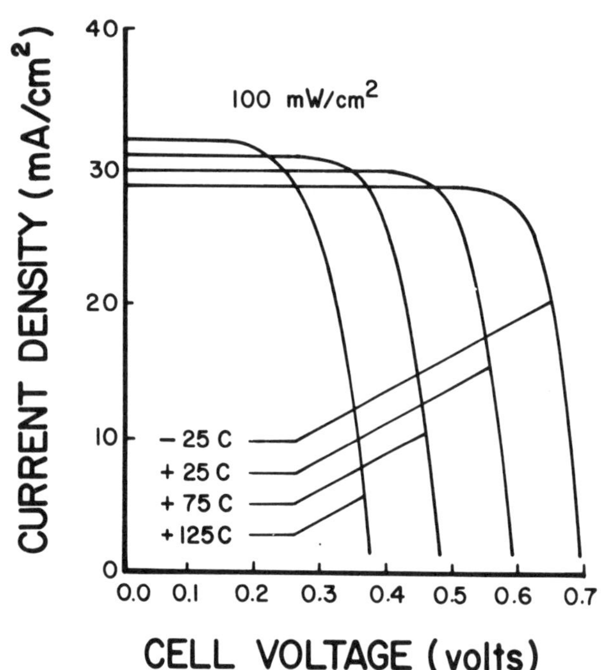

FIG. 2-6 Effects of temperature on a terrestrial solar cell I-V characteristic at 100 mW/cm^2 AM 1 conditions. From Kirpich [40, p1302].

It is easy to see that increasing the cell temperature results in a decrease in the cell output power. The characteristic shown in Fig. 2-6 is for an N on P cell, phosphorous diffused, 2000 Å junction depth, tantalum oxide anti-reflection (AR) coating, silver collector grid, Czochralski grown Si, boron doped, 2 ohm-cm material.

The fall-off of solar cell output power with temperature is known to be an important limitation of solar cells for some applications, and systems engineers should be generally aware of this characteristic. Heat sinking and other well known heat removal techniques can be used to minimize the problem. The effect is particularly acute where photovoltaics are being used with concentractors at many suns[1]; there the temperature can quickly rise to destructive ranges unless the heat is carried away. Cells specially designed for such high temperatures must usually be used in concentrating photovoltaics if the concentration ratio is high.

The Fill Factor, F, is another solar cell parameter of importance to a systems engineer comparing alternative solar cells. It is sometimes called the curve factor [41, p125] in the literature. The fill factor is found directly from the solar cell[2] I-V characteristic or its normalized equivalent, the J-V curve. Normalization is accomplished by dividing the actual cell output current by the cell exposed area to give typically milliamperes/cm^2 of cell area. Normalization to a standard 1 cm^2 area permits easy and quick comparisons to be made of different cells even though they are not all of the same area. Such comparisons can be a nuisance if one has to constantly take account of the differing cell areas. Normalization to J-V curves overcomes this difficulty.

The Fill Factor is based around a simple idea embodied in Fig. 2-7. It is simply the dimensionless ratio of the two areas shown. An ideal solar cell with a sharp knee would have a fill factor of 1.0. Fill factor may be 0.8 or higher for a well designed solar cell. Excessive series resistance and other effects can seriously lower the fill factor; thus fill factor is one measure of the quality of a solar cell--both of design and of manufacture.

Solar Cell Equivalent Circuit is shown in its approximate form in Fig. 2-8. [42] [22]. It is an approximate model because a solar cell is basically an area distributed kind of converter, and a more exacting model would account for this distributed nature of the cell. Wolf and Rauschenbach [43] investigated such distributed constant models and found that excellent agreement between measured and calculated I-V characteristics could indeed be achieved but that such models exhibit considerable complexity. The model in Fig. 2-8 is the

---

[1] 'Suns' refers to the solar concentration on the input optics to the cell. 1 sun = 100 mW/cm^2, 10 suns = 1000 mW/cm^2, etc.

[2] A similar 'fill factor' could also be found for a solar array of cells as an indicator of the goodness of the overall array; this use is not common apparently in the photovoltaic literature where chief interest has been in the cell.

generally accepted equivalent circuit model, though even it can be further simplified in normal power operation by deleting the R_{sh} element. Prince and Wolf analyzed [44] the effects of R_{sh} and found it became significant only when cells were made to operate under low illumination levels.

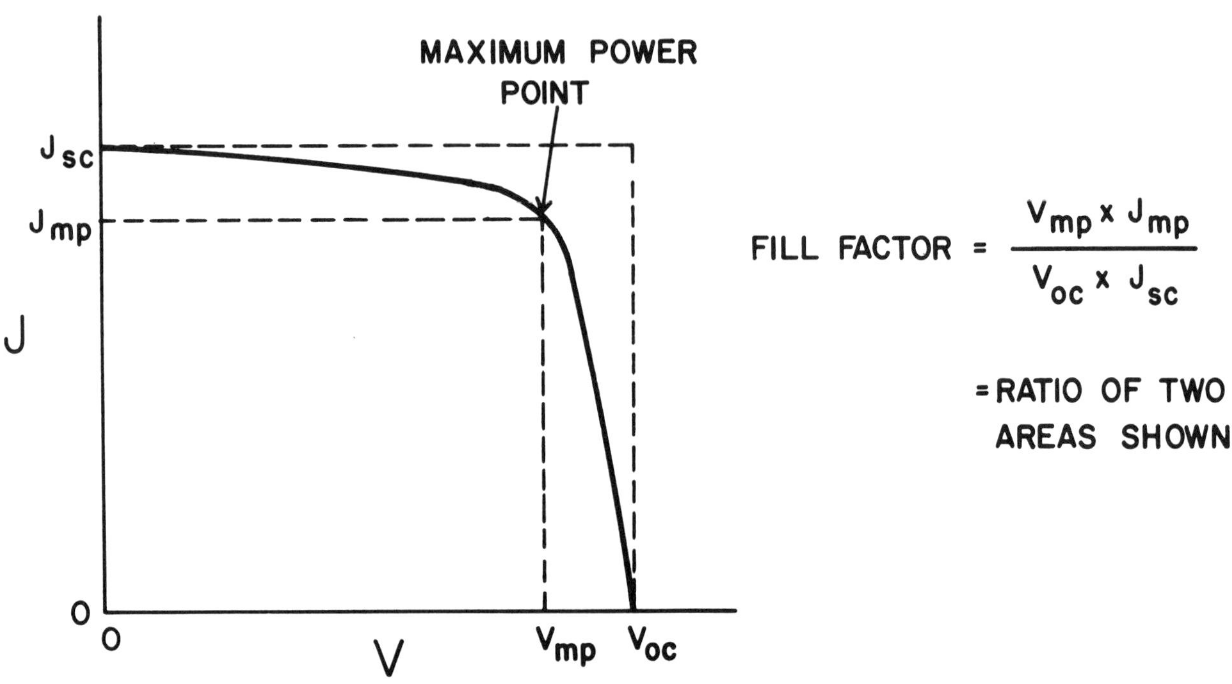

FIG. 2-7 Definition of the Fill Factor of a solar cell.

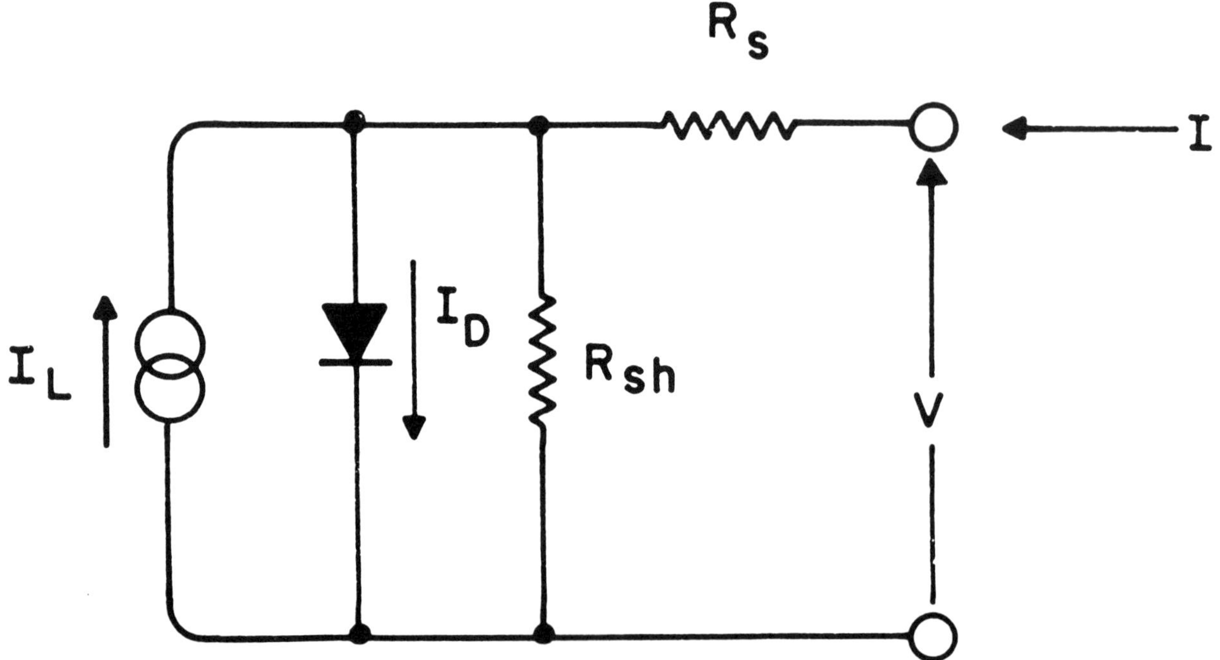

FIG. 2-8 Approximate lumped-constant equivalent circuit model for a solar cell. From Wolf [45, p131].

We note that the equivalent circuit model is in the form of a constant current generator, I_L, as the principal forcing function[1]. Spratt [46, p404] says it can be shown that

(2-1)

$$I_L = q\,\eta\,(\lambda)\,\phi\,(\lambda) \qquad \text{amps.}$$

where:

q = electronic charge

$\eta(\lambda)$ = quantum efficiency
= no. of hole-electron pairs released within the P-N junction per quantum of incident light energy at wavelength λ.

$\phi(\lambda)$ = photon flux incident on the cell junction
λ = wavelength of incident light on the cell junction

It is thus clear from (2-1) that the current generator magnitude is directly related to the quantum efficiency of the cell and the incident light photon flux. It is then easy to see why the solid-state physicist would place so much emphasis on trying to choose cell materials and geometry which maximize the quantum efficiency for all λ. Wysocki and Rappaport [47] give an alternate expression to our (2-1) which relates I_L and the illumination.

Finally, we observe that if R_s is sufficiently low to minimize the voltage drop across the diode, then $I_L \sim I$ short circuit of the cell. R_s is principally due to the resistance of the thin surface layer on the cell plus the resistance of the grid contacts to it.

The diode current, I_D, according to Spratt [46] and other previous researchers [26] [48] is:

(2-2)

$$I_D = I_o\,(e^{\alpha V_j} - 1) \qquad \text{amps}$$

where:

I_o = reverse leakage current of diode
$\alpha = q/\eta kT$
η = diode ideality factor (also designated A in some papers)
V_j = junction voltage

The load current is simply, assuming $R_{sh} = \infty$

---

[1] In the photovoltaic literature there is a tendency of authors to not always relate this current generator magnitude back to the incident light flux.

(2-3)

$$I = I_L - I_D$$

Eqs. (2-1), (2-2), and (2-3) may be combined in many useful forms.

If the load current is divided by the cell area to yield the cell current density, J, then the cell power generation density, P, is

(2-4)

$$P = V \times J \qquad \text{watts/cm}^2$$

Multiplication of (2-4) by the total cell area results in the absolute power output for an array of parallel cells ideally interconnected.

<u>Efficiency And Operation at Maximum Power Point</u> is a subject of substantial concern to the systems engineer. Consider a physical solar cell delivering useful power to a load as depicted in Fig. 2-2. We would like to know how the cell I-V characteristics relate to those of the load, how to find the operating point for the system, and what the conversion efficiency is.

We start with the plot of the load line shown in Fig. 2-9 (a).

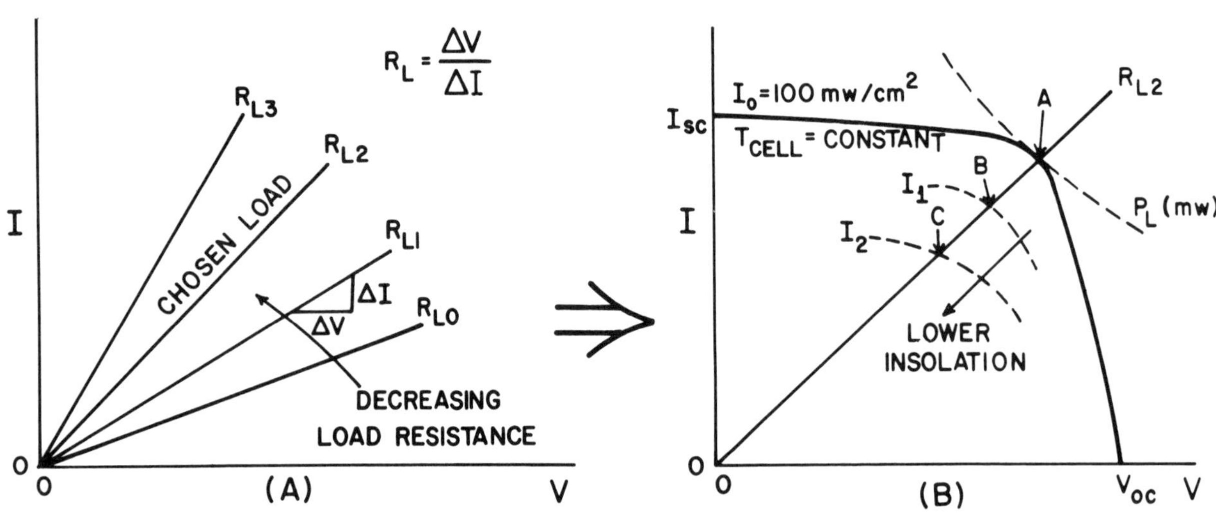

FIG. 2-9 Integrating load (A) and solar cell characteristics to find the system operating point and conversion efficiency (B).

Here a family of load lines is plotted, each of constant resistance over the range of interest. We choose I and V scales to be the same as those for our solar cell. This set of I-V characteristics for the load is entirely independent of solar cells, being governed solely by Ohm's law.

We now elect to superimpose our load characteristics on those for the solar cell as shown in (B). It is then evident wherever the load and cell characteristics intersect will be the operating point for the cell-load system. For the sketch shown the operating point is at A, the maximum power point, because we deliberately chose R_{L2} to fall on the cell's maximum power output point, delivering power P_L to the load. We've assumed the insolation is 100 mW/cm^2 and that the cell temperature in some way has been fixed at a constant known value.

What is the conversion efficiency, η, of the system corresponding to operation at point A? It is easily seen, from the definition of efficiency, eq. (1-8), to be

(2-5)
$$\eta = \frac{P_L}{I \times A} \qquad \frac{(\text{mW})}{(\text{mW})}$$

where

A = The cell area (cm^2)
I = Insolation[1] = 100 mW/cm^2 for illustration shown. (Assumed normal to the cell).

We now inquire about the loci of the system operation when the insolation is changed. If it is lowered to I_1, our operating point shifts to point B, again the intersection point of the cell and load characteristic curve at that insolation. If it is lowered further to I_2, operation is at point C, and so on.

The same general procedure would be followed had we initially chosen a non-optimum load resistance.

From the above discussion it is evident that, given a measured I-V curve for a solar cell, finding the operating point for a given load resistance is a straightforward procedure from which the conversion efficiency can also be easily found graphically.

Finally, using eq. (2-5) and the definition of the fill factor shown in Fig. 2-7, it is easy to see that the conversion effiency also is

(2-6)
$$\eta = \frac{V_{oc}\, I_{sc}}{I \times A} \times (\text{Fill Factor})$$

which explains why it is desirable for the solar cell designer to make the

[1]The symbol I here for solar insolation should not be confused with an electrical current any place in the system. The distinction is usually evident from the context.

fill factor as nearly 1.0 as possible. Eq. (2-6) also tells us of the desirability of making V_{oc} and I_{sc} as high as possible for the cell. These are largely fixed by the cell designer, but the system engineer has some control over them via the insolation and cell temperature as shown earlier in Figs. 2-5 and 2-6 respectively.

The general graphical methods elucidated above furnish the basis for experimental determination of the conversion efficiency. By placing the cell to be measured at the end of a long light collimating pipe and aiming the assembly directly at the sun, the I-V characteristic curve for the cell can be directly plotted on an x-y recorder by simply varying the load resistance from a low to a high value. Care must be exercised, however, to insure that the recorder is, in fact, measuring the cell terminal voltage and not the load voltage which may be occasioned by lead drops. Assuming a pyrheliometer was simultaneously used in the above measurement to measure the insolation, the measured maximum efficiency can easily be computed from the ratio of the measured maximum load power to the measured insolation times the cell area. This experimental method has the merits that the measurement can be made under terrestrial direct sunlight conditions and that it is quite simple to execute once the apparatus is set up. Skolnik [49] has described this method used many years in the author's laboratory; similar apparatus is installed at Goddard Space Flight Center and probably at other laboratories around the country. It has the disadvantage in some locations of having to wait for a cloudless day to perform the measurements.

We, once again, make the observation mentioned earlier that it is evident from the previous technical discussion and equations that the solar cell-load system is not limited by a thermodynamic Carnot efficiency. There are, however, other limitations of an electronic nature.

Photovoltaics And The Net Energy Issue An introduction to photovoltaics would not be complete without mentioning the net energy issue. It is embodied in the ratio of the net electrical energy produced over some specified time to the total energy required to manufacture the cell. This general kind of analysis appears to be in its infancy, and calculations made using the methodology vary widely at present. One criterion sifting out of this area is the so-called pay-back time defined as the time required for the cell to operate to produce as much energy as it took to build it initially. Beyond the pay-back time more energy would be produced than was used in its manufacture. This leads to the notion of 'photovoltaic breeders' where the excess energy theoretically could be used to produce more solar cells each of which could in turn become breeders and so on--all without consuming any nonrenewable resources once the breeder process was started.

Hayes [27, p27] points out that "from a 'net energy' perspective, photovoltaics are appealing. Detailed studies of the energy needed to manufacture such cells shows that the energy debt can be paid in less than two years of operation". He says that "The two-year payback period (for cells with an expected lifetime of more than 20 years) has become conventional wisdom among the silicon photovoltaic specialists". He cites Slesser and Hunam [50] as the source of the two year calculation.

More recently Wihl and Scheinine [51] have developed a model called the Solar Breeder which allows them "to carry out a simulation and analysis of a photovoltaic manufacturing process based on any technology employed, and it calculates its energy demand and energy return to society". They conclude that the payback period is typically five years and "the Solar Breeder can swing into the net energy mode earlier than generally expected and consequently represents one of the major potential energy sources of the future".

It is highly probable we shall see much more published in the future on the net energy issue for photovoltaics.

We close this introduction to photovoltaics to examine in the next section selected portions of the prior photovoltaic art of particular interest to systems engineers.

2-2 Prior Photovoltaic Methods

By 'prior' we mean here before about 1975, the arbitrary dividing date chosen in this research between prior and present art.

In this section we limit our view to that of the "devices", i.e. the solar cell only, for the solar cell is the basis of photovoltaic power producing arrays which are of most interest to the systems engineer.

The topics and materials in this section have been highly filtered from the extensive and very detailed R & D literature on photovoltaics. The intent was to select only those topics of concern to a systems engineer whose interest is to constructively use solar cell arrays for terrestrial purposes. It was therefore necessary to prune topics on the highly complex solid state physics and materials involved in virtually all photovoltaic converters, regardless of the type material. That art appears to also need integrating, but this task was beyond our present scope.

Also largely deleted here was the extensive photovoltaic art relating to spacecraft unless portions of that technology were more or less applicable to terrestrial use. The whole subject of high energy particle damage to photovoltaics--so important to space craft cells--is not relevant to terrestrial photovoltaics and was therefore deleted. Nor has the whole subject of array deployment mechanisms been included, as terrestrial arrays are mostly mounted in one location; there is no need for compact stowage and then deployment as for space.

We first briefly identify some publications summarizing the prior art in the complex photovoltaic R & D field--publications which may be of most value to the systems engineer previously inexperienced in photovoltaics and the electrophysics-materials inherently involved in this field.

In the latter part of the section we identify other topics of interest to the systems engineer and indicate where more information is to be found.

Helpful Review Publications on photovoltaics include the following references.

1961

- Wolf [52] in a significant paper summarized the state-of-the-art of photovoltaics about this time. Though he emphasized photovoltaics in the space program, the paper accurately mirrors the R & D then. He dealt with the silicon cell, the measurements problem including the need for creating sunlight simulators, improvements in the silicon cell made about 1960, and new configurations and approaches on silicon solar cells. The problems of preparing cells from grown crystals were dealt with. The early thin film solar cell work was summarized. Research work on non-silicon cells, specifically gallium arsenide, gallium telluride, and cadmium sulfide was summarized. These latter did not lead to any improvement in performance or economy compared to the approximately 11 percent (AMO) efficient silicon arrays then.

<u>1964</u>

- Rau [53, Ch. 8] in an interesting and easily readable style reviews the history of photovoltaics from the telegraph engineer Joseph May (1873) discovering light effects on his selenium resistors on the transatlantic cable, to Bell Laboratory's 1954 'solar battery' first installed at Americus, Georgia ($\eta \sim 6\%$). He also shows some interesting terrestrial applications of solar cells on low powered devices.

<u>1965</u>

- Cherry [54] sums up photovoltaics technology as of about this date. Though his principal outlook is for spacecraft applications, many of the current solar cell ideas were born in the space era. The superiority of silicon cells was evident in 1965, with Ga As cells then costing 30 to 60 times silicon cells. Shingle and flat mounting the cells was clearly in existence then with preference, overall, for flat mounted cells because of the ease of replacing broken cells. Cell interconnections using expanded metal were in use, permitting stress relief between cells and modules during temperature cycling.

  Vee ridge concentrators were beginning to be seriously examined. Concentrators were recognized to provide increased cell output and reduced array weight. Many areas then needing improvement were clearly identified: thinner cells, wrap-around cells, dopant impurities, cover glasses, lightweight substrates, materials other than silicon, stacked cells, and the possibility of tracking an array to always operate at its maximum power point.

<u>1966</u>

- Cherry and Zoutendyk [8] did a paper then on the state-of-the-art, somewhat elaborating on the ideas in Cherry's 1965 paper above but more orientated toward arrays. The problem of electrically insulating the cells from the mounting structure was dealt with along with the use of silicone elastomer adhesives to bond the cells to the support structure. Work was underway to develop modules that could be welded rather than soldered together to improve array reliability under thermal cycling conditions. Equations were given permitting solar cell temperature to be calculated based on measured absorptivities and emissivities of the solar cell. The R & D problems seen then in cells, arrays, and concentrators were clearly exposed. Solar cells in arrays at this time had efficiencies about 11 percent (AMO).

- Elliott [Ch. 1] in a book edited by Sutton [18] had a section in which he attempted to sum up the state-of-the-art of solar cells. His summary of developments in materials other than silicon is clear and easy to follow. Cells made of Ga As were reported to have η = 11 percent, Cd Te about 6 to 8 percent, and CdS up to 4 percent. The I-V curves were shown also.

1968

- Crossley,Noel and Wolf [45] issued a massive final report called "Review and Evaluation of Past Solar-Cell Development Effort". This towering work examines solar cells in great breadth as well as depth, starting with Becquerel's 1839 paper on liquid cells. Their report, in this author's opinion, is by far the most comprehensive state-of-the-art document generated to date in the photovoltaics field and is readable and interesting to both 'devices' and 'systems' engineers. A total of 528 papers and books were surveyed, and a bibliography arranged chronologically was accumulated. One can hardly leave this impressive piece of scholarly work without recognizing the large number of effects, materials, and configurations for solar cells which have been thought of, investigated and many discarded by the scientific R & D community.

  This volume is a highly recommended entrée to newcomers to the photovoltaic field desiring perspectives on the prior art.

1972

- This year saw the formation of the NSF/NASA Solar Energy Panel. Their final report [7, p52] succinctly summarized the photovoltaic art as of that time, though in no detail. The use of photovoltaics on buildings was stressed along with ground central stations of photovoltaics. Key problem areas were identified, development goals set, and some appraisal of success likelihood made.

  This historic management type report later formed the broad framework on which the nation's terrestrial photovoltaic program was based.

1973

- This year a very important national conference, Photovoltaic Conversion of Solar Energy for Terrestrial Applications, was held in Cherry Hill, N. J. It was sponsored by NSF and organized by JPL. Virtually all those actively involved in photovoltaic research in the United States attended. The published proceedings [55] of this conference, in toto, constitutes the state-of-the-art at that time. Virtually every known aspect of photovoltaics was presented and extensively discussed.

  It was at this meeting that the important national goal of producing solar cells for \$500/kW (\$0.50/peak Watt) was set by the photovoltaic community after extensive and sometimes heated discussion.

- This year saw a growing global interest in solar cells for terrestrial applications as manifested by the Paris, France International Congress by CNES [56]. World-wide developments in solar cells and their applications were presented and discussed.

1974

- JPL issued a report [57] assessing the technology required to develop photovoltaic power systems for large scale use. Alternative candidate photovoltaic systems and development tasks were analyzed. It pointed up the need for an early emphasis on the development of the single crystal Si photovoltaic system for commercial utilization; a production goal of 5×10^8 peak W/year of $0.50/peak W cells was projected for 1985. Envisioned applications ranged from houses to central power plants. The report included a plan, milestones, schedules, and funding requirements based on a technical analysis and evaluation by JPL.

1975

- Interest in thin film solar cells appeared to be strong this year. Li [58] summarized the state-of-the-art in this area, discussing polycrystalline silicon thin film and Cu_2 S-CdS thin film cell operation, construction, and performance.

1976

- Backus [14] edited his book, Solar Cells, in which he collected the classic papers in the photovoltaic field. He also assembled what was probably the most complete bibliography to then on photovoltaics, arranged chronologically. His information was limited to through 1974.

  In it, Wolf [p38] in an interesting paper titled "Historical Development of Solar Cells", attempted to integrate the complex history of solar cell R & D. He prepared a chart (his Table 1) showing the timing of the various key events from the early 1800's discovery of selenium to the 1967 model of Cu_2-CdS solar cell. He also showed a graph of silicon solar cell costs ($/W) from 1956 through 1972, ending at about $50/W then. His paper brought severely needed perspectives to an otherwise highly diversified photovoltaic devices field.

  Backus' bringing together of these important technical papers rendered a much needed integrative service to the photovoltaic community.

1978

- Palz [25] published a book in which he gave a succinct history (p191) of the photovoltaic effect up through about 1975. Those systems engineers preferring a quick glimpse of the prior art would do well to examine this source.

It is thus evident from the above citations that there are several papers and books attempting to capture the prior art of photovoltaics and to permit one to see the larger patterns within this highly sophisticated R & D field.

We now leave the review subject and focus on several selected topics taken from the prior art literature on photovoltaics, all of importance to the systems engineer.

Photovoltaics And Carnot Efficiency It is of considerable interest to the systems engineer to know whether, fundamentally, the photovoltaic converter is limited by the Carnot efficiency. In the 1957-1967 literature there were conflicting statements on this question.

Essigman [59] has examined this question in some detail to resolve the conflict. He makes the important point that "The Carnot limit applies only to a heat engine, which is a system operating in a cycle while only heat and work cross its boundaries. Therefore, a device must fit this definition in order to be subject to the limitation." He argues that the only way a photovoltaic system fits the definition is to define "the system" as consisting of the sun as the hot source at about 6000 K and the temperature of the cell and its surroundings at about 300 K, thereby establishing a Carnot limit of (6000 - 300) /6000 = 95%. Thus he points out that "whether a solar cell is so limited or not is academic for practical purposes", as solar cell efficiency is limited by the mechanisms of the energy conversion process.

But if "the system" is defined only as solar radiation impinging on a solar cell on the earth, then it does not fit the Carnot definition of a heat engine and the Carnot efficiency limitation does not apply.

Avoiding the Carnot efficiency limitation, as pointed out in Sec. 2-1, is one of the salient advantages of photovoltaics.

Choosing The Type Cell to use in a given terrestrial array is an important decision for a systems engineer. The choice of cell types is, as with many engineering decisions, the result of conflicting factors making trade-offs necessary; the systems engineer is well advised to study this whole area carefully before making his choice.

There is firstly the practical consideration of what type cells are commercially available at the time the choice is made. One would of course seek to obtain the latest commercial literature, samples, and data in addition to being generally familiar with the R & D of each type cell.

The choices are basically those revealed in Fig. 2-1. Silicon has a long well proven history; cadmium sulfide is more recent. Mytton [60] in a helpful review paper compares the relative merits of these two candidates on the basis of costs and efficiency. He points out that the prime attraction of CdS "is their lower cost which, for medium scale production to meet terrestrial markets from a few kilowatts to a few megawatts annually, is expected to be less than half that of silicon."

The systems engineer should be aware that silicon cells have had a long history of stable operation under both space and terrestrial environments. Cadmium sulphide cells, on the other hand, have had a record of problems with stability throughout life, particularly from moisture; encapsulation or protection of some kind is necessary. Also the systems engineer should know that, according to Mytten's data in 1972, silicon cells had a mean conversion efficiency of about 12 percent while CdS was only 6.5 percent. This implies that if a given electrical power is to be delivered the cell array area required will be almost twice as large for CdS cells than for silicon.

The systems engineer would also want to consider the merits of any new material cells which might have recently been commercially introduced before making a responsible choice of cell type.

Solar Cell Models were mentioned earlier in Sec. 2-1. There has been on-going work in the prior art toward developing mathematical models that will accurately describe solar cell performance. In addition to the modeling efforts cited earlier, Barrett [61] investigated (1970) a solar cell performance mathematical model based on a computer program. Though his model was intended for space solar cells, it may be possible to modify it to be suitable for the terrestrial solar spectrum; this is an unknown.

Experimental data points he obtained appear to agree reasonably close to his computer program predicted results, though he recognized additional refinements were needed to narrow the errors even more.

Systems engineers interested in pursuing the cell modeling issue further may find Barrett's report a useful starting point.

Cell Contacts And Interconnects Systems engineers should be aware of the fact that the problem of how solar cells are to be reliably physically connected in an array is a nontrivial technical design problem. Interconnecting wires or straps tend to fatigue crack, welds break, solder interfaces to cells peel off, solder that wasn't supposed to crack does, and other types of interconnecting failures have been observed on solar arrays--all leading to open circuits and partial or total loss of array power. Hence the array reliability is in no small part determined by the security of the cell interconnection means chosen by the array design engineer.

Rauschenbach and Gaylard [62] nicely state the nature of the problem. "Cyclic variations in temperature cause alternating thermal stresses in solar cell interconnecting systems. These stresses may lead to the fracture of soldered or welded joints or the interconnector or the silicon solar cell itself. This failure mechanism, called fatigue, occurs by the formation of minute cracks which gradually grow until the strength of the member is so reduced that the member ruptures completely." Their paper analyzes this important problem.

Garner and Rauschenbach [63] showed how statistical analysis can be applied to the interconnect failure problem. Eakins [64] showed how one company dealt with this problem by welding. Curtin and Billerbeck [65] studied advanced interconnects and showed that use of silver plated INVAR can result in an order of magnitude improvement in fatigue life compared with that of materials used previously. Reference [39, Ch. 5] summarizes JPL's space-related experience on interconnections.

Whether interconnections tend to fail as soon in a terrestrial environment as in space appears an unknown. In space the depth of the thermal cycles is much greater than on the earth's surface, and one would therefore suspect less of a problem terrestrially on this issue; but apparently no one has shown whether this consideration is in fact valid.

Solar Cell Encapsulants to protect it from the various deteriorating environmental influences is another subject of concern to the systems engineer. The protection of solar cells on space vehicles from high energy proton, electron, and micrometeorite bombardments has a long history which will not be reviewed here. That work principally resulted in the addition of special inorganic 'cover glasses' adhesively mounted to the cells [39, Ch. 4].

In about 1973 attention started to be directed toward using one of the various polymeric plastics for cell protection. A sample of the considerations involved in choosing the plastics is in Fayet's [66] paper. He was principally interested in protecting Si and CdS cells for space use, but some of the considerations may have a carry-over value to protecting terrestrial arrays.

More recently (1975) Carmichael, et al, [67] were researching the general encapsulation problem from the standpoint of terrestrial arrays having a useful life of 20 or more years. One of their first tasks was to assemble and evaluate world experience and properties of encapsulating materials. They quickly found most of the previous R & D had been space orientated and that the classes of materials which had been used as protective covers on silicon solar cells included halocarbons, polycarbonates, polymides, acrylics, epoxies, polyesters, and silicones. They found the adhesives and primers fell into the classes of silicones, epoxies, and elastomers. They pointed out the important conclusion to the systems engineer that "Only limited life data on encapsulant materials for terrestrial environments is available, and the actual experience to date is much shorter than 20 years, with service experience for less than 2 years being typical."

Solar Cell and Array Manufacturing is of some interest to the systems engineer.

First, we must know that prior to about 1975 only a small number of firms were actually manufacturing solar cells or arrays or both. They principally serviced the spacecraft industry with its erratic non-constant demand for solar cells. Because of the erratic market nature several firms who had been in the business left; a few have recently returned, and several new small manufacturers of various solar cells have been created within the past few years. All these firms--old and new--have been reluctant to commit the necessary large capital needed to create modern highly automated solar cell or array manufacturing capabilities, especially in light of an uncertain and unproven terrestrial solar cell market.

Methods of producing solar cells and arrays in 1974 were very expensive because the production rates were small and the processes were not mechanized.

The 1972 Solar Panel Report [7, p65] envisioned the national need for photovoltaic manufacturing capability when it said "The main development required is the production of very low cost arrays, regardless of what ultimate photovoltaic system is used". It proposed the construction of a low cost solar array pilot plant and laid out the expected timing for realizing it.

One step in the direction of a mass production solar array plant was taken in a 1974 report to NASA by Bartels [68] of Spectrolab who concluded that "The cost of solar cell arrays without concentration can be reduced to approximately $250/m^2 (or $2/peak Watt) through the utilization of known technologies and mass production equipment. A detailed cost analysis showed that this objective could be reached utilizing process technology that is known today and conventional crystal growing and slicing methods. No major technological breakthrough would be required, and the effort would essentially involve production engineering activities rather than a basic research and development effort. There would need to be a growth in production rate to about 2.5 MW per year to justify the mass production equipment."

As an apparent follow-up to Bartel's above study, Taylor and Schwartz, also of Spectrolab, in a 1975 report [69] undertook "to determine the process steps and design requirements of an automated silicon solar cell production facility. Identification of the key process steps was made, and a laboratory model was conceptually designed to demonstrate the feasibility of automating the silicon solar cell fabrication process. ---The study--established the technical feasibility of automating the solar cell manufacturing process--. Estimates predict--an estimated manufacturing cost of $0.866 per cell or $1.22 per watt."

Graham, et al [70] in 1976 also investigated a program to study process routes for silicon solar cells.

It was thus evident in 1975 that it was technically possible to create a mass production manufacturing facility to produce silicon solar cells at a price reasonably close to the $0.50/Watt cost goal. Seemingly only the capital and the management decisiveness needed to do the job were missing before such a mass production factory could become a reality.

A number of researchers have also investigated the possibilities of creating ribbons of single crystal silicon aimed eventually at mass production. One sample of such process-orientated research can be found in Hilborn and Faust [71] who set out "to develop web-dendritic process methods that will (1) minimize the cost of processing silicon into ribbons of solar cell quality with a terrestrial energy conversion efficiency greater than 10 percent, and (2) be suitable for large quantity production".

Testing Procedures for cells and arrays is a subject of importance to a systems engineer.

It is well known that there are two ways to test solar cells or solar arrays.

- Use Natural Sunlight (AM1). This method has the merit that the cell or panel is irradiated with the actual serrated terrestrial solar spectrum earlier shown in Fig. 1-21 which is the same as the cell or panel would later see in actual power producing service. Skolnik [49] described an example of a test set of this type used at the University of Florida for testing cells from 1971 on.

Ray [72, pl] succinctly pinpoints the disadvantage of this method. "The sunlight method for testing solar panels has always presented a problem. There are few cloud-free days with uniform light intensity; therefore, the panels must be tested with existing sunlight conditions, and then the resulting data must be normalized. In the past, many tests on solar panels and the satellite were delayed or not run because of insufficient sunlight at the scheduled test time. Also, control of the temperature of solar panels being tested in sunlight is difficult, if not impossible, whereas the artificial sunlight of the test set permits the temperature of the solar panel to be measured and controlled."

- Use Sunlight Simulator to overcome the above difficulty. Ray [72] in 1970 developed such an apparatus at Johns Hopkins APL to permit testing of spacecraft solar panels. It used iodine-quartz (tungsten) lamps as the radiant energy source. Others are known to have developed sunlight simulators also; this area was not pursued further in this research.

  One of the major problems with the sunlight simulator is to get the spectrum of the illuminating lamps to match that of natural sunlight. This is a nontrivial physics type problem difficult enough for space AM0 conditions but even more difficult for terrestrial AM1 conditions to match the sun's serrated spectrum shown in Fig. 1-21. The problem is further compounded by shifts in the spectrum at Air Masses higher than 1 which are encountered for all sun positions except directly overhead.

Many systems engineers will undoubtedly argue that since the terrestrial array is to be used in natural sunlight it should thus be tested in natural sunlight, thereby avoiding the spectral matching errors of sunlight simulators in addition to their initial cost.

The whole question of terrestrial photovoltaic measurements was studied and brought into focus at a significant 1975 workshop held at NASA Lewis Research Center. A workshop Proceedings [73] was published containing much helpful technical information on the various aspects of solar cell and array testing.

Later Brandhorst et al [74] published a report laying out interim solar cell testing procedures for terrestrial applications based on the results of the above workshop.

Testing is a complex technical area beyond our in-depth scope here, but the systems engineer involved in test planning or actual tests on photovoltaics is well advised to consult the last two cited information sources.

Standardization of Solar Cells as an important issue has not gone unnoticed by the photovoltaic R&D community. Standardization when implemented offers the major advantage of cost savings. If not initiated too early, i.e. while a given cell type is in the early R&D phases, standardization can also greatly stimulate the large-scale commercial introduction of photovoltaics. It has other advantages also, as minimizing supply and stocking problems.

In 1974 NASA initiated a standardization program for silicon solar cells. By then silicon cells had experienced a long successful history. NASA's overall goal was to develop standard specifications for solar cells and covers; it was envisioned that later arrays might also be standardized. This program is described by Bifano and Forestieri [75]. While this effort was space directed, it has obvious carry-over value to terrestrial cells and arrays.

Note that standardization is predicated on agreed-upon testing procedures, an issue treated in the previous subsection.

Solar Cell Performance Degradation is another issue of substantial importance to a systems engineer planning a solar-electric facility to last 20 or more years.

Solar cell degradation, as measured by a fall-off during actual service life of the electrical power output may be caused by dust, dirt, bird droppings or ice accumulations on the array; deterioration in the transmissivity of the encapsulant or protective glass due to ultraviolet radiation; cracking or leakage of the encapsulant allowing moisture to enter and deteriorate cell operation, corrode contacts, etc.; long-term chemical actions and reactions between the various elements used to make the cell initially; delamination of the contact grid; failure of one or more cell interconnects, as discussed earlier; cell hot spots in an array; cell-substrate thermal bond failure with resulting high cell temperature; and other causes--individually or collectively can cause cell and array performance degradation.

In face of the known facts from the space program and limited terrestrial experience about cell and array wear out and failure modes and mechanisms it is clear that:

- Solar Cells and arrays do not have infinite life as some early optimistic researchers claimed. They just did not specifically look for wear out mechanisms in many cases, relying on short-term test results.

- There are various wear out mechanisms of a physical-materials-thermal-electrical-chemical nature operative in real world terrestrial solar cells and arrays. In general these wear out or degradation mechanisms are little understood at this time; the reason is that so much of the R&D has necessarily had to be devoted to researching the various photovoltaic candidate materials and cells that relatively little has been done to study degradation except on CdS cells where it was a major problem.

A sample of the pre 1975 work on degradation is Kennerud's paper [76] in which he showed for CdS cells how it is possible to infer from the I-V characteristics before and after degradation the probable exact degradation cause within the cell, e.g. an increase in R_S suggests that the substrate delaminated from the active material or the grid separated from the cell in places. This information can then be used to effect design or process improvements within the cell.

It is likely we shall see a great deal more published on the general subject of degradation of cells and arrays as photovoltaic technology matures. The subject is inherently linked to testing procedures and standardization. The photovoltaic literature prior to about 1975 appears relatively silent on degradation as an important topic relevant to cells and arrays to be used terrestrially. One can speculate that by 1985 a much larger body of engineering factual knowledge about degradation will exist.

Concentration is a topic the systems engineer should know about. The use of optical concentration permits a larger power output to be obtained from a group of solar cells; conversely if a given power output must be delivered the use of concentration generally implies fewer cells would be required than if no concentration had been used. Clearly in the latter case the cost/KW has been decreased, a desirable result.

Berman [77] was among the earliest (1967) to have recognized that "the most logical approach to the use of silicon solar cells for terrestrial applications is to utilize solar concentrators in conjunction with cells especially designed for such operations." He clearly showed cell efficiency varies with light intensity, peaking at some optimal value; thus he concluded that it is necessary to optimize the cell for an intensity. He investigated the range from 1 to 4 suns, and the work resulted in 1 x 2 cm silicon cells optimized for 400 mW/cm^2 light intensity where they gave 102 mW electrical power output.

A sample of later (1975) concentration work is that done by Spectrolab [78] where solar cells were investigated when driven by Fresnel lens, mirrors, and Winston type concentrators. This work was recognized to be of a preliminary kind. Bell, et al [79] of Mobil Tyco Solar Energy Corp. also investigated concentration in 1975. They found that the compound parabolic concentrator[1] "system will produce power at very much lower cost than will flat panel solar cell arrays."

The concentration subject was not investigated in-depth in this research; those interested in this facet of photovoltaic R&D will find an entrée to the prior art in the above references and in those to be cited in Sec. 2-3 on this subject. It is evident that the researchers who had studied concentration prior to 1975 held high hopes for its cost reduction potentialities. They do not always point out, however, that in general it is necessary to use tracking apparatus (at increased cost) with concentration type photovoltaics whereas tracking is not necessary with a flat plate array of solar cells.

This completes the photovoltaic issues of principal interest to the systems engineer gleaned from a sifting of the prior art. We now turn to summarize the present art in the next Section.

2-3 Present Art And Problems Being Addressed

By 'present art' we mean from about 1975 to early 1978, i.e. a continuation beyond the 'prior art' of Sec. 2-2. The introductory comments of that Sec.

---

[1] Also known in the literature as the Winston collector, after its inventor.

also apply here. The principal thrust here is highly selected information of potential value to a systems engineer planning to use photovoltaics.

Before treating the detailed technical topics some general matters deserve mentioning about the 'present art.'

General Matters

- The DOE Photovoltaic Program The present-art documents defining it were cited early in Sec. 2-1. The nation's photovoltaic program is a dynamic one, continually changing. Several newsletters[1] record these changes on a timely basis, and it is incumbent on a well versed systems engineer to be knowledgeable of developments within DOE and other government agencies active in photovoltaic R&D.

  One change in DOE goals was the reported [80, p 175] setting of a new cost goal[2] of $0.10-0.30/peak watt (electrical) with the production of 50 peak gigawatts (electrical) capacity[3] for the photovoltaic industry by the year 2000. Three major photovoltaic options were being pursued: flat plate, single-crystal silicon arrays; concentrating arrays; and advanced material thin-film arrays.

  A succinct summary of the state-of-the-art as of about 1977 for silicon cells, cadmium sulfide cells and gallium arsenide cells appeared in reference [37] where potential material constraining factors on each technology were also examined quantitatively. Gallium arsenide was seen as being the only relatively scarce material for future U. S. use.

  Again we remind that two of the best entrées to the 'present art' of photovoltaics government sponsored R&D are in subscriptions to references [9] and [10] which abstract hundreds of technical reports on virtually all phases of photovoltaic R&D work.

- The Electric Power Research Institute (EPRI) Photovoltaic Program is summarized briefly by Balzhiser in reference [81] and deWinter in reference [82, p249]. Their program is "aimed primarily at assessment of future potential prospects for photovoltaic conversion devices"; the intended application would be for potential use in electric power systems as fossil fuel savers. They try to design

---

[1] Solar Energy Intelligence Report (Business Publishers, Inc., P. O. Box 1067, Silver Springs, Md. 20910) and Energy Digest (Scope Publications, Inc., 6425 Maplewood Dr., Falls Church, Va. 22041).

[2] The earlier goal had been $0.50/pWe (in 1975 dollars) by 1985.

[3] It is interesting to compare this figure against Fig. 1-26 for perspective.

their R&D program to complement DOE's rather than to try to compete. The DOE, overall, appears considerably deeper involved with the forcing of photovoltaic devices-orientated R&D progress than EPRI whereas EPRI's basic interest is in photovoltaic systems aspects; there is, naturally, some overlapping in the photovoltaic systems area.

- Foreign Photovoltaic Programs may be of some interest to the systems engineer. An overview is presented in reference [82] (photovoltaics is only a small part of other solar programs here) along with U. S. programs. German activities are in [83], and Australian activities are in [84, Sec. 5]. Photovoltaic R&D progress in other countries was not exhaustively pursued in this research. Many foreign countries appear strongly interested in the application and systems aspects of photovoltaics--more so than in the detailed device physics and the R&D needed to realize low cost solar cells.

- A Much Needed Present Art Review Paper appeared in 1977 by Hovel [85]. This excellent paper summarizes the art as of then on Si, CdS, concentration cells and thin-film cells; many detailed references are cited where more information can be found.

- Photovoltaics Specialist Conferences constitute an important source of technical R&D information on the 'present art' in photovoltaics. The 13th such conference was held June 5-8, 1978 in Washington, D. C., and abstracts of papers are in reference [86].

  This important annual conference mirrors state-of-the-art technology in photovoltaics. Several papers of interest to systems engineers were presented at the 1978 conference; these will be cited herein at appropriate places.

We now leave these general items to address some selected 'present art' specific issues of interest to the systems engineer, as we did in Sec. 2-2 for the 'prior art.'

Solar Cell Efficiency Limits concerned several researchers in the 1975-1978 period. This old problem was reexamined with new insights and new results by new people. Previously it had been thought that the approximate upper limit solar cell efficiency was about 25 percent.[1]

Dunbar and Hauser [87] in 1976 examined "the various mechanisms which limit the conversion efficiency of silicon solar cells. This is being accomplished by means of a computerized semiconductor device analysis program which gives a complete numerical solution to the general semiconductor device equations including excess carrier generation due to the full spectrum solar irradiance." The program lead to various cell modification possibilities

---

[1]The exact value varies with the type cell material.

with the best predicted efficiency being approximately 20 percent for a back surface field (BSF)[1] structure.

In this same time period Lindholm published several papers broadly aimed at reexamining the fundamental device physics affecting conversion efficiency, the aim being to better understand the theory, to resolve certain discrepancies between measured and predicted efficiency, and to determine fundamentally what must be done within the cell to improve its efficiency [88] [89].

A number of papers dealt with efficiency, its calculation and improvement in an important special issue on photovoltaic devices edited by Brandhorst in reference [91].

The relationship between the incident solar spectrum and cell efficiency was examined by Boes and Hall [91A]. They concluded that "the errors in solar cell performance predictions which result from neglecting the variations in the terrestrial solar spectrum are generally less than 5%" and that "there does not appear to be a need for spectral data measurements on a widespread basis." Ertel [92] [93] also investigated spectral effects and efficiency and claims efficiency increases linearly with wavelength. An efficiency of 40 percent at 0.8μm is predicted. He envisions such a cell should "be combined with a solar thermal mechanism through which the excess heat could be trapped for other uses."

Advanced High Efficiency Cells A recent development of potential importance to the systems engineer on a longer range basis was the 1978 reporting of research results on tandem cells expected to have an efficiency in the 27-35 percent range [28][29][30][31][32][33][34][35]. These, however, are still very much in the R&D phase.

In 1977 Brandhorst [94] filed an invention disclosure on an improved back wall cell. No claims were made for it, however, in the reference. Also in this year Thornhill [95] investigated the possibility of automated fabrication of back surface field solar cells in a preliminary way. A total of 1852 cells were produced. They had an average efficiency of 11% at AMO and 28°C. The "program --- was successful in demonstrating the feasibility of automated solar cell production." While his efficiencies are somewhat below expected efficiencies for the BSF cell, they undoubtedly will rise as more is learned about the manufacturing process. No cost estimates were made for BSF cells; these data will probably appear in the near future as the BSF development matures.

Another recent development has been the vertical multijunction solar cell. This is a special cell built to receive the insolation on the edges of the PN junctions. Such cells have been reported or studied by Sodha and Agarwal [96]

---

[1] A back surface field cell has a built in electric field at the back of the cell. It "exhibits enhanced short circuit current density, and most significantly, unusually high open-circuit voltage," according to Fossum who explains the cell operation in reference [90]. He analyzed its performance in [97].

and Wohlgemuth and Wrigley [98]. They are definitely in the 'advanced' class in 1978.

Advancements also occurred in GaAs cells which will be treated shortly under the title of Concentration.

Polycrystalline Solar Cells have the potential major advantage of being much cheaper than the high purity but expensive single crystal material required for the normal solar cell. Polycrystalline materials have therefore been investigated for use in solar cells by Chu et al [99], Feldman et al [100] and at a 1976 National Workshop [101]. All these polycrystalline cells had efficiencies below about 5 percent.

More recently (1978) polycrystalline research results have been reported by Belouet (France) [102] who reports about 6 percent conversion efficiency, Hanoka [103] who reports 8.5 percent on his best cell made, Daud [104], Dapkus [105], Yeh [106], Kuper [107], Lindmayer [108], Fabre [109], and Chu [110]. These 1978 R&D results indicate polycrystalline cells have been made in laboratories with up to about 10 percent conversion efficiencies; thus polycrystalline technology appears to have progressed significantly since 1975.

Thin Film Solar Cells[1] are also being researched; one motivation is to reduce the amount of expensive crystalline material in the cell and thereby lower its cost. Representative present art in this facet of photovoltaics include works by Lindmayer [111], Chu [112], Iles [113], Kirkpatrick, et al [114], Chu [115], and Shirland [116].

Thin film cells are in a comparatively early stage of development, and efficiencies are only in the 4-6 percent range in 1978. Efficiency may rise as more detailed knowledge about the processes to make them is acquired.

Cu_xS/CdS Photovoltaic Cells were some years ago thought to be another low-cost solar electric alternative, but many technical difficulties caused interest to wane. Recently interest in this type cell has been rekindled. A sample of present art researchers in this facet of photovoltaics include Chin [117], Böer [118] (he claimed possible conversion efficiency in excess of 15 percent in 1975, life expectancies in excess of 20 years, and production yields in excess of 90 percent), Windawi [119], Burton and Haacke [120], Trivich [121], De Meo [122],[2] Olsen [123] [124], Loferski [125] Shewchun et al [126], Armantrout [127], Shea and Partain [128] Rothwarf et al [129], Baron et al [130], Brogagnolo [131], Singh [132], and Jones [133].

---

[1] Roughly, those thinner than about 0.2 mm (8 mils).

[2] This assessment of CdS arrays was done for EPRI with large-scale utility applications in mind. The reference contains many items of interest to utility system engineers.

An important perspective on these devices is nicely stated by De Meo. "Recent results strongly suggest that these devices can be stable for long periods, but their large-scale utilization will require conclusive verification of both cell stability and process reproducibility."

Optical Concentration combined with solar cells in various forms is a subject which many researchers are exploring in the present art. This work is a continuation of the pre 1975 research in this facet of photovoltaics which we mentioned at the end of Sec. 2-2. Interest in this area appears accelerating, judging by the large number of recent papers found in the literature.

Optical concentration technology in photovoltaics appears to be growing rapidly, and as cost effective combinations of optics and photovoltaics evolve, this field may be of increasing importance to the systems engineer planning advanced power systems. Accordingly we summarize here--in the briefest possible form--some of the significant present art. An in-depth examination is beyond our scope.

1978

Edenburn [134] summarized the DOE/Sandia Concentrator development project, and Burgess [135] described a Sandia program aimed at applications in the 20 to 400 KWE range. Watkins [136] described Sandia's 1 KW Fresnel lens concentrating photovoltaic system along with some of the failure problems experienced. Burgess [137] describes the performance of it. Lowe [138] preliminarily evaluated a Ga Al As concentrator system for space use, finding a 24 percent cell efficiency. Wilkening [139] described the plans for a 10 KWE photovoltaic 200/1 concentrator. James and Williams [140] treated Fresnel optics for photovoltaic cells, pointing out that the cost cutting emphasis should be on the optics rather than the cell. Zimmerman [141] described a low cost lightweight 100:1 concentration ratio parabolic concentrator, and Neuner [142] showed a sealed beam headlight type concentration technology providing concentration ratios in the 100 to 500 range.

Rapp and Boling [143] described a novel design based on sunlight incident on the face of a sheet of luminescent fluorescent glass or plastic with photocells coupled to the edges. They claim the advantages "of high concentration ratios without tracking the sun and efficient collection of diffuse light." A solar-electric 5 percent efficiency is expected, and the cost estimate is $0.54/pWe. Passive cooling for concentrated (25 suns) solar cell arrays was demonstrated by Williams et al [144]. Swart and van Wyk [145] investigated tracking and its influence on the power conditioning in order to match the load to this varying source.

Lindmayer and Wrigley [146] presented "some theoretical and experimental factors concerning the photovoltage of silicon solar cells in a range of magnitudes of illumination intensity," and Hu [147] analyzed the open circuit voltage of high intensity silicon solar cells. Chappell [148] investigated a new type of silicon cell with the potential for 24 percent conversion efficiency in 50 suns.

Masden [149] investigated geometrical parameters on tracking arrays with particular attention given to the shadowing problem and land cost.

Robb et al [150] compared the performances of silicon ($\eta_{peak} \sim$ 15-18%) and gallium arsenide ($\eta_{peak} \sim$ 22%) in the range up to 1000 suns, and Castle [151] evaluated Si cells under concentrated conditions. Van der Plas, et al [152] created cells 21 percent efficient in AM 1.5 at 1000 suns concentration. Ewan [153] examines the principles and technology for making (Al Ga)As-GaAs solar cells under 1000 suns concentration. Sahai et al [154] claim to have achieved 24.6 percent efficiency at 180 suns and 20.5 percent at 440 suns, the highest reported values thus far for concentrator cells they say. Dalal [155] presented ideas on the design, technology, and cost considerations for high intensity silicon solar cells. Kemthong [156] revealed fabrication experience with high efficiency silicon concentrator cells. Reliability of cell contacts on silicon cells for concentrator use was studied by Selikson and Mickelsen [157].

Donovan, et al [158] describes a 10 KWe concentrating array developed by Martin Marietta Corp. which operated at 40-50 suns and used Fresnel lenses. Castle [159] also describes a 10 KWe system developed by Spectrolab, Inc. The application of concentrator technology to a rural village in Spain has been revealed by Luque et al [160].

Evans and Florschuetz [161] [162] studied concentrating photovoltaic systems by means of a computerized model. One of their results of interest to systems engineers is "that concentrating systems should not be ruled out relative to fixed flat unconcentrated arrays even for low solar irradiation locations such as Cleveland."

In summary, the present art of concentrating photovoltaics in 1978 is that hardware for these systems has been created which is about 24 percent efficient under terrestrial conditions. This is a sizeable advance over prior art flat plate arrays which, practically, have efficiencies in the 10-15 percent range. Also there is much current R&D active in this facet of photovoltaics.

<u>1977</u>

Mollé [163] investigated a concentration unit for residential use using silicon cells. Gorski et al [164] describe the design, construction, and testing of a small array (~ 100 WE) of compound parabolic concentrators and photovoltaic cells. They also analyzed a two-dimensional version which they believed had the potential of achieving $0.15 to $0.50 per peak watt.

Nine papers by various authors dealing with high intensity cell design and performance were presented early in 1977 at an ERDA photovoltaic program review meeting [165].

Heinbockel and Roberts [166] analyzed various silicon hybrid systems and compared them with gallium arsenide hybrid systems. The hybrid system also produces useful thermal power. They also found that concentrator hybrid systems produce a distinct cost advantage over flat plate hybrid systems.

Backus et al [167] laid out some general considerations for using photovoltaic concentration systems. Their work showed that "concentrators in photovoltaic systems offer a viable alternative to flat, fixed arrays.

Concentrating systems offer a better potential for reduction in the cost of energy in the near term." Yekutieli, et al [168] reveal a small laboratory model two axis tracker based on use of Fresnel lenses and solar cells.

O'Donnell, et al [169] described the characteristics of experimental solar cells for concentration and revealed experimental results.

Finally, in 1977 IBM researchers Woodall and Hovel announced [170] the achievement of a 22 percent efficient GaAs cell at 1 sun and 26-27 percent at 1735 suns.

1976

Kaplow of MIT was reported by McClintock [171] experimenting with two foot diameter mirrors of 500-1000 concentration focussed on solar cells.

Burgess [172] described the Sandia concentrator program based on single crystal silicon and gallium arsenide cells combined with various kinds of concentrators. Fossum [173] studied the optimization of silicon cells under concentrated conditions.

1975

Researchers at Spectrolab were engaged in an in-depth study of photovoltaics under 3-100 suns [162]. Their analytical modeling of silicon solar cells under high illumination was completed indicating "they can be designed to operate at over 100 suns with conventional processing and with characteristics comparable with conventional space cells." They also found that "Even at the location of the lowest direct insolation (Omaha) several of the concentrator systems collected more energy than flat, fixed tilted arrays." They were also studying comparisons between silicon and GaAs cells.

Davis and Knight (England) [174] briefly reported their results of testing GaAs cells up to 2000 suns and intermittently up to 5000 suns. They showed a measured I-V curve for the cell showing about $10A/cm^2$ at about 1.05 volts for 1000 suns.

Spratt and Schwartz [46] of GE reported their research on concentrated photovoltaic power systems, assuming that concentrators can be built at a lower cost per unit area than solar cells. They found that the cell series resistance plays the greatest role in determining the performance of concentrating systems and that a new type of structure is needed. They proposed one on which they were working.

It is obvious from the above citations that there was almost a frenzy of activity in concentration photovoltaics in the 1975-1978 period, all of it predicated on the belief that electrical energy produced this way will be substantially cheaper than with the more conventional flat plate photovoltaic arrays.

Finally, it is abundantly clear research developments in the concentration area are of major concern to the systems engineer planning advanced solar-electric power systems.

Thermophotovoltaics is a natural outgrowth of the concentration work and that of conventional photovoltaics. The key idea is described by Horne [175] in a most significant paper of interest to systems people. "The concept consists of a solar concentrator which focuses solar energy onto an absorber which is heated to incandescent temperatures (2500 K to 3000 K operating range). The radiation emitted from the incandescent source has a peak energy spectrum that closely matches the bandgap energy of silicon. The emitted long wavelength radiation is directed onto photovoltaic cells which absorb that portion of the spectrum which has photon energies greater than, or equal to, the bandgap energy of silicon. The absorbed energy is then converted to electrical energy and heat in the cells. The photons having energies less than the bandgap energy of silicon pass through the silicon and are reflected back to the source where they are reabsorbed. -- Theoretically such a device could reach efficiencies of the order of seventy percent." Horne goes on to state that "A computer model of the generator system shows that overall conversion efficiencies for the generator of twenty-five percent to be practical with current technology."

Swanson and Bracewell [176] preliminarily investigated this concept for EPRI, including experimentally evaluating thermophotovoltaic (TPV) cells, achieving overall efficiencies of only 7-12 percent. They recognized with improved cells efficiencies in excess of 30 percent could be achieved.

Thermophotovoltaics would appear to be an interesting and promising means of achieving high conversion efficiency solar-electric conversion, and the systems engineer would do well to be sensitive to present art R&D in this area. Such devices would appear to be inherently Carnot limited in efficiency, however. This technology is much newer and relatively less explored than other subfields of photovoltaics.

Solar Cell Interconnects for arrays were discussed in the prior art of Sec. 2-2. The only more recent document found in this research was reference [177] which is a study of interconnect and bonding technologies for flexible arrays for spacecraft. Nine different conductive epoxies were investigated and 13 different interconnect materials were compared. The study was intended to determine whether thermocompression or conductive adhesive bonding could substitute for welding on flexible solar arrays. This space-orientated research appears to have a high carry-over value to terrestrial solar arrays, although, as earlier pointed out, the technical requirements are somewhat different.

Solar Cell Encapsulants were treated in the prior art of Sec. 2-2. Samples of the present art (1975-1978) work include:

1977

Anagnostou et al [178] invented an embossed encapsulating film geometry into which solar cells would be placed during assembly. This, combined with use of printed circuitry for the array interconnects, results in an encapsulated assembly believed suitable for automatic processing.

Sierawski and Currin [179, pp. 22-14 through 22-17] investigated the use of silicone gel for potting photovoltaic arrays.

1978

This year saw the following reports generated for JPL for the well-known Low Cost Silicon Solar Array Project (LCSSAP).

- De Bell & Richardson [180] [181] investigated test methods, material properties and processes for solar cell encapsulants. They studied 24 encapsulant materials under accelerated aging conditions.

- Kolyer and Mann [182] investigated accelerated test methods on solar cell encapsulants along with basic data and a statistical model. Their accelerated program included weathering tests at Phoenix and Miami. They were evaluating the materials on the bases of temperature, humidity, and insolation effects.

- Carmichael et al [183] [184] [185] [186] of Battelle investigated encapsulation materials specifically for terrestrial photovoltaic arrays. The aim was to be able to predict performance of encapsulation materials over a 20 year lifetime.

Also in 1976 Broder [187] at NASA Lewis Research Center investigated two types of fluorinated ethylene propylenes (FEP) for possible use as solar cell cover materials. The motivation was a low cost substitute for quartz covers. FEP-A appears superior when exposed to the earth's atmospheric conditions.

It is evident that solar cell encapsulation R&D is ongoing. Undoubtedly this type work will have to continue for many years into the future before it can be reliably known that certain encapsulant materials are in fact suitable for 20 or more years of life on actual solar arrays under terrestrial conditions.

Present Art of Solar Cell Manufacturing In face of a bewilderingly large number of candidate solar cells of various designs and variations which have been created and studied by the photovoltaic R&D community it is incumbent on the systems engineer to know--before power system planning proceeds too far--what plans, if any, have been made for manufacturing a given cell design. It would be easy, for example, to lay plans for erecting a large power array by a certain date based on a cell type we were impressed with from the R&D history only to discover that absolutely no tangible plans had been generated to mass produce the cell or the cell-concentrator system combined.

Thus the systems engineer should have some elementary sense of overview of the present art of the manufacturing aspects of solar cells. Some overview prior art material was presented in Sec. 2-2, and we now slightly update that information.

A List of Photovoltaic Manufacturers names and addresses as of about 1977 has been compiled by Bereny and de Winter [188, p. 137]. These are all 'small businesses,' and as of 1978 there is no single large-scale mass production manufacturer or solar cells in the United States.

Studies Directed At A Solar Cell Manufacturing Facility are a part of the present art scene. The following serious studies, presented at the 13th IEEE Photovoltaics Specialist Conference, were aimed in the direction of eventually creating a mass production facility for photovoltaics:

- Bickler et al [189] of JPL envision a factory producing about 20 MW/yr. of silicon photovoltaics at $2.00/peak watt employing near-term technology and employing about 546 people. Their proposal is a result of JPL's Low-Cost Silicon Solar Array (LSSA) project.

- Grenon and Coleman [190] of Motorola envision the most likely near-term process as utilizing Czochralski ingot silicon crystal technology. Round cells of 7.6 cm and of 12 cm diameter would be produced at about a 20 MW/yr. rate. The plant capital cost would be about $20 million. Some of this work was reported earlier by Coleman [191] [192].

- Carbajal's [193] [194] of Texas Instruments idealized plant proposal is for a tandem cell, polycrystalline silicon, Czochralski grown crystals. Their cell would be encased in a module of 13 percent efficiency. The module would be of glass-steel construction. He estimated the manufacturing cost would be $1.31 to $1.51/peak watt. The plant capital cost was estimated at about $20-25 million and is envisioned for the 1982 time frame.

- Kran [195] of IBM presented the analysis and outlook for a plant based on silicon ribbon technology in the 1978-1986 time frame. It would be about a 10 MW peak/yr. production rate beginning production in 1982. He felt they could meet the $50/m^2$ price goal by 1986. No figure was presented of estimated plant capital cost.

- Breneman, et al [196] of Union Carbide Corp. described a silane plant for producing silicon. They estimate it will cost about $6 million and is planned for the early 1980's.

- Buhs [197] of AEG-Telefunken described the German program, their Phase I of which includes building "a complete production line for terrestrial silicon solar cells and generators." The plant capacity would be up to 300 peak KW/yr. by about 1985.

An important 1976 report by Thornhill and Taylor [198] describes the demonstration at Spectrolab of the feasibility of automated silicon solar cell fabrication. A plant of 3.37 MW/yr. production rate was envisioned producing cells at $1.22 per peak watt. These would be hexagonal cells.

While studies by others aim at some sub-part(s) of the manufacturing process for solar cells, the above are the only ones found in this research which appear to have integrated the technology into a complete manufacturing plant proposal involving processes, production rates, yields, cost estimates, and all the other variables involved in planning a large complex manufacturing operation.

The problems in the manufacturing area are important and nontrivial. Much present art R&D is underway on the various sub-steps involved in manufacturing, e.g. growing silicon ingots, ingot analysis, impurity studies, and production, slicing into thin wafers, growing silicon ribbon and sheets, hot forming silicon sheets, various process methods for producing the cells--these and many more technical areas relevant to eventually manufacturing silicon on a large-scale have been researched and are being researched as parts of the JPL/DOE Low Cost Silicon Solar Array program. This program involves many companies in the R&D. Technical reports on these sub-areas were accumulated in this research; but no attempt was made here to integrate this complex process-oriented area, both because of its complexity as well as its lesser relevance to systems engineers.

The systems engineer should, however, know the above process orientated activity exists, that the R&D work going on in this area is responsible creative activity, and that the photovoltaic technical community has not been deficient in recognizing and researching this important area. Engineers going into solar photovoltaics manufacturing R&D can explore this area on their own.

Unfortunately, there appears to be no single overview document which integrates this LCSSA program,[1] and one is forced to initially enter it at the detailed level.

Automated Array Assembly is of more interest to the systems engineer. This work springs from the JPL/DOE Low Cost Silicon Solar Array Program and is interrelated with the manufacturing process R&D mentioned above which in turn is interrelated to the solar cell design.

Only the following report relevant to automated array assembly was uncovered in this research:

- Williams [199] of RCA investigated arrays composed of modules, gas diffusion junction formation technology, screen-printed Ag contacts, spin-on AR coating, gap welding for interconnects, and a glass plate-conformal coating-metal foil moisture barrier panel. His cost estimates indicate such arrays could be produced for $0.90/peak watt. His lowest cost projected to that date for material and expense items for array module manufacturing was $0.30/peak watt.

Perhaps other present art work is also aimed at automation of array production, but the rather extensive literature surveyed in this research gave little such indication. By far most of the automation work appears directed toward the solar cell itself rather than the automated assembly of

---

[1] Someone could render a useful service, it appears, to the photovoltaic community and those outside it but interested in it to generate such an integrative document which would authoritatively summarize the state-of-the art of the various manufacturing processes in a coherent form readable by nonspecialists in this area.

them into a finished array package. The latter is largely presently done manually.

Solar Cell and Array Testing is a subject bound to be of great interest to the systems engineer, for cell and array specifications must be written clearly and with knowledge of the present art. Such specifications may also define the tests and test procedures for cells, modules, arrays, etc. along with acceptable performance limits. Such engineering specifications are ordinarily a part of the purchasing contract with a cell supplier, and hence they are used as the reference document upon receipt of the cells to tell good ones from unacceptable ones.

Solar Cell Testing Procedures for terrestrial use were proposed by Magid and Brandhorst [74] on an interim basis in 1975 with a final version planned for the summer or 1976. The final version was not available to this researcher. The interim document contains much helpful information on definitions, standard test conditions, testing in natural sunlight, testing indoors, supplemental test procedures, and air mass two spectral data.

It may be of interest to systems engineers to know that the European Space Agency specification for silicon solar cells in reference [200] contains similar but more detailed information; but remember it defines cell specifications for space use. We have repeatedly indicated throughout this state-of-the-art review that space requirements are somewhat different from terrestrial requirements on solar cells.

Solar Array Testing There exists in 1978 a DOE/LeRc Photovoltaic Systems Test Facilty located at Lewis Research Center near Cleveland, Ohio. It was built "as a national facilty to serve the needs of the entire DOE National Solar Photovoltaic Program." It provides a place "where photovoltaic systems may be assembled and electrically configured, without specific physical configuration, for operation and testing to evaluate their performance and characteristics. The facility as a 'breadboard' system allows investigation of operational characteristics and checkout of components, subsystems, and systems before they are mounted in field experiments or demonstrations," according to Cull and Forestieri [201]. The system in 1978 had "two inverter test stations, a battery storage system, interface with the local load and the utility grid, and instrumentation and controls --." An earlier (1976) report on this project was made by Forestieri, et al [202].

Kolecki and Gooder [203] also describe an indoor high voltage solar array test facility at Lewis Research Center permitting testing of up to nine arrays yielding a net voltage of 15 KV maximum.

Array testing and technical problems therein are treated in a collected group of 11 papers in reference [165, pp. 200-311]. Many of these were preparatory to the Test Facility described above. This series of papers is highly recommended study for systems engineers planning large-scale photovoltaic installations.

Solar Array Applications constitute a growing part of the present art in 1978 photovoltaics. Surveying all of these is not possible here; we therefore only briefly present a few selected applications of possible interest to the systems engineer.

Shepard and Sanchez [204] have described a novel hexagonal photovoltaic module formed into the shape of a shingle for use in place of shingles on the roof of residential or commercial buildings as shown in Fig. 2-10. The cells

FIG. 2-10 Shingle photovoltaic module. Courtesy of General Electric Co., Valley Forge Space Center.

are individually bonded to the embossed surface of a 3 mm thick thermally tempered hexagon-shaped piece of ASG SUNADEX glass. The average power output per module was measured at 5.80 watts for 1 KW/m^2 insolation at 28°C. This translates to 98 $watts/m^2$ of module area under their standard operating conditions of 1 KW/m^2 and a wind speed of 1 m/s.

The idea of using solar cells as 'shingles' is very old, springing from the space solar paddles. More recently the invention of another form of solar cell shingles has been disclosed by inventors Forestieri, Ratajczak, and Sidorak in reference [205].

The largest terrestrial solar array installation to 1977 was made on an experimental farm near Mead, Nebraska for irrigation purposes. The array was "composed of 28 individual panels each capable of generating approximately 1 KW of peak power output," according to Romaine, et al [206] who described this important system. Its peak output is 25 KW. It drives a 10 Hp pump for irrigating 80 acres of corn and soy beans, pumping 1,000 gallons of water per minute. The average purchase price was $15.50 per peak watt [207]. A DC to 3-phase AC inverter is between the array and the pump motor. The system also supplies auxiliary power.

A final application, probably destined to be of historic importance in photovoltaic technology, is the world's largest photovoltaic installation underway in 1978 at the Mississippi County Community College, Blytheville, Arkansas. The heart of the system, manufactured by Solarex, is the solar receivers, each containing 100 glass-faced 20 sun silicon cells that can achieve 14 percent conversion efficiency at 55°C. There are 600 such receivers. The expected power output is 250 KW peak. The expected $6-$7/peak watt cost of these concentrator-receivers is lower than that of any photovoltaic system installed to date. The photovoltaic arrays are mounted at ground level and are installed in a fenced in enclosed area adjacent to the electrical load.

It is the expectation of the photovoltaic community that in future years many thousands of arrays will be applied to the many and varied government, residential, commercial, and industrial uses we have for electricity. We shall glimpse another view of the application of solar arrays in Sec. 2-5 where we address the problems of integrating them with existing electric systems. First, however, we should address in the next Sec. the issue of sites for them.

2-4 Site Selection Considerations

In contrast to the wind-electric literature (Chap. 3) which deals extensively with the importance of siting, the photovoltaic literature is nearly silent on the siting subject. Of the many hundreds of photovoltaic references searched only Imamura [208, Ch. 2] deals with siting as a specific subject; even his viewpoint is severly limited to that of the siting for NASA's Residential Prototype System Test (PST). Generalized siting methods applicable to a utility sytems engineer appear to not exist in 1978; at least if they do,a diligent search of the photovoltaic literature did not reveal them. A few other references briefly mention the subject as an aside or minor part of the main thrust of a paper. As siting principally concerns the first step in the application or deployment of solar arrays and since such wide-scale terrestrial deployment of photovoltaics is only now in its infancy, perhaps the near silence of the literature is understandable.

From the systems engineering viewpoint, it is believed that where a given photovoltaic array is to be located may have considerable effect on its electric energy produced over its lifetime, substantial effect on the array lifetime and maintenance, and in general be a determining factor in the suitability of the solar-electric array for the intended application.

We limit our discussion in this section solely to the kinds of factors which might enter into initially choosing the specific site where the photovoltaic array is to be located. We shall treat the subject in a very preliminary way only, as no distinct body of engineering knowledge about siting has to date been created.

The Kind of Application strongly influences the siting considerations. From the systems engineer's viewpoint and excluding a whole host of small-scale special purpose photovoltaic applications, there are two main types of applications currently envisioned for photovoltaics:

- Centralized Photovoltaic power plants of a high power class. (We shall briefly describe these in section 2-5.)

- Decentralized Photovoltaics where perhaps eventually millions of arrays of only a few kilowatts capacity each would be deployed on residential rooftops.

The siting factors are substantially different for the two applications.

The General Solar Insolation, as revealed in solar radiation maps of the type shown earlier in Figs. 1-19 and 1-20, is the beginning point for both the above applications;[1] but at the practical level such maps are of limited usefulness. A utility company has a service territory over a limited area, and the territory is not movable. Florida Power Corporation (FPC) located in an area of about 6.0 KWH/m^2 mean direct normal solar radiation in June cannot seriously consider erecting a centralized photovoltaic plant in Utah just because the insolation there is 10.0 KWH/m^2. It is of interest to the FPC planning engineer to know this fact, but basically he has to use the available solar insolation within his firm's authorized territory. Thus his siting problem is one of "Where within our territory are the best sites?"

The insolation may vary greatly within a given utility territory if clouds tend to be present in some locations and not at others as in peninsular and mountainous terrain.

In contrast, the typical residential user of photovoltaic arrays finds his siting problem to be more of a microscopic nature, i.e. where to locate the array on his property. Again knowing the insolation is higher in some other part of the country is of little practical help--unless he is willing to move there. It appears doubtful even a small percentage of the population would elect to move solely on such energy grounds.

But for both applications the insolation maps could tell them whether to even consider photovoltaics; additionally they give a sense of needed perspective.

Some General Siting Guidelines can be identified for photovoltaic arrays. They should be located:

- Southerly Facing to collect the maximum solar energy. For a centralized plant a southerly facing hill might be available on which the large array could be deployed. Sites with northerly facing hills or cliffs would be generally unsatisfactory sites.

  On a residence the preferred site is on a southerly section of the roof, i.e. the roof ridge runs east-west. If the roof ridge

---

[1] Bartels [209] has pointed out that errors of over 10 percent in the published U. S. insolation data are not uncommon.

runs north-south, it poses special problems for siting the array on the roof.

- To Avoid Shadowing caused by nearby buildings, trees, or other objects or apparatus. Avoiding shadows on the array is particularly more difficult in general in the urban environment. Photovoltaic arrays sited where the array is partially shadowed are known to suffer from 'hot spots' and possible failure of the shaded cells caused by forcing the shaded cells to operate in the third quadrant, i.e. the power dissipating mode, of their characteristics as Partain [210] and Palz [25, p. 197] have shown. The shadowing problem was earlier encountered on spacecraft where parts of it could shade the solar array. While there are ways of minimizing cell burnout possibilities caused by shadowing, e.g. the addition of auxiliary diodes, the solution adds to an already high initial cost. A better solution is to simply avoid the shadowing in the first place.

  Due thought should be given to the fact that the shadow patterns at a given site may significantly change with the seasons.

  The shadowing problem may be of lesser importance to a very large utility array which would in all likelihood be sited in an open area free of nearby shadowing objects.

- To Minimize Visual Pollution wherever possible. For residential application, Böer [211, p. 380] suggests that "By planning properly and using attractive architecture, visual pollution can be kept to a minimum."

  For a centralized plant, utilities may find it necessary to plant trees to hide such a facility, using basically the same techniques used so successfully to hide unsightly conventional transmission lines and sub stations.

- To Minimize Dust And Dirt Accumulations on the array panel. At dusty sites it is preferable to mount the array above the dust, perhaps on a building or tower. Some thought should also be given during the siting process toward making the array accessible for cleaning a few times a year to remove the dust and dirt.

- To Minimize Snow And Ice covering the array as much as possible [212]. If the array is to be mounted on a mountain, perhaps it can be installed at a level below the ice. If it is to be installed on an urban building in the snow belt, then the array should be sited where daily access can be had to it during the winter to scrape off the snow and ice; in the latter situation siting the array on a tall tower is thus precluded.

- To Avoid A High SO_2 Area if at all possible, as Thomas [213] has some evidence in accelerated life tests on solar cells that SO_2 is the number two cause of cell deterioration--behind temperature--though the testing is not yet complete. This guideline suggests that it is less desirable to site a photovoltaic array near a coal burning power plant than farther away.

- Reasonably Near The User Load Center as is possible consistent with meeting other siting criteria. For a centralized photovoltaic plant sites in or near the suburbs would be preferred over one, say, a hundred miles away. For a decentralized residential application the preferred siting should be such as to minimize the connecting wire length between array and power conditioning or household load.

- Away From Known High Lightning Areas inasmuch as possible. The effects of lightning on photovoltaic arrays is a large unknown in 1978,[1] and research on this aspect of photovoltaics must be initiated. In the meantime wisdom dictates suitable lightning protection be installed on large, expensive, critical arrays, especially if they are to be sited out in open flat terrain or high on a building, tower or other structure.

- Away From Known High Bird Concentrations insofar as possible. Their droppings tend to cut off the insolation and decrease array efficiency. If an array site must be installed in or near such bird roosting concentrations, suitable spikes or other anti-landing deterrents should be installed on the upper edge of the tilted array.

- To Avoid Severe Windloads On The Array where possible to do so. Photovoltaic arrays typically are not presently made to withstand high wind loads; Also if an array is mounted out in the open and a northerly wind comes the array, being mounted at an angle, is a natural airfoil and could be lifted up and destroyed by the wind.

  Some thought to prevailing winds, if any, should be given in siting a critical array. Perhaps suitable spoilers on the north side of the array, as trees or bushes, could be installed to minimize array damage possibilities at the proposed site.

- To Minimize Vandalism Possibilities insofar as possible. Photovoltaic arrays are expensive items; already in the technology there have been several known instances of vandalism or complete stealing of the array. Ravin [214, Ch. 2] gives the following suggestions

---

[1] It is interesting to observe that of the many hundreds of photographs of actual solar array installations examined for this research not a single one made use of Franklin's well-known lightning rod! It would appear that as photovoltaic applications grow in number this unprotected state must be reversed, for there will surely arise cases where very expensive arrays are irreversibly damaged by lightning.

to minimize vandalism and other hazards:

▲ "In order not to attract attention to the equipment do not use bright colors. In some cases consider the use of camouflage

▲ Mount the equipment on a pole or as high off the ground as possible.

▲ In general, do not post "do not tamper" or "keep off" signs except on building type installations which have alarms installed. These signs are invitations to vandals. "Scientific Ecological Experiment in Progress" signs have been found to be generally more effective than warning signs.

▲ Depending on the location, both domesticated and wild animals can cause damage to and be injured by outside equipment. The equipment should be placed off known animal paths wherever possible.

▲ Secure the equipment with as heavy a mount as feasible. Permanently weld the mount to the equipment in order to make carrying of the unit more difficult.

▲ Permanently engrave an identifying serial number in at least two locations on the equipment. One number should be visible, the second number should be concealed.

▲ In applications where there is constant activity such as maintenance, or in marine applications, the solar arrays should be protected against being stepped on, stood upon, and against small hand tools being dropped on the array."

It is thus evident that many factors must be appraised by the systems engineer in siting a photovoltaic solar-electric system. It appears probable that knowledge about photovoltaic siting will grow as future applications become more widespread. As stated earlier, the above is only the beginning in an area on which the photovoltaic literature is mostly silent.

There appears to be no substitute for good engineering judgment by the systems engineer in thoroughly thinking through a given site recommendation.

2-5 System Integration of Photovoltaics

The Meaning of 'Integration' is treated in Sec. 3-8. There it is pointed out there are two facets of integration:

- Electrical Integration
- Technology Integration

Emphasis here will be on electrical integration. We further limit our considerations to integrating a single photovoltaic system with the electric grid. The problems of multiple photovoltaic systems feeding the power grid will be treated in Chapter 5. Also the various aspects of energy storage, while briefly mentioned here, are principally treated in Chap. 4 as a separate topic.

In photovoltaics, as in Wind Energy Conversion Systems (WECS), the system must be designed and engineered:

- To provide acceptable power to the utility grid.
- With associated equipment which is capable of providing full protection for both the photovoltaic generator and the grid.

It is intuitively obvious that the voltage, frequency, and phase of the photovoltaic generator power output should meet the utility's requirements. In photovoltaics, as we shall see, there is an additional concern over the current and voltage waveforms and the extent they can depart from a sinusoid and still be acceptable. This concern does not arise in a WECS system because there the power is being supplied by an alternator instead of a DC-AC inverter as in photovoltaics.

Integration, as used here, is the electrical connection of a photovoltaic solar-electric system into a localized small-scale or a large-scale electric energy system so that the photovoltaic system either delivers acceptable useful electric power locally or into the system grid or some of both.

In this section our view is principally directed toward the photovoltaic-electric grid system; we leave the photovoltaic cell treated so extensively in earlier sections behind now, considering it as a 'component' in the larger photovoltaic system which also has other important components necessary for the successful operation of the system.

We shall first briefly remark about the decentralized vs centralized philosophic controversy and then about the systems aspects of The Department of Energy's (DOE) National Photovoltaic Program as well as the Electric Power Research Institute's (EPRI) program. Then we shall examine the technical details of a basic photovoltaic system, including examining in some detail photovoltaic arrays, inverters, a system with energy storage, and the optimization of photovoltaic systems. We shall also treat large (utility orientated) and small (residential orientated) photovoltaic systems. Finally we close the section by examining a few of the known special system problems unique to photovoltaics. These topics are of major importance to systems engineers and managers and should give us a reasonably accurate picture of the state-of-the-art of photovoltaic systems up to 1978.

As we shall see, there are many aspects to the successful integration of a photovoltaic generator into existing electric power grids, and we necessarily must treat these aspects one at a time. The reader is cautioned to not attempt solving all the many integration problems simultaneously if maximum progress per time increment is to be made.

The Decentralized Vs Centralized system philosophic controversary repeatedly surfaces in the photovoltaic literature, in research discussions, and at photovoltaic conferences. The kernel of the argument is whether photovoltaic systems should be implemented via millions of low power spatially dispersed systems, as on residential rooftops, supplying a few KW to the homeowner, or whether fewer photovoltaic systems should be created in larger sizes, perhaps megawatts, at a relatively few centralized locations. The first decentralized approach more nearly matches the inherently dispersed nature of solar energy and would result in 'on-site' photovoltaic systems. It has much merit. The second centralized approach is more along the lines of conventional utility system planning where a few large generating stations are located in the territory served by the utility. The issue is, fundamentally, one of an appropriate scale for deploying future photovoltaic systems.

Recall we treated this issue rather extensively in Sec. 1-6, and the merits of each approach need not be repeated here.

The decentralized vs centralized issue is of the highest importance to the future growth of solar-electrics; photovoltaic systems in particular offer the technical opportunity for the nation to go either route or, more probably, both routes. Either route, if implemented on a large-scale, clearly would save fossil fuels compared to existing fossil fueled conventional electric generating plants. It is very clear that were the national decision made to exclusively create small-scale residential photovoltaic systems we would <u>not</u> have 10 years later large centralized photovoltaic systems; conversely were the decision made to exclusively create large-scale centralized photovoltaic systems, then 10 years later we would <u>not</u> have widely dispersed decentralized photovoltaic systems in the same way we do not have today a system of dispersed on-site small-scale conventional electric generators.

In an era of rising electric energy costs (see Fig. 1-28) many utility customers appear to see small decentralized photovoltaic systems as, at last, an opportunity to experience relief from increasing electric energy rates charged by utility companies; thus the 'little people' seem to favor the decentralization approach. They feel as Vincze [215] that "solar-electricity should be generated--preferably by solar cells--where it is used, and all effort should be directed to reducing the cost of such cells."

In contrast there seems to be an undercurrent of feeling among some utilities to view the creation of dispersed photovoltaic systems as a threat, tending to decrease their saleable load and hence revenue, and hence they react by favoring creation of large-scale photovoltaic systems under utility direct control. The development of large-scale photovoltaic systems also involves large dollar value R&D contracts whereas the converse is true for the small-scale systems. Large-scale systems may have reliability and accountability advantages, however.

This issue cannot be settled here; rather it is simply pointed up here as a current issue vitally affecting future technical and business growth directions in photovoltaic systems. It is an issue Keith of Alabama Power Co. summed up in his remarks during the plenary session of the 1978 13th IEEE Photovoltaics Specialist Conference when he said, "The dispersed nature of solar energy poses a problem the utilities will have to cope with."

<u>Systems Aspects of National Photovoltaic Program</u> have been an integral part of the program since its inception. Systems thinking permeated the founding 1972 Solar Energy Panel report [7]. Photovoltaic systems on residential

buildings--including combining photovoltaics with thermal heating--were clearly envisioned as were large-scale ground central stations of photovoltaic arrays "to feed power into a distribution grid in the manner of a utility power plant."

Systems thinking was also present at the 1973 Cherry Hill Workshop Conference [55] [216] sponsored by NSF and recorded by JPL as the national photovoltaic program was starting to be implemented. There a systems panel[1] looked at the evolving photovoltaics R&D technology from the systems viewpoint, in sharp contrast to the 'devices' viewpoint largely dominating the conference. A resume of that panel's conclusions are of sufficient importance to merit repeating here. Quotation is from reference [216].

> "The difficulty of fashioning meaningful studies of solar photovoltaic conversion installations in an operating electric power generating plant without having the use of cell costs as a basis was cited in a conclusion that such system studies are really not appropriate until cell costs have been defined. In turn, the availability of sufficiently cheap cells means that the photovoltaic system can be used directly for peaking, eliminating the need for energy storage, although the properties of adequate conversion efficiency and lifetime must be additional attributes of a photovoltaic device to make it acceptable for installation in large-scale generating plants.
>
> The conclusion that, although solar cells have been associated mainly with electrical systems for individual homes, the use of solar cells for utility power applications should be the major objective of a development program was derived from several considerations. It was pointed out that the present restricted view of terrestrial applications, which made use of limited evaluations of photovoltaic conversion devices for large-scale electric utility uses, resulted from three faults: (1) the utilities did not seriously evaluate the prospects of solar energy because of its intermittent character; (2) in turn, solar cell experts have remained unfamiliar with the concepts and economics of electric utilities; and (3) those involved in implementing photovoltaic conversion systems for terrestrial applications have not considered utilities as a reasonable market. As support for the argument that the utility market is better than the residental market, the following discussion was presented: (1) In contrast to a single residence, which has a poor load factor, a utility has a diversified load which facilitates amortization of equipment. (2) Because of the load mismatch, single residences would require energy storage devices. This is not the case for utilities. (3) A utility has alternative power generating capabilities and would need no energy storage in case of bad weather. In contrast, a residence would not usually have an alternative source as a part of its system. (4) Photovoltaic systems would fit in nicely with the projected use of battery storage by utilities. (5) Centralized photovoltaic device plants can be installed and maintained more inexpensively than individual units on homes. The recommendation

---

[1] Composed of knowledgeable people in photovoltaics as well as utilities.

derived from this discussion was that utility applications should dominate the program to use photovoltaic devices, and the recommendation was bolstered by the statement that the utilities should participate early so that the overall program would have the advantage of their experience in the stability and synchronization problems of large-scale power generation and distribution and in the effects of these on component and subsystem developments.

The issue of the involvement of the electric utilities in the program to establish the commercial practicability of photovoltaic conversion systems for large-scale terrestrial power generation was the subject of many remarks during the discussions of Panel II, User Requirements for Photovoltaic Systems. That the government should control and direct the overall program at this stage of development was a view that was not contested. However, a dichotomy was revealed in the statements from utility representatives on the one hand and representatives of solar cell manufacturers on the other. The position declared by the first group was that the utilities will become involved at the point that production of an acceptably low-cost cell has been demonstrated. The point of view of the solar cell industry was that if large purchases of solar cells were not made during the program before the time for demonstrating the production of the final version, low-cost cell, the industry would be unable to achieve the final price objectives, since the pricing would be based on a volume-unit price learning curve. The only response to this enigma was that the demonstration of the practical, low-cost solar cell would probably result from a government-funded program.

During this discussion, a reference was made to a joint venture by the fuel cell industry and a group of utilities, and this was used to illustrate the point that a joint solar photovoltaic development program might depend upon the willingness of the solar cell industry to back up technology claims with investment money and in this way to convince the utilities of the correspondence between production and price capabilities.

The involvement of the Electric Power Research Institute (EPRI) was suggested as a means of establishing a cooperative, mutually advantageous program. It was indicated that the program planning unit of EPRI could function as an information bank and as the agency for interacting the solar photovoltaic program with the utilities."

From that Conference forward a few utilities, along with EPRI involved themselves in photovoltaic systems R&D.

More recently (1977) at a semiannual review meeting of ERDA's silicon technology programs [217] about half the papers presented dealt with systems issues: Photovoltaic systems analyses (5 papers), power conditioning and storage subsystems (5 papers), and photovoltaic tests and applications (6 papers), and this volume is highly recommended study for systems engineers. Samples of the kinds of systems activities then (1977) being planned at Sandia (Schueler and Marshall's paper, p 392) include: (1) assessment of the utility/customer interface for on-site systems, (2) conceptual design of photovoltaic/thermal electric systems, (3) reliability and capacity displacement for PV systems in a utility grid, (4) transient effects and reliability of photovoltaic array systems, (5) and power conditioning studies. Many significant systems questions are raised in a paper by Watkins and Forestieri, e.g. regarding system vulnerability (lightning strikes, shadowing, aging effects), system safety, system performance, and system efficiency. Most of these questions are still unanswered. One also finds in this volume some of the early cost work showing concentrator photovoltaics produce the lowest busbar electric energy cost.

Most recent (1978) is the reporting by Schueler [218] [219] of the work being done by Sandia Laboratories for DOE on the photovoltaic systems engineering and analysis project. Schueler is studying many issues of interest to power systems engineers.

EPRI Systems Activities In PV's were referred to earlier. EPRI has been somewhat active in the systems area of photovoltaics both in sponsoring R&D and in disseminating information.[1] Reference [220] defines 1976 utility involvement in solar energy activities; five utilities out of 116 surveyed were cited as being involved in photovoltaics. They were:

- Arkansas-Missouri Power Co.: Consulting time and monitoring equipment for Mississippi County Community College photovoltaics installation.
- Burlington Electric Co.: Submitting a proposal to ERDA to study the feasibility of installing a 5-8 MW photovoltaic demonstration plant on the Burlington Municipal system.
- Florida Power And Light Co.: Studying inverter designs, safety locks, and other interconnect equipment for conversion of electricity from photocells.
- Pennsylvania Power And Light Co.: Provided data and consulting assistance for an ERDA-funded study conducted by GE on the conceptual design and system analysis of residential photovoltaic power systems.
- Salt River Project Agricultrual Improvement And Power District: Providing operational and forecasting data to Westinghouse for a utility impact study of conceptual photovoltaic power plant systems.

It is thus evident, from EPRI's survey data, that only about 4.3 percent of the utility companies responding have any kind of photovoltaic activity.

In reference [221] EPRI's total solar program is reviewed as of October 1976, their second such review meeting. Of the 10 solar contracts

---

[1] It is interesting to know that in the conduct of the author's research involving hundreds of reports and research documents, research reports ordered from EPRI arrived promptly in a few weeks at most; in contrast reports ordered from the government's NTIS required many months for some to arrive, many were lost, and others were needlessly duplicated.

reported therein only one deals exclusively with photovoltaics (contract RP651-1, Section 10 of the reference) and one (contract RP551-1, Section 5 of the reference) partly deals with environmental aspects of photovoltaics. Section 10, written by Marsh of GE, deals with the requirements assessment of photovoltaic electric power systems.

It is clear from the last reference document that EPRI is emphasizing solar-thermal power generation rather than photovoltaics. Perusal of their first solar program review meeting document [222] reveals the same emphasis. The last reference also reveals thermophotovoltaic contract work which apparently was terminated, as no reference is made to it in their second review meeting.

One must conclude, in face of the evidence, that utilities in toto, while beginning to become involved in photovoltaic systems on an individual utility company basis--an encouraging sign--both the utilities collectively and their principal R&D organization, EPRI, are only minimally involved in advancing photovoltaic systems compared to the R&D activities of DOE, NASA, Sandia and other government R&D systems groups. Thus the state-of-the-art in 1978 is a clear lead by the government in photovoltaic systems R&D--exactly as envisioned at the 1973 Cherry Hill meeting quoted above.

It is not hard to see that significant greater future involvement by the utilities in photovoltaic systems technology will be essential if the potential large fossil fuel saving potentialities of photovoltaics are ever to be realized on a grand scale.

We now leave the more general present art sytems aspects of photovoltaics to focus on the known technical aspects.

A Basic Photovoltaic System integrated with the utility grid is shown in Fig. 2-11. It permits solarly generated electric power to be delivered to a local load. It consists of:

- Solar Array, large or small, which converts the insolation to useful DC electrical power.

  Arrays will be treated in more detail later in this Section.

- A Blocking Diode which lets the array-generated power flow only toward the battery or grid. Without a blocking diode the battery would discharge back through the solar array during times of no insolation (recall from Fig. 2-8 that the cell equivalent circuit has a forward biased diode in it).

- Battery Storage in which the solarly generated electric energy may be stored.

  Energy storage is treated as a separate topic in Chapter 4.

- Inverter/Converter, usually solid state, which converts the battery bus voltage to AC of frequency and phase to match that needed to

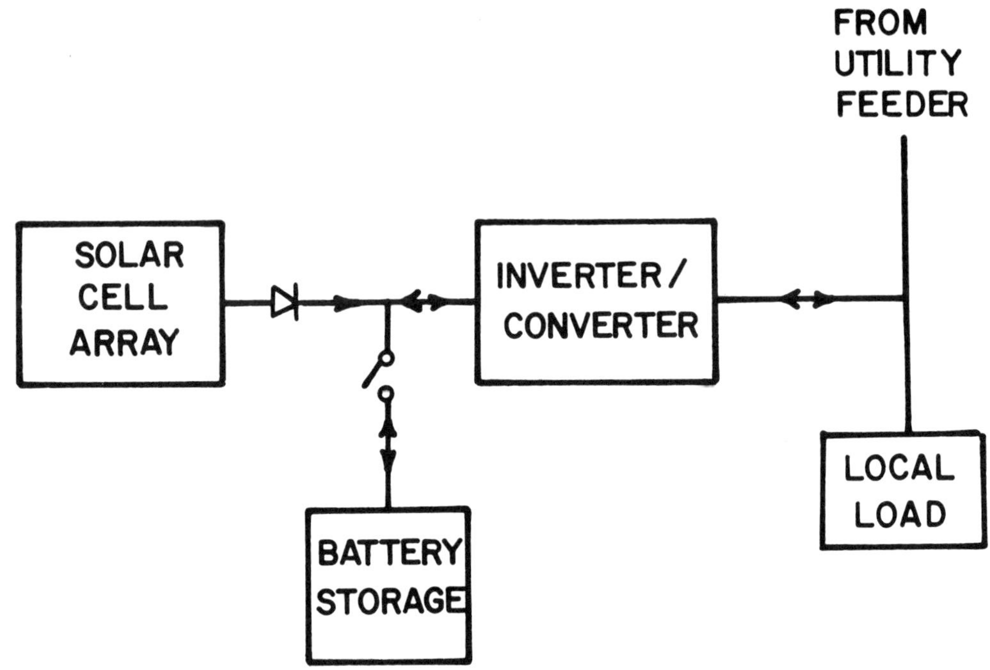

FIG. 2-11 Basic photovoltaic system integrated with the power grid. The system shown was studied by Pittman and Chowaniec [223] for peak powers in the 100 KW to 10 MW range (industrial and commercial customers).

integrate with the utility grid. Thus it is typically a DC-AC inverter. It may also contain a suitable output step-up transformer, perhaps some filtering and power factor correction circuits and perhaps some power conditioning, i.e. circuitry to initiate battery charging and to prevent overcharging. Some researchers prefer to show power conditioning as a separate system functional block. In the system shown in Fig. 2-11 Pittman and Chowaniec [223] also used this block to function as a rectifier to charge the battery from the utility feeder when needed and when no insolation was present.

Inverters will be treated in more detail later in this Section.

- <u>Appropriate Switches/Circuit Breakers</u> to permit isolating parts of the system, as the battery. One would also want to include breakers and fusing protection, not shown, between the inverter output and the utility grid to protect both the photovoltaic system and the grid.

Variations, usually more complex, of the Fig. 2-11 system have been used on spacecraft for photovoltaic systems, though naturally there was no utility feeder for such applications, i.e. it was operating as an isolated system. The same basic system configuration has also been used on many special purpose remote applications of photovoltaics where the installation is far removed from a utility line. Many of the latter applications only need DC in which case the inverter is deleted, leaving a simpler system.

Systems of the type shown in Fig. 2-11 may be designed to integrate with either single-phase or three-phase utility lines. The system studied by Pittman and Chowaniec [223] had a three-phase 440 volt output.

It is interesting to observe from Fig. 2-11 that photovoltaic power systems have no inherent slow reacting inertial components therein, as the wind-electric systems shown in Chap. 3 inherently have because of the rotor inertia. Photovoltaic systems can therefore be expected to be comparatively more stable, easier to control, and react quicker to load changes.

Pittman and Chowaniec [223] investigated operating strategies for the above system driving commercial loads, as shopping centers. They also investigated proprietor ownership vs utility ownership for various solar array costs/KW and for silicon and cadmium sulphide cells. The systems engineer should study their significant paper carefully.

Biran and Erlicki [224] have modeled essentially the system shown in Fig. 2-11 and created a highly flexible computer program in terms of loads. It permits computation of the required charging voltage, computation of the slant angle at which the solar array must be installed, computation of the working cycle of the system, retrieval of all radiation data for the location, computation of the array size, computation of battery capacity and computation of the system cost. They claim their program may be adapted for any location for which climate data are available.

Bartels and Moffett [225] analyzed a similar photovoltaic system to Fig. 2-11, but without the inverter, for use on low power terrestrial applications. They also developed a computer program.

Photovoltaic systems of the kind shown in Fig. 2-11 are normally designed to operate at the maximum power point of the array; but as Landsman [226] points out "the solar array maximum power point changes because insolation and ambient conditions continually change. Battery voltage varies with state of charge and ambient conditions. As a result, the system rarely operates at the maximum power point of the solar array. A maximum power tracker can increase the system energy output thus enhancing its economic value." He described such a tracker and demonstrated it on a 300 watt version. He says, however, that "there is disagreement and uncertainty as to whether a maximum power tracking scheme can increase the utility of a solar array. The answer would seem to be site dependent." Maximum power tracking apparatus appears to be used only on a minuscule number of photovoltaic systems in 1978. The subject is included here for completeness and to indicate that photovoltaic system analysts have not been unaware of this problem.

In the next several subsections we separately treat the various subsystems indicated in Fig. 2-11.

<u>Solar Photovoltaic Arrays</u> is, in itself, a rather involved subject. We therefore only briefly treat each of the subtopics the system engineer should know about and indicate where more information can be found.

- <u>Tracking Arrays vs Fixed Arrays</u> is an early issue which must be faced by the system design engineer. A tracking array is defined as one which is always kept mechanically perpendicular to the sun-array line so that at all times it intercepts the maximum insolation. Such arrays must be physically mechanically movable by a suitable prime mover and are generally considerably more complex than fixed arrays. A fixed array is usually orientated east-west and tilted up at an angle approximately equal to the latitude of the site. Fixed arrays are mechanically simpler than tracking arrays.

Mosher, et al [227] have examined this issue in some depth, finding that "the tracking cell will produce over 30 percent more electrical energy in the course of a relatively clear day than will the stationary cell." Their experiments were done on only a small number of cells rather than a large array, but the result is probably independent of array size.

Thus one of the initial choices a systems engineer has is whether to track his array and pick up a little more energy or whether to fix it and simply increase the array size slightly to make up for the extra energy he could have received had he tracked. One would expect that eventually when array prices come down significantly that there would exist a point where it would be cheaper to simply add on more array area than it would cost to add tracking; but the present art is not at that state yet, it appears.

- Solar Cell Connecting Arrangements are shown in Fig. 2-12. These arrangements are a result of the distillation of space-related solar array failures where the failure mechanisms and corrective

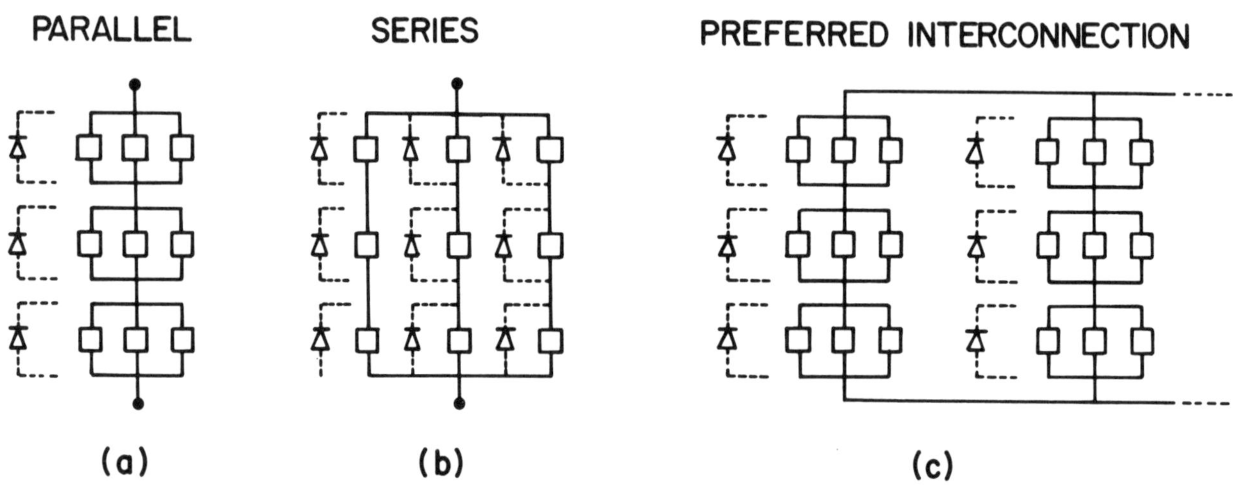

FIG. 2-12 Solar cell connection schemes in large arrays. Adapted from Hovel [85].

means were extensively studied. Hovel [85] summarizes that experience in his significant review paper.

Cells may be connected in parallel as in (a) to achieve the desired current and then stacked in series to achieve the desired voltage. Alternatively the series connection method shown in (b) may be used. Under ideal conditions they would behave the same; their behavior is not the same under abnormal conditions. If a cell shorts in (a) the output of a whole row of cells tend to feed current into the shorted cell; but a shorted cell in the series string of (b) has only a minimal effect. An open cell in (a) has a minimal effect while in (b) an open cell results in power loss from that entire series string of cells.

Shading can seriously affect array operation, leading to 'hot spots' and possible burnout of cells in the shade as Blake and Hanson [228] and Sullivan have shown [229]. Placing diodes across the cells as shown greatly minimizes the hot spot problem by forcing current through the externally added diodes rather than through the shaded cells.

The connection scheme in (c) has been found optimal with respect to shading, cell shorting, unequal cell I-V characteristics, and nonuniform illumination. It also has the fewest number of shunt diodes. In toto, it results in the highest array reliability.

Solar arrays can also be regulated by shorting out appropriate combinations of series and parallel segments, as Gooder has shown [230] on a 40 watt experimental array.

Lahri et al [231] have recently experimentally studied series/parallel configurations of solar cells for power output optimization.

From the brief remarks it is evident there are subtleties involved in how the solar array cells are to be connected to result in optimal system reliability. These facts are known, however, and additional research on them appears not needed.

- Array Design information for terrestrial arrays is notable by its paucity in the rather extensive literature surveyed in this research. Much of this kind of information is probably considered proprietary by the various few fabricators of solar arrays existing in 1978.

The engineering literature on space solar arrays is rather extensive and is recorded in the various Proceedings of the Intersociety Energy Conversion Engineering Conference (IECEC) and elsewhere. Some of that technology is undoubtedly transferrable to the design of terrestrial solar arrays; but the requirements for space arrays are much different from terrestrial design requirements. In space power/weight ratio is important, and that led to developing lightweight honeycomb substrate support structures on which the cells were mounted. Weight is of far lesser importance for terrestrial arrays. For space, cost was important but performance was dominant.

For terrestrial use cost/KWe is a dominating influence. For space use large arrays generally had to be deployed which lead to complex and expensive deployment mechanisms. For terrestrial use such mechanisms are not needed. Finally, for space use the rejected thermal heat from a solar array had to be radiated off into outer space, as there was no atmosphere around the array to carry off the heat. For terrestrial use the rejected heat can be carried off by a combination of radiation, conduction, and convection; in some cases, e.g. combined with solar air heating, in residences, the rejected heat can be used. In the latter case the physical array design would be substantially different from an ordinary flat plate array where the heat was carried off by radiation and convection.

Array designs fall into two broad classes:

- Flat Plate Arrays wherein solar cells are attached with a suitable adhesive to some kind of substrate structure, usually semi-rigid to prevent cells being cracked.

  This technology springs from the space-related photovoltaic technology, and many such arrays have been built in various power sizes.

- Concentrating Arrays wherein suitable optics, e.g. Fresnel lenses, parabolic mirrors, Winston type concentrators, and others, are combined with photovoltaic cells in an array fashion.

  This technology is relatively new to photovoltaics in terms of hardware development, and comparatively fewer such arrays have actually been built as of 1978.

Flat plate arrays by far dominate the present state-of-the-art of photovoltaic arrays. Such arrays are usually custom engineered for a specific application, and only comparatively recently are small arrays--or modules--being offered for sale commercially. Palz [25, pp. 209-214] shows photographs of sample flat plate photovoltaic arrays, though no details are given of the engineering considerations involved in the physical design of such arrays.

JPL published [232] a Solar Cell Array Design Handbook, Vol. 2 (1976) which contains a wealth of information on solar cells and material properties; unfortunately its title is misleading, as no place therein does it show how to physically design a terrestrial photovoltaic array. Nor can such needed design information be gleaned from progress reports on the Low Cost Silicon Solar Array (LSSA) Project [233] [234] [235], though they contain a great number of Gantt Charts and other management control devices.

It seems clear that from the systems engineer's viewpoint a great deal more work needs doing in integrating present art knowledge about the physical engineering design of photovoltaic arrays. Such work should seek to record what is known about array physical design, to organize it in a sensible way, and to present it in a form and language useful to system engineers actually designing terrestrial arrays. There simply was not time to achieve this task in the present research. Also it appears of sufficient importance to the future use of photovoltaics to be handled as a separate matter.

- Solar Array Encapsulation was treated in the prior art of Sec. 2-2 and in the present art of Sec. 2-3.

  Encapsulation is a part of the ongoing R&D currently underway on JPL's Low Cost Silicon Solar Array Project as recorded in reference [233]. Recent research on encapsulation at Battelle has been reported by Carmichael, et al [236] who examined all polymeric encapsulants as well as glass encapsulants.

- Array Support Structures is another facet of photovoltaic arrays which appears to suffer from a dearth of information. While the subject of support structures is known to have been investigated extensively by mechanical engineers in the growing solar heating field, unfortunately almost none of this technology appears to have been transferred to the photovoltaic community.

  In this research the only publication turned up dealing with support structures was a paper by Keeling, et al [237] of Motorola. They surveyed currently available units for Sandia Laboratories to determine the most cost effective approaches. These structural support concepts include residential, intermediate, and large-scale arrays, but are limited to fixed angle and seasonal adjustment capability. They also discuss the interesting new concept of combining array supports with high tension power line supports and/or array placement on the adjacent right-of-way.

- Array Testing And Degradation are topics of substantial importance to the systems engineer. Solar cell and array testing were briefly treated in Sec. 2-3.

  We update that information slightly here. Forman [238], of M.I.T. Lincoln Labs in an outstanding paper of interest to systems engineers, describes field testing and evaluation of photovoltaic module performance. His report summarizes actual field testing experience with a total of 41.5 KW of photovoltaic modules at various experimental test sites in the United States. The two largest arrays evaluated were the 25 KW array in Mead, Nebraska and a 9 KW rooftop test bed in Lexington, Mass.

  Forman's detailed evaluation of realistic degradation mechanisms is all the more significant in light of early R&D reports some years ago indicating solar cells had infinite life. While it is now known this is not true, it is still too early to say how long a life an array will have. Forman indicates that overall the cell-arrays gave "excellent performance for 1 year of service, but we can't yet predict whether they will last 20 years."

Using the Sherlock Holmes inspection approach of a hand held magnifying glass, Forman found evidences of degradation in:

- ▲ Cracked cells
- ▲ Broken interconnects
- ▲ RTV delamination
- ▲ Dirt and sand creeping into cells
- ▲ A 30 percent loss of array power in 6 months due to air pollution (dirt accumulation).

In spite of the observed failures only modules amounting to 0.13 KW (out of 41.5 KW) actually failed, he said.

Forman's paper is probably the most complete up-to-date study of degradation in solar arrays and is must study for systems engineers. He emphasizes that "It is important to determine the nature of changes that occur in modules that may or may not affect the 20 year-lifetime DOE goal."

Watkins and Pritchard [239] of Sandia Laboratories have examined degradation in concentrating photovoltaic arrays, including the effects of various numbers of thermal cycles.

Anagnostou and Forestieri [240] have summarized testing and degradation experiences of modules exposed in Puerto Rico, Florida, Arizona, and Ohio. Exposure periods were up to one year. They found that "In all cases, there was a loss of performance of the modules with outdoor exposure. The loss was dependent not only on the exposure site but also on the brand of the module. --- For all modules, a large percentage of the performance loss could be recovered by washing. As time progressed, this percentage was reduced, except for the glass-covered modules. All other brands showed a permanent loss in performance." They conclude that delamination and dirt retention are likely trouble spots in solar cell modules.

Pope and Matlin [241] summarize field tests on 4 arrays in the 10-100 KW range scattered around the country. Based on their limited tests to 1978, they found only 10 failures out of 3651 modules.

The problems JPL has encountered in environmental qualification testing of terrestrial solar cell modules are revealed in a paper by Hoffman and Ross [242]. They are examining hail hardness, UV weathering ability, factors affecting dust attraction, and humidity effects on modules. JPL's program of development, procurement and evaluation of solar arrays is laid out in reference [233].

- Standardization efforts are currently under way in Europe for measuring the electrical performance of solar cells, modules and arrays, as reported by Treble [243].

  The U. S. standardization effort was described in Sec. 2-2. It appears, in light of the heated discussion over standards at the 13th IEEE Photovoltaics Specialist Conference, that standards which are widely accepted and followed by the growing photovoltaics industry are some years away.

- Lightning Protection is believed by this researcher to be of substantial importance to the protection and safety of future photovoltaic installations. In all the hundreds of photovoltaic papers examined in this research only a paper by Bond [244] deals with the subject. Unfortunately it was withdrawn at the last minute and not presented at the 13th IEEE Photovoltaics Specialist Conference, so its contents are unknown.

  Of all the photographs and actual installations examined of photovoltaic arrays only a single one shown by Palz [25, p. 210] has what appears to be Franklin lightning rods actually installed.[1] It was on a navigational lighthouse in Indonesia.

  Research in the area of lightning protection of photovoltaic arrays of various sizes and locations obviously needs to be initiated by researchers knowledgeable of lightning effects on power apparatus.

  We now leave the subject of Solar Photovoltaic arrays to address another major component in the system of Fig. 2-11.

Inverters are devices, usually solid state, which change the array DC output to AC of suitable voltage, frequency, and phase to feed photovoltaically generated power into the power grid or local load, as shown in Fig. 2-11. We lump the subject of the control of it here also. These functional blocks are sometimes referred to in the literature as power conditioning.

The overall state-of-the-art of inverter technology has been succinctly summarized by Kroeger [245] of Sandia Laboratories: "Many manufacturers market hardware below 25 KW, transistor inverters dominating, and a significant number build SCR inverters at powers to 1 MW or so. At intermediate power levels, the use of transistors or SCR switching devices depends on the manufacturer's history, philosophy, and product line. Most applications are of the standby or uninterruptible power source genre, in high reliability power supplies for critical equipment like computers. These inverters operate from battery or battery-like DC sources and are not designed to operate with their outputs parallel-tied to the utility grid." A representative sample of this type of isolated DC-AC inverter is the work of Corry [246] [247] for inverters in the 10-100 KW range for Army use. Such inverters, he claims,

---

[1] It is possible they are bird arrestors though.

are about 88-93 percent efficient in converting DC to AC.

Inverters and their many facets have been discussed at the IEEE Power Electronics Specialists Conference (PESC) Palo Alto, California, June 14-16, 1977 along with many other power electronics applications. A Workshop on Power Electronics For Photovoltaic Power Systems was held at that meeting, though the results were not available to this researcher.

At least three books have been written about inverters; see Bedford and Hoft [248], Gyugyi and Pelly [249], and Pelly [250].

Kroeger [245] indicates that a Photovoltaics Power Conditioning Workshop was held Oct. 13-14, 1976 and that he was preparing a Proceedings of this conference attended by over 50 representatives of government, private R&D organizations, power conditioning hardware manufacturers, electric power utilities, and universities. Based on the workshop discussions and preliminary survey results, Kroeger sees the types of power conditioning equipment for photovoltaic systems classed as:

- ▲ Special and Remote (100 W - 100 KW) - Transistor switching regulators.
- ▲ On-Site Residential (2-10 KW, 1ϕ) - Voltage-fed transistor inverters, transformer output.
- ▲ On-Site Load Center (100 KW - 10 MW, 3ϕ) - Voltage-fed 6 or 12 pulse inverters, transistors slowly replacing force-commutated SCR's at higher power levels, transformer output, peak power tracking, excess power fed into utility grid.
- ▲ Central Station Plant (1 - 20 MW Modules) - Current-fed, line commutated, SCR-type 12 pulse inverters, transformer output, peak power tracking.

In 1977 Kroeger was (1) evaluating the power conditioning industry, (2) formulating specifications for power conditioning equipment in photovoltaic applications, (3) making recommendations to ERDA for power conditioning program, and (4) contracting for innovative hardware development.

Pickrell [251] of NASA LeRC, in 1977, was in the process of testing three types of inverters for photovoltaic power systems: an 8 KW line commutated inverter, a 5 KVA stand-alone unit, and a 10 KVA new design unit. The first two were off-the-shelf designs, and the latter one was in process of being custom tailored. The block diagram of the 8 KW inverter is shown in Fig. 2-13. This single-phase inverter was originally designed for windmill operation and modified for photovoltaic system application. It has the major advantage according to Pickrell, of simple design, low cost, and delivers power over a wide range of input voltages. Its disadvantages include: low power factor, requires KVAR correction, the output has high harmonic content, and the series inductor significantly lowers efficiency. The peak measured efficiency was 89 percent at about 2 KW power output, falling off for power levels below and above that.

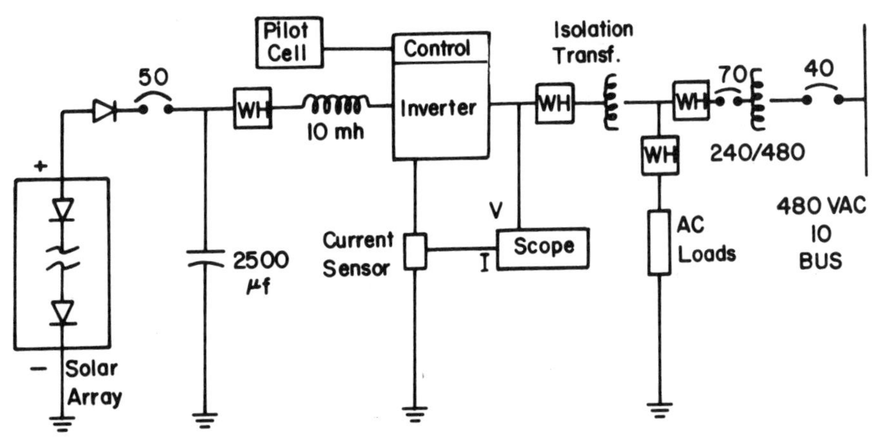

FIG. 2-13 Block diagram of 8 KW line commutated inverter evaluated by Pickerell [251].

Pickrell later (1978) described [252] in detail the 10 KVA inverter which had been optimized for photovoltaic applications. This single-phase DC-AC inverter is designed to operate directly connected to the utility line, feeding solarly generated power therein. It also was designed to continuously control the solar array operating point at the maximum power level based on variable solar insolation and cell temperatures. A second requirement was that the inverter be designed for high efficiency at rated load. The basic control system was simulated by a hybrid analog-digital computer and studied thereon before constructing the inverter hardware. Special attention was paid to the transient response stability; this is important in system operation, as fast moving clouds passing in the sun-array line would cause the array operating point to quickly shift. The inverter and its control system must follow this. A very thorough engineering analysis, synthesis, and follow-up appears to have been made on this inverter, and the reader will find the details in Pickrell's paper. Inverters of this general kind may have wide applicability. particularly for on-site residential applications.

Roesler [253] in 1978 is in the process of creating a 60 KW photovoltaic power system for a typical military utility network. Much attention is focussed on the DC-AC inverter with peak power tracking to permit the system to feed a 227-480 V 60 Hz utility network. Urish (Westinghouse) and Watkins (Sandia Laboratories) [254] have described the design and construction of a 50 KVA photovoltaic power conditioning unit, though details are missing.

Wood and Pelly [255] of Westinghouse on an EPRI contract have done an impressive piece of in-depth research on inverters integrated with utility networks in a broad general sense. Their work is aimed at a much higher power level--in the megawatt range--than for inverters cited above. Their work also appears to be orientated toward the move within the present utility industry of eventually going to large-scale battery energy storage at various points throughout the utility distribution network; such a development implies two-way exchanges of power between the utility grid and a localized battery bank. Though their work is not specifically directed toward photovoltaic power systems, it is believed relevant to them, and some of it undoubtedly may have a carry-over value when photovoltaic systems evolve to a higher power level.

Wood and Pelly's objectives were to:

▲ "Establish the best power conversion technology (or technologies) applicable to a number of advanced energy storage and conversion technologies.

▲ Determine whether a common technology can be applied in all cases.

▲ Determine whether adopting a modular approach to the construction of such a common conversion technology would lower costs.

▲ Establish areas of power conversion and related technology wherein research and/or development is needed before practical embodiments of the technology (or technologies) selected can be effected for application in the utility environment.

▲ Study a preliminary design of the best technology."

Their major conclusions, after a very in-depth examination of many inverter circuits, were:

▲ "A common technology can be applied to all the sources considered (Battery, Fuel Cell, Superconducting Magnetic Energy Storage, Magnetohydrodynamic generation, and Flywheels) excepting flywheels.

▲ The current-fed, naturally commutated inverter (essentially as used in HVDC transmission) is unquestionably the best short-term candidate. No other scheme has significantly lower estimated costs, nor do any possess lower full or part load losses."

They examined in-depth virtually every known type of inverter circuit which might be considered as a candidate for the utility application. From the many circuits contained in their report, the basic Graetz circuit shown in Fig. 2-14 is apparently[1] the general type of circuit found best. The waveforms are shown also. Note that the circuit can be used in two modes: (1) as an inverter

---

[1] 'Apparently' because it is somewhat difficult to tell from their report precisely <u>what</u> the preferred circuit is, it being mixed in with all the many candidate circuits and not being distinguished except by name only. The serious systems engineer will want to study their otherwise excellent original reports.

BASIC CIRCUIT OF 3-PHASE BRIDGE NATURALLY (LINE) COMMUTATED CONVERTER

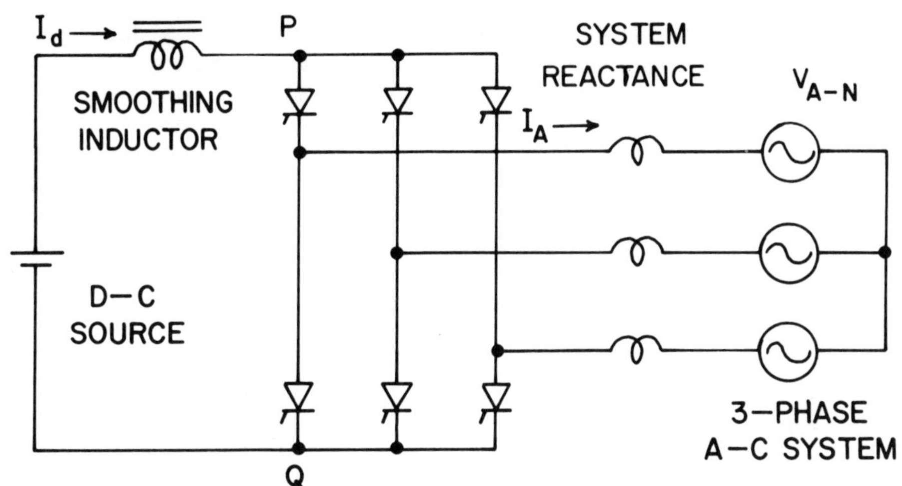

(a) INVERTER OPERATION

VOLTAGE ACROSS D-C TERMINALS V_{P-Q} ← mean value ≏ D-C source voltage

0

A-C VOLTAGE and CURRENT

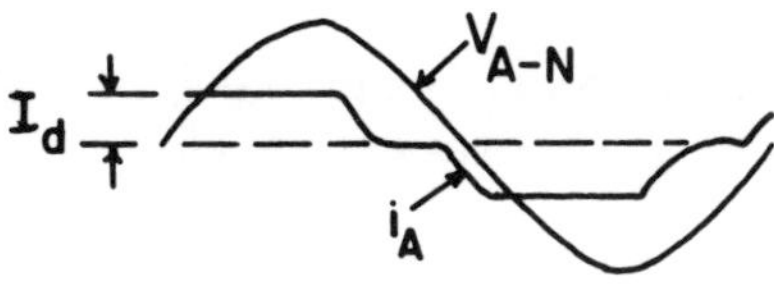

(b) RECTIFIER OPERATION

0

VOLTAGE ACROSS D-C TERMINALS ← mean value

A-C VOLTAGE and CURRENT

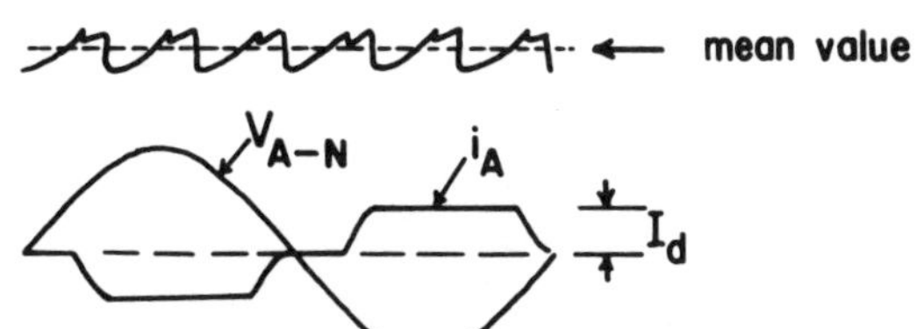

FIG. 2-14 Current-fed line commutated inverter. Circuitry for varying the thyristors' conduction periods is omitted. The basic waveforms for inverter and rectifier modes of operation are shown. From Wood [255, p. 168, 169].

changing DC to AC or (2) as a rectifier changing AC to DC, thus charging the battery.

It should be noted that when operated in the inverter mode the resulting line current is nonsinusoidal, containing harmonics.

Photovoltaic System With Energy Storage offer systems designers many options not ordinarily considered in conventional power systems. We have already considered in Fig. 2-11 the basic system with battery storage, and we subsequently treated the inverter needed to get the power into the AC line.

Chowaniec et al [256] investigated the operation of combined photovoltaic/battery systems in a total utility context. Their analysis made use of the conventional economic dispatching philosophy used in examining the day-to-day operation of utility systems. Their key issue is the need to assess combined photovoltaic/battery systems as a part of the total utility system rather than as individual units.

A considerably broader systems view of advanced photovoltaic systems with energy storage integrated with other energy systems is shown in Fig. 2-15 after the work of Eldridge, et al [257] [258] [259] of Mitre Corporation. Eldridge envisions large-scale photovoltaic arrays integrated with utilities via inverters, as we've treated here earlier, but also with the interesting possibility of using the array DC output directly to electrolyze water and produce hydrogen and oxygen which could then be stored and transmitted via gas pipelines to the point of utilization; there it could be used directly or run into fuel cells and a DC-AC inverter to produce electricity.

It is clear that with this system photovoltaics offers the options of DC power, AC power, hydrogen and oxygen fuels in either gas or liquid forms from which electricity can be generated.[1] The system has many advantages and few disadvantages [257]. It is also clear that with energy systems of this kind the term 'integration' takes on a much broader meaning than the narrower interfacing with an electrical network used heretofore. Time was not available, unfortunately, to research the 1978 status of this interesting advanced systems development.

Photovoltaic System Optimization is of concern to a systems engineer who would like to plan for, design and install nearly optimal apparatus. From the systems point of view, photovoltaic power system optimization studies appear presently in their infancy. This is understandable because of the enormous R&D effort heretofore necessarily devoted to creating solar cells; thus relatively little systems optimization analysis effort has been applied to this technology to date. Samples of the optimization work that has been done will now be summarized.

---

[1] See Fig. 1-32 for how this basic system might also be applied to other solar-electric systems. In this system, as many other advanced energy systems note we again encounter multiple energy conversions, a topic treated in Sec. 1-7.

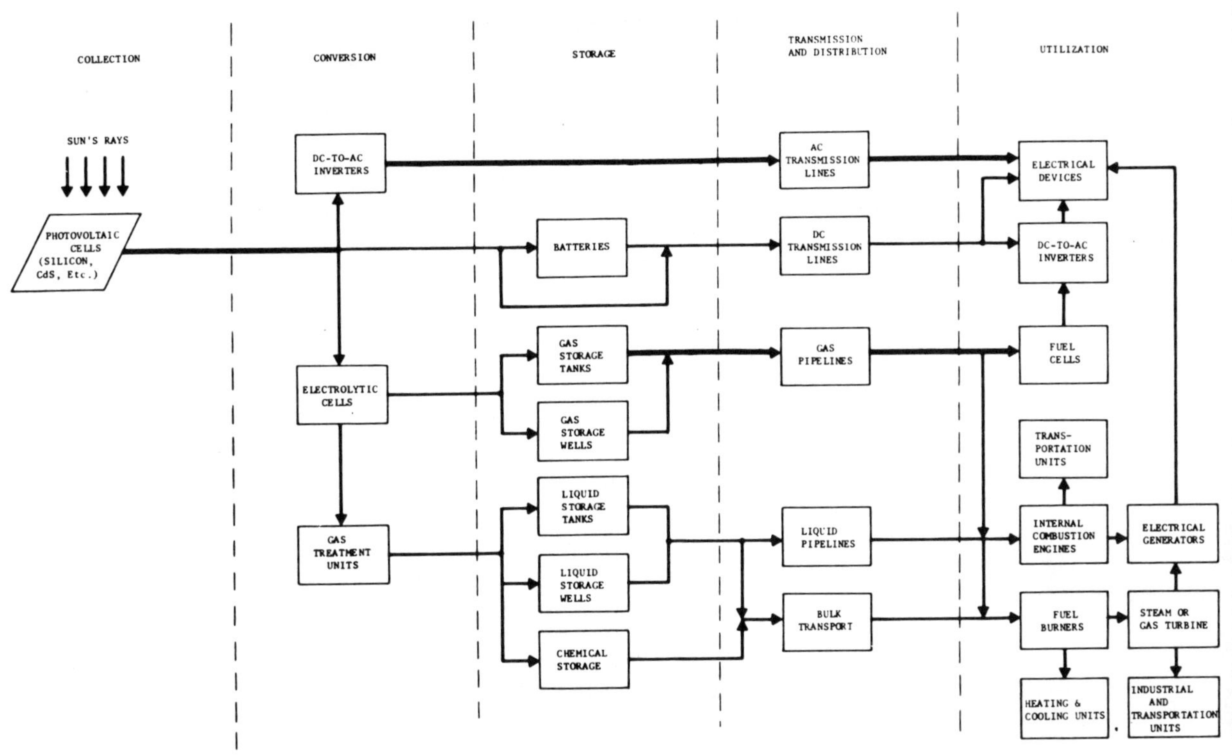

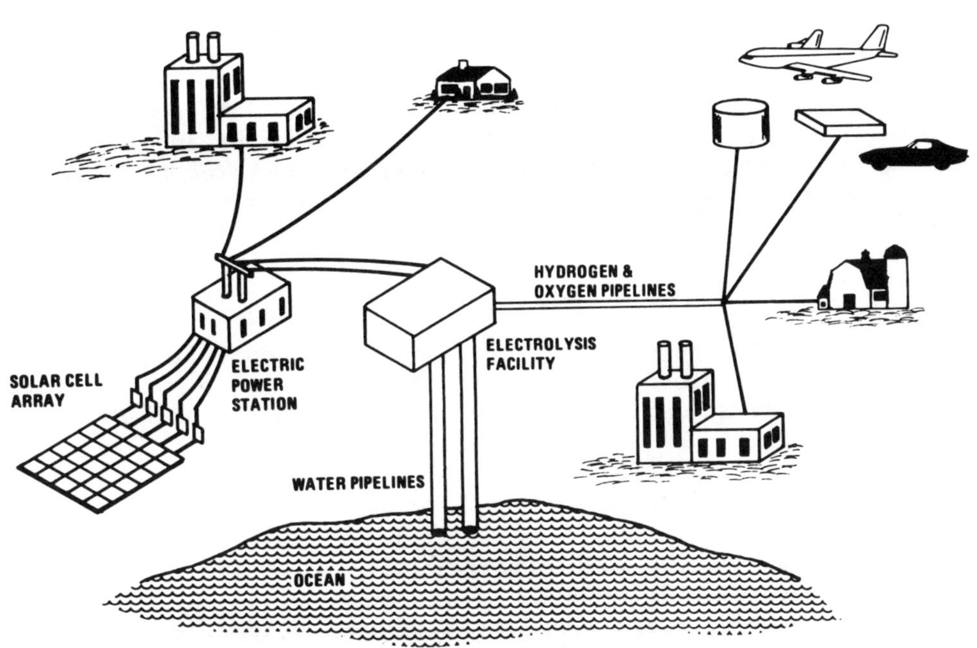

FIG. 2-15 Photovoltaic energy system under study at the MITRE Corporation. From Eldridge [257].

Linn [2 6 0] has done a computer optimization study of terrestrial photovoltaic power systems, addressing the "combinations of photovoltaic subsystems and component sizes which offer promise of producing economically attractive electrical power. --- Real electrical load data from residential, commercial, and industrial applications is combined with true meteorological data to evaluate system performance." Linn's "Results clearly show the superiority of certain concentrating collectors for ranges of solar cell costs. Areas where system sizes are not economically justifiable for certain load characteristics are identified."

Masden [2 6 1] has examined optimum values of collector and array geometry as a function of land cost relative to collector cost. He also considers the important shadowing problem, recognizing that realistic arrays will operate with shadows for a significant period of the day.

Ross [2 6 2] of JPL is studying module cost optimization including structural design optimization, solar cell geometry optimization, optimization of encapsulant optical qualities, thermal design optimization, optimization of environmental protection features, and others. His studies include consideration of the total system life-cycle costs as well as site variability.

If all solar cells in an array could be made precisely alike, then one would not have to address the optimization question of their variability. Watkins and Burgess [2 6 3] have addressed the question of the effect of solar cell parameter variations on array power output. Their "results indicate that for series connected arrays variations around a mean of cell series resistance and cell operating temperature insignificantly affect array maximum power. However, variations in cell photon generated current significantly affects power output as array current is limited by the weakest cell."

Samples of the early spacecraft power system optimization work are papers by Bauer [2 6 4] and Schwartzburg [2 6 5], the latter of which felt that "The use of computer simulation early in the design of an electric power subsystem not only establishes the sensitivity of subsystem performance and load capability to parameter changes, but also helps to define the required external interfaces, especially in the area of thermal control." Though spacecraft power system models are different from the terrestrial model of Fig. 2-11, the optimization philosophy is obviously similar.

Undoubtedly the future will reveal many more optimization analyses of various facets of photovoltaic systems, particularly as photovoltaics moves toward commercialization: necessarily optimization is implied in well designed commercial products.

The cost facet of optimization will be treated in Sec. 2-6.

Having presented the basic photovoltaic system and its subsystems, we now enlarge that view by examining (1) large photovoltaic systems and (2) residential photovoltaic systems, again embodiments of the centralized and decentralized philosophy mentioned several times.

Large Photovoltaic Systems One of the first proposals for large-scale centralized photovoltaic stations was made by Cherry [2 6 6] in 1971. The station he envisioned is shown in Fig. 2-16. He explains that "a square mile of solar array, during the summer months (at Washington, D. C.), assuming 60% sunshine hours at 7% conversion efficiency, could produce enough power to accommodate about 18,000 homes. This same power station during the winter months with about 50% sunshine hours and lower intensities could accommodate about 10,000 homes.

By providing an electrical storage system for the station as illustrated (in Fig. 2-16) around the clock power would be available. Using lead acid storage batteries similar to those installed in telephone exchanges, a storage

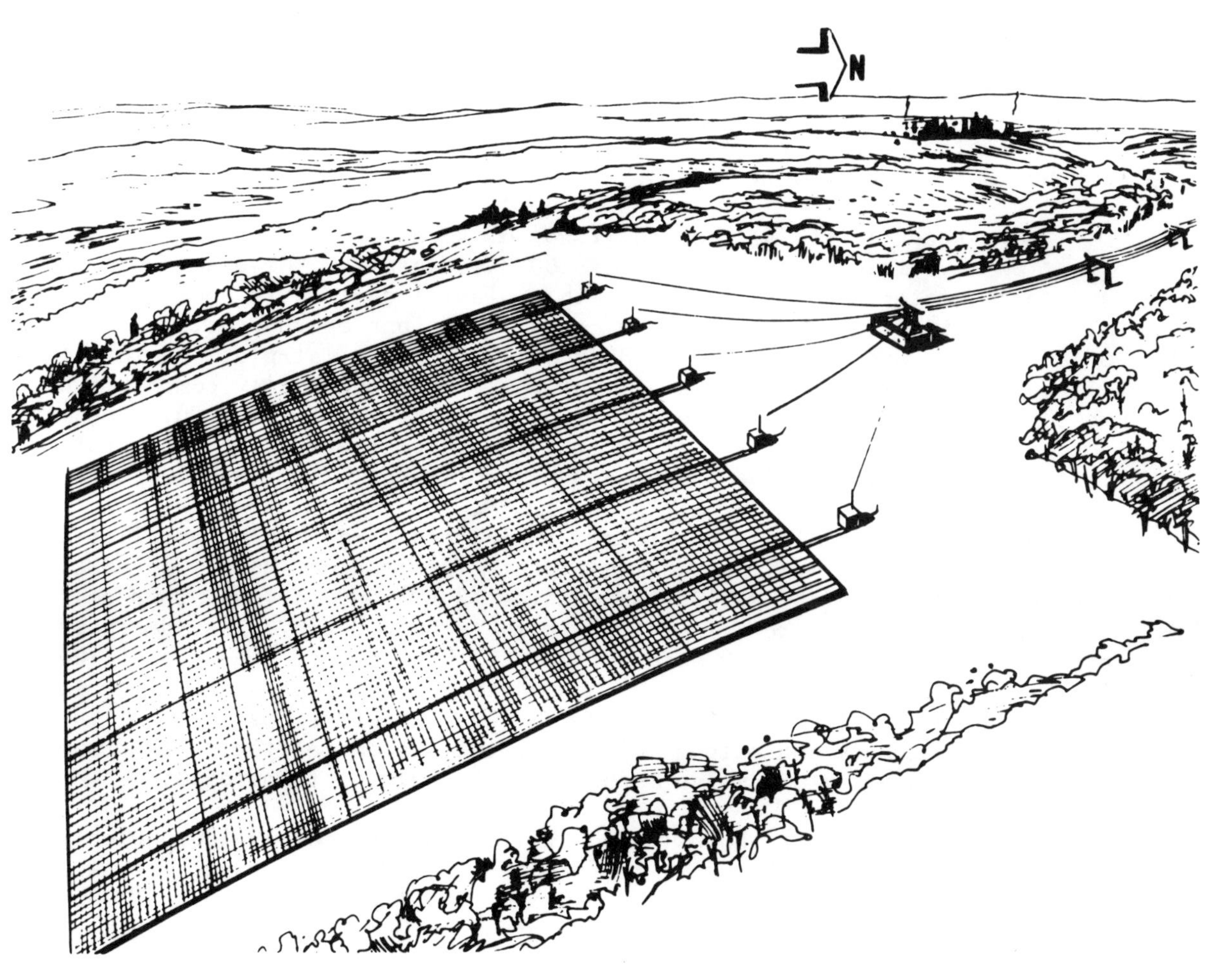

FIG. 2-16 One square mile terrestrial solar plant envisioned by Cherry [266] circa 1971.

capacity of 2.6 KWH/ft^3 can be obtained. A 1 million KWH storage capacity would require about 400,000 ft^3 or a building approximately 115 ft. x 115 ft. x 30 ft. high. This storage could provide around the clock power to the 10,000 homes in the wintertime for 4 full days, should this be necessary, or it could be used to handle peak power demands.

The entire electrical power requirements for Washington, D. C. and Prince Georges County, Maryland, (PEPCO) for 1969 were 1.1×10^{10} KWH/yr."

Cherry's imaginative solar-electric system has been much discussed in the advanced energy conversion community.

About the same time (1971) Ralph [2 6 7] made some preliminary calculations for a 1 $mi.^2$ solar power station which possibly could be located in the Phoenix, Arizona area. His station would have had a peak capacity of 0.259×10^6KW, transform the DC to AC and switch it onto the existing distribution grids. He found that such a station would require only about 5 percent of the U. S. production of metallurgical grade silicon, and he correctly recognized to make such a station a reality "efforts must be directed toward a substantial reduction in the cost of solar cell arrays."

More recently (1978) Leonard [2 6 8] of Aerospace Corp. has analyzed central plant photovoltaic power stations of 100 MWe capacity. He examined both flat-plate and concentrator collectors for possible plant locations in Phoenix, Dodge City, Miami, and Sterling, Virginia. He computed bus bar energy costs as a function of array area and found minima. He "concluded that flat-plate photovoltaic plants with $100 - 300/peak KW arrays will probably be economically competitive in most parts of the U. S. by the year 2000 but that systems with high or medium concentration will require that most of the thermal energy be used profitably in order to be competitive anywhere in the U. S. prior to the 2000 time period."

An interesting widely publicized real-world large-scale installation is the 25 kilowatt photovoltaic agriculatural experiment at Mead, Nebraska. This system was described by Matlin and Romaine [2 6 9] and various other contemporary researchers. This system, installed in the summer of 1977, was designed to power a 10 Hp pump to irrigate 80 acres of corn 12 hrs./day at the rate of 1000 gals/min. It was also used in crop drying. The system is an isolated one, i.e. it is not integrated with a local utility grid. It was therefore possible to linearly vary both frequency and output voltage of the system to start the AC pump motor without encountering excessive motor starting current transients.

The Mead System is shown in Fig. 2-17. "Approximately 100,000 cells are connected together on 28 panels to form the 25,000 watt array. The panels are wired to produce 6.3 amps at a maximum power voltage of 160 volts." Batteries store up to 85 KWH of energy for off-peak power use. Power is provided at either DC at 120 volts or 3-phase AC, 60 Hz, 240 V line-to-line.

The Mead plant was important in actually demonstrating the application of solar-electric power to practical farm needs at a power level significantly higher than previous terrestrial photovoltaic arrays.

Another even larger (60 KW) photovoltaic system is under construction in late 1978 at Mt. Laguna Air Force Station in California. It is expected to

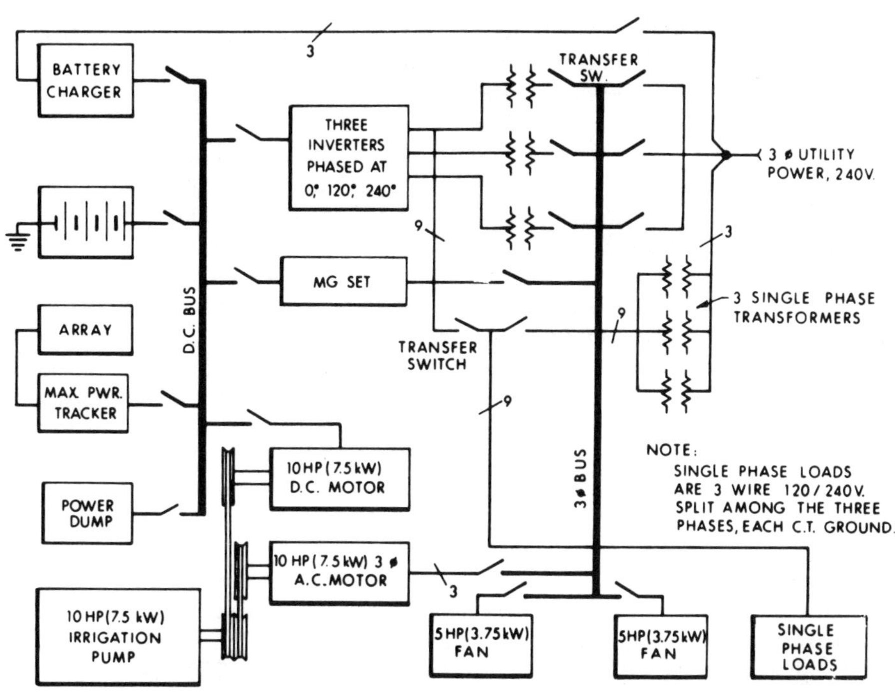

FIG. 2-17 Block diagram for Mead, Nebraska 25 KW photovoltaic systems. From Martin and Romaine [269].

be operational early in 1979. The system has been preliminarily described by Roesler [270]. It features peak power tracking and an utility interface. "The objective --- is to augment a typical utility network with solar cell power and thereby reduce the fuel consumption of the utility power plant. --- this system will be one of the first attempts to actually connect a solar cell power system to an utility network, and the findings of this project will have an important influence on future solar cell power systems."

The Laguna system is basically that shown in our Fig. 2-11 with the addition of some impedance cushioning between the inverter and the power grid.

Roesler describes the integration of this sytem with the utility network:

> "The operation of the solar array and power conversion system is automatic. As the array output power increases and the output voltage of the inverter reaches that of the utility line, the output from the inverter will automatically be phase locked to the utility line and the inverter will be connected to the net. At this point there is no power transmitted to the net. As the available power from the array increases further, power will be delivered to the net and the solar cell power system will augment the net. If the power output from the array decreases sufficiently, there will eventually be a negative power flow, i.e. from the net into the inverter. The design of the system limits this negative flow to the power losses (less than 3000 W) of the inverter. If the negative flow persists for more than a few minutes, the inverter will be isolated from the net until the output voltage of the inverter again equals the utility line voltage.

> The utility interface utilizes three main parameters: impedance cushioning, and phase and voltage control of the inverter output. By controlling these parameters, a smooth power transfer is ensured."

This inverter uses 12 self-commutated SCR bridge circuits to generate a 12 step sine wave approximation. It is expected that "the lowest order harmonic generated is the 11th, and minimal filtering is required to ensure total harmonic distortion of less than 3%."

Roesler's system appears to be of major significance to power systems engineers. Undoubtedly operating experience with it will be described in the future after construction is completed. Its significance appears to be, as he claims, that, it "will uncover problems connected with operation and maintenance and it will provide the required feedback for other, larger systems planned in the future." It will furnish valuable experimental data in real-world utility operation as opposed to the popular but usually incomplete computer analysis of such systems, an important point he does not mention.

An even larger photovoltaic system (100 KW) has been described by Lyon [271]. This system "will be installed by MIT/Lincoln Laboratory at the Natural Bridges National Monument (Utah) during the spring and early summer of 1979." The "array will contain approximately 1560 m^2 of flat plate solar modules consisting of 266,000 individual solar cells, and it will utilize 1.4 acres. --- a combined battery/switchgear/power conditioning building will house 750 KWH of lead acid batteries, the control subsystem, and a 40 KVA main inverter." Additional details of this system were not available to this researcher, and it could not be determined whether this is to be an isolated or an integrated system.

A still larger system (250 KW), the Mississippi County Community College system located in Blytheville, Arkansas, has earlier been mentioned in this report. It is scheduled for operation in mid-1979. This is the largest photovoltaic system ever assembled [272], and it appears destined to be another important benchmark in the evolution of photovoltaic systems to higher power levels.

The arrival of such systems as described above make it clearer and clearer that the 1971 dreams of Cherry and Ralph cited earlier in this section are in fact becoming reality in the late 1970's. Technical details on the electrical integration of this system were not available to this researcher. They can be expected to appear in the future technical literature as this development is completed.

From the above specific examples it is evident that the progression toward large-scale terrestrial photovoltaic power systems development is already underway in the United States in 1978. This research is slowly demonstrating that larger and larger photovoltaic systems <u>can</u> be built and integrated with the utility grid, obtaining needed electric energy directly from the sun. Only a few years ago such systems were thought by many to be utterly ridiculous; now, however, they are near realities. But a great task lies ahead to make such large-scale systems cost effective.

The above approximately summarizes the state-of-the-art of large-scale direct, centralized, solar-electric systems as of late 1978. We now leave this approach and focus on the other end of the application spectrum, i.e. dispersed, small-scale, residential systems.

Residential Photovoltaic Systems were clearly envisioned by the 1972 Solar Energy Panel [7, p. 55-58] where the benefits were seen to be:

- "Minimal effect on the ecology through use of land areas already being used for other purposes.
- About three times the present average household consumption of electric power can be collected from average-size family residences, even in the northeastern U. S. Assuming the energy storage problem solved, this energy surplus may make the electric automobile feasible.
- Invulnerability to breakdowns in energy distribution or centralized generation systems.
- The small size of the individual unit makes prototype testing and demonstration relatively inexpensive, and will help to attract consumer oriented industries."

The report showed [p. 57] a schematic of a generalized energy system for buildings, a major part of which was a solar-electric system. It envisioned supplying electric power to the home as well as to the battery of an electric car.

The idea of rooftop solar-electric panels to supply electric power to a residence is very old, and this ultimate application permeates photovoltaic literature. The idea is at least as old as 1929, as Palz [25, p. 230] shows a Paris house roof envisioned covered with "solar electricity panels." The idea is undoubtedly even older. No attempt was made in this research to further explore the origins of this idea. The idea has long had potential appeal to the world's individual homeowners. Conversely the idea appears to have had little general appeal to electric utility companies who generally disdain the small on-site approach to electricity generation in preference to large centralized systems, principally because of the so-called economy-of-scale argument for conventional power plants that the cost of power decreases with increasing plant power and size. The idea has also not been popular with utility companies probably because it has been seen as a threat to existing electric energy systems. In idealized theory, at least, were the idea of residential solar-electric systems to become a practical reality on a large-scale, then the entire utility system would have been bypassed and its energy output and profits would drop substantially, for the individual homeowner then would be receiving all his electric energy from the sun and would no longer need the utility. It is therefore easy to see why historically the utilities have collectively been substantially less than enthusiastic about residential photovoltaic systems. Their development has therefore sprung up outside the utility industry, principally within the photovoltaic community and the advanced energy systems community.

Residential photovoltaic systems are the principal and classic example of the decentralized philosophy for deploying solar-electrics as was thoroughly treated in Sec. 1-6 of this report. Maycock [273] of DOE in 1978 sees "The real reason for photovoltaics is for residential use!" The reader should be aware, however, that there is substantial argument within the 1978 photovoltaic community over whether the potential residential or utility markets are largest for photovoltaics. It is well-known that both are enormous potential markets. The issue is beyond our scope here.

In this Sec. we shall, in an all too brief a manner, simply sketch some of the state-of-the-art technical work that has been done on integrating photovoltaic systems into residences. The previous work is largely of a theoretical nature, as the cost of solar arrays has heretofore been too high to permit very many residences to be so equipped. A few such experimental houses were so equipped, however, notably at the University of Florida and the University of Delaware in the late 1960's and early 1970's respectively. The reader desiring more detailed information on photovoltaic systems applied to residences will find it in the references cited. The references cited here are by no means exhaustive.

1972

Backus [273A] studied "the technical feasibility of using a solar cell covered roof on a private residence to provide all of the energy requirements ---. The energy storage was by electrolysis of water coupled with a fuel cell." His reference design was based on 1972 technology, and the house was to be located in the Phoenix area. His electrical aystem was based on a 2150 ft^2 rooftop array driving a 75 percent efficient electrolysis unit followed by a 65 percent efficient fuel cell, thence to a 95 percent efficient DC-AC inverter[1] coupled to the house main breaker box. The power conditioning was quite complex because sometimes "the inverter is fed directly from the solar cells and at other times is fed from the fuel cell."

Finally, he opines that "The installation, maintenance, and operation of the rather sophisticated system, such as the solar-fuel cell system, should be the responsibility of a utility company. It would seem reasonable that the company would own the system and have a leasing arrangement with the home owner that would be a fixed fee per month minus a credit for the excess gases which would be metered into the utility company's lines."

His paper was principally a systems study and showed no details, as of the inverter design.

1976

Imamura, et al, [208] [274] of Martin Marietta Corp., completed an impressively thorough systems type study for NASA LeRC called "Definition study for Photovoltaic Residential Prototype System." They thoroughly studied several options for the electrical configuration, finally choosing the best baseline electrical configuration shown in Fig. 2-18, a configuration similar

[1] We are, once again, reminded of cascaded energy conversions treated in Sec. 1-7.

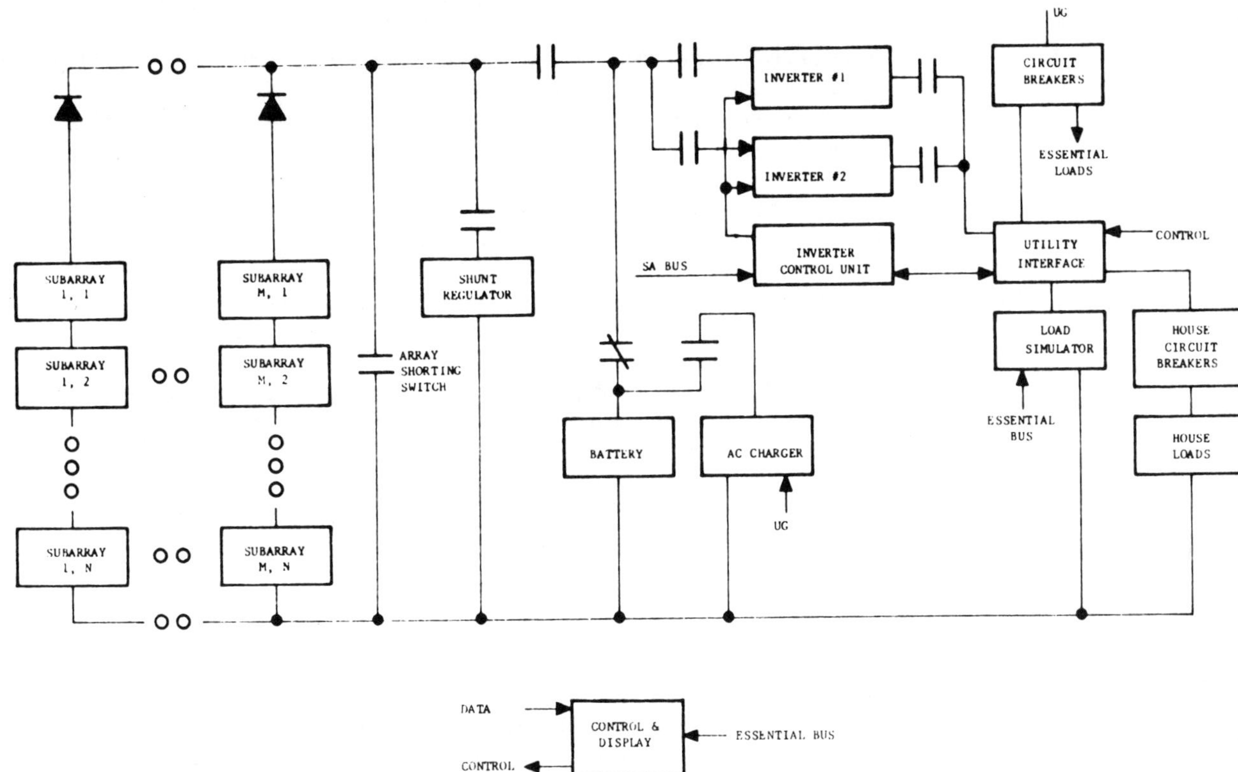

FIG. 2-18 Functional block diagram of baseline power system for residence. From Imamura, et al [208].

to our earlier Fig. 2-11. They performed a sensitivity analysis and found "the most sensitive parameters are the solar isolation and the inverter efficiency. A key result of the sensitivity analysis was that a significant increase in the overall (Prototype System Test) PST efficiency can be obtained by eliminating the battery." The system proposed was for 115/230 V three wire, single-phase, 60 hz output at 13 KVA at 0.85 lagging power factor. The inverter input voltage would range 105-140 VDC. The inverter efficiency would be 90% at full load. The array was to have an area of 133.8 M^2 with a total of 31,680 cells resulting in a power output of 9183 watts at 28°C and 100 mW/CM^2 insolation.

They concluded, among other things, that:

- "A power system configuration, containing only the solar array, inverter, shunt regulator, and associated control circuitry provides the best overall performance and is the most cost-effective arrangement for the PST program.

- The configuration without the battery as compared to that with the battery yields a substantially higher energy displacement (ratio of energy supplied by the PST power system to the total energy used by the residential electrical loads) computed over a one-year period.

- The solar array size can be significantly reduced by eliminating the battery (i.e. eliminating the energy storage requirement and using the utility source for night power demand). However, the use of a battery is recommended as part of the PST program for experimental purposes."

Of possible interest to the systems engineer are their equations [p. 3-81] for sizing the battery, if one is to be used.

The systems engineer involved in integrating photovoltaics into residences would be well advised to study Imamura's reports in detail.

Federman, et al [275], Westinghouse Electric Corp., have made a preliminary examination of photovoltaic residential systems in the range of 1-10 KW, including attempting to estimate the cumulative number of such residences vs time for market estimation purposes. They also examined intermediate power systems (100 KW - 10 MW) and central power systems (50 to 100 MW). Few technical details are provided in this paper, however.

Sater and Goradia [276] described a 36 M^2 high intensity 3 KW solar energy system "which can supply almost the entire energy for the heating, air conditioning, hot water and electrical needs of an 1800 ft^2 model home in Albuquerque, N. M." They estimate their system would cost less than $4,500 and would supply 45,000 KWH/yr. thermal energy at $0.0072/KWH and over 10,000 KWH/yr. of electrical energy at $0.02/KWH. They estimate that widespread implementation of such houses could save the nation over 30 million barrels of oil total per year. The solar cell assumed was a high efficiency edge illuminated one. Theirs is a concentrating tracking type system.

Kirpich, et al [40], General Electric Co., have examined "the performance of an on-site residential photovoltaic power system, both with and without energy storage, --- for site locations in Cleveland, Ohio; Phoenix, Arizona; and Miami, Florida." Their system analyzed was similar to our Fig. 2-11. The results of their sensitivity analysis yielded a preferred system sizing associated with the minimum cost of electric energy supplied. They also found that "For a 30-year levelized cost of $0.05/KWH, the maximum allowable capital cost of solar arrays varies from $0.90/peak watt in Phoenix down to $0.45/peak watt in Cleveland for the case of the no energy storage system."

1977

Eltimsahy and Delombard [277] performed an experimental investigation on a low powered solar cell/inverter system for coupling to a conventional three-phase network. Power outputs in the 0-400 watt range were experimentally studied. Their system was shown "to deliver over 80% of the optimal solar cell array power to the electric network." Their inverter "used a 6 pulse SCR bridge inverter similar to those used in DC transmission.

White, et al [278], Mobil Research and Dev. Corp., offer a few perspectives on solar electric power for small homes, making the important point that "solar-electric homes will require supplemental power from the utilities only in off-peak hours, thus offering an incentive to distribute such power at reasonable rates."

1978

Federman, et al [279] examined stand-alone photovoltaic residential systems. They found "that at least three solar energy options are available to new single-family residences. The utility can be used to provide back-up energy with the residential on-site electrical and thermal storage systems eliminating daily peaks for the utility; the utility can own and maintain a part of all of the system with either on-site or central back-up; or the residence can stand alone. Likely, all options will be used--the choice depending upon location and circumstances."

Kirpich and Tully [280], General Electric Co. are studying residential photovoltaic system concepts in a regional context. Twelve U. S. regions representing a broad spectrum of climatic characteristics are being examined. Their systems are roof-mounted photovoltaic systems in conjunction with solar thermal collection systems to provide space heating and domestic hot water.

This completes the state-of-the art review on strictly solar-photovoltaics for residential use. It is clear all the above systems type studies are of a preliminary nature, all awaiting the arrival of a practical $500/KWe or lower solar array which in turn hinges on cost reductions in solar cell technology, a need this review has repeatedly made. The entire photovoltaic community is well aware of this need and is diligently trying to reduce cell costs.

Photovoltaics and Total Energy Systems are of some interest to the systems engineer. A total energy system is one engineered to permit almost all the available energy to be usefully employed. Typically 'total energy', in this photovoltaic context, means use of both electric and thermal outputs.

We only make a partial survey here of the literature on this subject turned up during this research.

The idea of combining photovoltaic cells with thermal systems originated in the 1972 Solar Energy Panel [7, p. 56]. A practical embodiment for use on a house was described in 1973 by Böer [281]. The rejected heat from the solar cells was used to heat the household air. Samara [282] described Sandia's 1975 program in photovoltaic-thermal solar systems.

In 1978, Chobotov and Siegel [283] analyzed total energy systems for single-family residences. Smith, et al [284] and Kern, et al [285] have also preliminarily examined photovoltaic total energy systems.

Special Systems Problems unique to photovoltaic systems will round out our treatment of this rather involved subject of 'systems integration'. We treat each subject only in the briefest possible way to alert the systems engineer to possible problems. More information will be found in the references cited.

- Motor Starting With Photovoltaic Systems, as with an array without storage driving one or more pumps, poses special system problems because of the starting transient current demanded by the motor. Landsman [286] has researched this problem in connection with the 25 KW Mead, Nebraska photovoltaic system. He succinctly states the problem: "The starting transient could be as large as 80 KVA (motor code J) at full voltage starting, even though running power is 10 KVA. Use of an 80 KVA inverter is not acceptable ---." For this isolated system he found "the best procedure for motor starting is to use a combination variable voltage/frequency scheme. The motor is started at low frequency, and a correspondingly scaled voltage ---. The voltage and frequency are both slowly increased to the nominal running values." His motor was a 10 Hp 3-phase shown in our Fig. 2-17.

Obviously Landsman's system would not be applicable to a photovoltaic system integrated with a local utility; but it does point up one of the technical problems of an isolated system having no storage or 'stiffness' to withstand the starting current transient. An advantage of an integrated system is that starting transients can be drawn from the power grid which is a 'stiff' supply.

- Specialty System Applications, usually at low power levels, are increasingly being made of photovoltaic systems. Many of these involve only DC power distribution, and no inverter is used. An example is the 3 KW (peak) photovoltaic array being engineered in 1978 for the Papago Indian Village of Schuchuli about 120 miles west of Tuscon, Arizona, as reported by Bifano, et al [287]. The system will supply electricity for village water pumping and basic domestic needs. A 120 V DC distribution system is to be used. Battery charge control is accomplished "by switching solar cell array strings in or out of the charging circuit. Load management (shedding) provisions are incorporated to protect against excessive battery discharge."

Numerous investigators of such small photovoltaic systems have indicated the systems must be very carefully thought through to integrate with the local village life-style. This facet of socio-engineering may become all the more important in the future if photovoltaics develops into a large export market for the third world, as many believe is an important major future market for photovoltaic systems.

Ratajczak [288] has described 16 geographically dispersed specialty photovoltaic systems including powering refrigerators in the National Park Service, powering a remote Indian village, powering forest lookout towers, an interstate dust warning system, insect survey traps, remote weather stations, and other applications. These ranged from 12 V DC to 120 V DC buses at powers up to 446 watts.

Weiner, et al [289] have generated a computer program for analyzing navigation-aid photovoltaic power systems.

Medler [290] describes the special design problems for creating lightweight photovoltaic power unit for military applications.

The above represent only a sampling of the known large number or specialty applications of photovoltaics. No attempt was made in this research to exhaustively explore these interesting, but low power level, applications.

This completes our rather extended treatment of system integration of photovoltaics. System integration was seen to have very many facets indeed. We now turn to another important aspect of photovoltaics, economics.

2-6 Economic Aspects of Photovoltaics

Economic aspects are obviously of great importance to the eventual commercialization and widespread use of photovoltaic systems. The economics have been given much attention by the photovoltaic community.

We make several observations:

- Photovoltaic Systems Must Eventually Compete Economically with other electric energy systems.

- The Key Parameter insofar as the customer is concerned is the busbar energy cost measured in \$/KWH. It is generally recognized within the photovoltaic community to be higher than for conventional electric energy costs; but the photovoltaic figure is generally falling as cheaper solar cells become available.

- The Viewpoint Must Be A Systems One and not that of the cost of solar cells only.

- Photovoltaics Is Rich In Design Options, and it is very important to know precisely the set of assumptions being dealt with in any photovoltaic economic analysis. Many cost figures exist because of this richness of options resulting in confusion and difficulty of making valid comparisons.

- Two Viewpoints exist within the photovoltaic community:

  - ▲ The Acceptable Price per KW of array or for a given photovoltaic system. This is the estimated price at which photovoltaic systems could become competitive with conventional electric energy systems; therefore it is the estimated price at which the mass market in photovoltaics could be expected to rapidly grow. This is presently believed to be in the range of \$100-\$500/KW.

  - ▲ The Actual Price, usually on a \$/watt or \$/KW peak basis for the array. It is well recognized within the photovoltaic community that the actual price in 1978 is substantially higher than the acceptable price, current lowest array prices running of the order of \$6000-\$7000/KW.

- One Must Distinguish between:

  - ▲ Solar Cell Costs
  - ▲ Array Costs
  - ▲ System Costs.

  Unfortunately, within the literature such clear distinctions are not always made, and some papers appear to equate cell costs with array or system costs, ignoring storage (if used), power conditioning, wiring, support structure, and other costs. It is not generally easy to find such responsible data.

- The Complexity of the whole field of economic calculations, assumptions, estimates, incomplete or partial data based on an unproven R&D concept, etc. are generally recognized. Part of the complexity is a result of the fact that there is in 1978 no large-scale solar cell manufacturer producing a single design cell under known controlled and repeatable conditions.

In this section, we can only outline the general first-order approach to costing a photovoltaic system, give a rough feel for current solar cell costs and trends and indicate sketchily where more information can be found in the references. Do recognize this is a highly dynamic nonstationary area, and cost figures can and do rapidly change as the technology develops.

System Costing Approach for photovoltaic systems, ignoring inflation, money interest, depreciation, taxes, land, and other costs, i.e. the simplest case, has been clearly laid out by De Meo [291] of Brown University whose method we include here. The model is basically that of our Fig. 2-11.

Let: PPC = Power plant cost in dollars
PPO = Power plant rated output in KW
AA = Array area
AC = Array cost/unit area installed
CC = Cell cost/unit array area installed
SWC = Structure and wiring cost/unit array area installed
SGC = Storage cost/KW installed
PC = Power conditioning cost/KW rated

Then

(2-7)

$$AC = CC + SWC$$

And

(2-8)

$$PPC = (AC)(AA) + (PPO)[(SGC) + (PC)]$$

Also, letting I_{Ave} = Average daily insolation in KW/unit area
CF = Plant capacity factor
η_{cell} = Cell conversion efficiency
η_{area} = Cell area/array area
η_{wire} = Wiring efficiency (accounts for losses in intra-and inter-array wiring)
η_{sg} = Storage efficiency
η_{PC} = Power conditioning efficiency
η_{plant} = Power plant efficiency

Then

(2-9)

$$\eta_{plant} = \eta_{cell}\ \eta_{area}\ \eta_{wire}\ \eta_{sg}\ \eta_{pc}$$

and

(2-10)

$$AA = PPO/\left(\eta_{plant}\right)\left(I_{Ave.}/CF\right)$$

Equation (2-8) is basic and the others feed into it. Equation (2-9), again reminds us of our earlier treatment in Sec. 1-7. The η_{cell} is presently the lowest factor in Equation (2-9), and, according to Corollary 1 in Sec. 1-7, here is where the R&D should be focussed; and that is precisely where it is being focussed as evidenced by our earlier Sections in this Chapter.

Numbers used in the above equations, naturally, depend on the assumptions made by the analyst as regards what is realistic. De Meo used the above method to study the allowable economics of cadmium sulphide arrays, deriving therefrom several useful graphs. He studied power plant costs in the range of \$800-\$1400/rated KW. He concluded, based on central station use of CdS cells, that:

- "Unless power plant costs significantly greater than \$800/KW rated can be tolerated, cell efficiencies greater than 12% will be required for non-concentrating, central station photovoltaic power plants operating in high insolation locations in the Southwestern United States.
- Allowable cell cost depends very heavily on structure and wiring costs; therefore valid estimates of allowed cell costs or required efficiencies will require very accurate estimates of structure and wiring costs.
- The smaller the allowed cell costs in relation to the remainder of the plant, the more sensitive are those cell costs to variations in plant parameters.
- A relative increase in cell efficiency may justify a relative increase in cell cost greater by a factor of 2 or 3."

Finally, he concluded that installed CdS cells would have to cost less than \$1/ft.^{2}, and this value has been projected in large-scale production.

A more complex costing approach has been described by Bradley and Costello [292]. They account for fuel savings, operation and maintenance costs, taxes of various kinds, cost of gas turbine capacity, cost of coal capacity and other factors. They clearly link the acceptable photovoltaic system cost (\$/KW) to the costs of current fuels (their Fig. 6). As conventional fuel costs go up, the acceptable price of photovoltaic systems also increases, making photovoltaics somewhat easier to attain. They have many interesting graphs containing acceptable cost information which the reader necessarily must find from the reference.

System Cost Estimates Chowaniec et al [293] of Westinghouse have investigated costing a solar photovoltaic power station. They assemble an interesting table [p. 22-11] showing the capital cost of arrays for both $100/KW cells and $500/KW cells were these available. The array cost ranges from $205/KW to $770/KW, depending on the type module and whether fixed or tracked. They also show an interesting graph [their Fig. 6, p. 22-13] plotting busbar energy cost vs solar cell efficiency for various solar cell costs ($/KW). As expected, increases in cell efficiency markedly decrease the busbar energy cost. They do not give the details for computing the costs, however. They conclude that "For less expensive solar cells (less than $400/KW), fixed tilt collector/reflector modules are more cost effective than those which track."

Cole, et al [294] have estimated the cost of electricity produced by four types of compound parabolic concentrators. Their main conclusion is that "photovoltaic compound parabolic concentrators can produce electricity at significantly lower cost than flat photovoltaic arrays. Their best conical

concentrator cost is in the $1000-$3000/average KW range.[1] They have many interesting graphs showing how cost is related to concentration ratio. There is an optimal minimum cost at some concentration ratio.

Evans and Florschuetz [295] have also examined costs of concentrator photovoltaics.

Böer [296] has some helpful advice on costing as applied to residential applications of CdS combined with air heating. He estimates the panel cost to be $1.86/ft^2$ (1972 dollars). His system is based on a 7 percent system efficiency.

Wolf [297] has also excruciated over the cost issue for silicon solar arrays for large terrestrial applications. He concluded that "a reduction in solar array costs by a factor of at least 200 from the (1972) levels" will be required.

Merrigan [22, Ch. 5] examines the many facets of photovoltaic economics, including marketing aspects.

Tsou and Stolte [298] have studied the balance of system costs. They found that "the array structures and foundations are the highest cost items outside of module costs."

Ross [299] of JPL "describes one of the methods being used for module optimization, including the derivation of specific equations which allow the optimization of various module design features. The method is based on

[1] They take great pains to point out that cost comparisons on the basis of average rather than peak KW are more realistic.

minimizing the life-cycle cost of energy for the complete system." His methods appear to be lending a clearer analytical base and rationale to the costing than in many papers examined.

Cell And Array Costing, as the above system citations indicate, constitute what is presently the most important economic limitation on the widespread deployment of photovoltaics.

Hayes [300] succinctly sizes up the key issue of cell costs: "Solar cells cost about $200,000 a peak kilowatt in the late fifties. By early 1975, the cost had dropped to $31,000; by September 1976, the figure was $15,500. In early 1977, the cost of solar cells fell to $11,750 a peak kilowatt. And in December 1977, an Arkansas Community College contracted for a photovoltaic system for $6,000 per peak kilowatt." Maycock [273] of DOE in August of 1978 provides the most recent data point that "Eight manufacturers are saying we can sell concentrating photovoltaics for $2,000/peak KW now!" Though such a curve is by no means a firm thing, the evidence strongly suggests that photovoltaic prices have dropped dramatically in recent years as a direct result of constant striving by the photovoltaic community.[1] Actually achieving the national goal of $500/peak KW appears only a short time away; but the difficulty of actually achieving it should not be underestimated.

We have earlier on this Chap. cited the vigorous work now under way within the photovoltaic community toward studying the feasibility of automated factories for producing solar cells. When realized one or more such factories will surely have the effect of further substantial cost reduction and greatly hasten the day of practical photovoltaic electric systems.

Progress in reducing the costs of silicon solar arrays will be found in progress reports by Williams [301], RCA, and Carbajal [302] [303] of Texas Instrument, Inc.

Iles of Centralab, in 1975, [304] was working on three solar cell options giving promise of costs below $10,000/KW.

It should be abundantly clear by now to the systems engineer that the photovoltaic 'devices' community has had and continues to have a dynamic effort aimed at reducing the eventual cost of solar cells. This is, by nature an ongoing activity; on their shoulders hinges the future success of the entire photovoltaic systems movement, a not unexpected conclusion.

---

[1] The magnitude of this accomplishment cannot be overemphasized. Once again the dreams of creative people are being realized while the pessimists are being left behind. This researcher recalls that after returning from one of the 1972 solar panel meetings a colleague chided him for recommending the further development of then $400,000/KW solar arrays. "You might be able to reduce cost by one order of magnitude, but your group will never be able to reduce it by three orders of magnitude!" Clearly in 1978 we are within sight of the goal, however.

2-7 Barriers To Photovoltaic Integration

By barriers we mean here those problems or issues standing in the way of photovoltaic significant progress. By integration we mean here both our electrical integration and technology integration of Sec. 3-8. Including the technology integration is necessary as one looks ahead a few years to the possibility of photovoltaics becoming a common reality on the nation's residences and in the utility industry. We are also considering here photovoltaics as an entity, i.e. a potential whole new industry beneficial to the nation, as opposed to considering a single photovoltaic generator located in a specific part of the country.

After surveying an appreciable fractional part of the published literature on photovoltaic energy conversion systems, it is this researcher's opinion that the following items appear to be the principal 1978 barriers to photovoltaic integration:

- High Cost of Arrays is the principal factor holding back photovoltaic integration. The DOE goal of $500/peak KW has not yet been achieved commercially. The lowest figure heard on the 1978 scene was of the order of $2000-$3000/peak KW for concentrated arrays in large quantities.

- Breaking The Market Barrier Costs are high because the market is presently small. Photovoltaic manufacturers are therefore reluctant to invest in an uncertain market. The market is small because the costs are high. This cycle must be broken for major progress to occur toward low cost mass produced photovoltaics.

- The Relatively Small Number of Visible Photovoltaic Applications constitutes a barrier. Many utility technical people don't even know what a photovoltaic cell is; they've never seen one. The average American home owner is totally unfamiliar with photovoltaics as a potential unlimited source of his electric power for his residence. As array prices continue to drop, increased visibility of photovoltaics will be essential to the industry's growth.

- Relatively Small Government Agency Involvement in photovoltaic system applications exclusive of DOE. The DOD[1], Air Force, Navy, Embassies, Commerce Dept., Interior Dept. and many other government agencies could help catalyze the civil market for photovoltaics with vigorous, informed, well-conceived photovoltaic application programs. The price of photovoltaics would thereby be lowered and photovoltaic systems would become more generally available at growing fossil fuel savings to the nation.

- No Mass Production Photovoltaic Array Factory exists in 1978 in the United States, though plans are underway to that end for realizing such a factory in the early 1980's.

- An Information Barrier exists. People--even well informed technical people--simply don't know about photovoltaic systems. Only one Solar-Electric Engineering course is known to exist in the United

---

[1] Medler [305] has preliminarily explored the DOD application possibilities along with the barriers.

States,[1] and no textbook exists for it. Further there are almost no well-written integrated pamphlets on photovoltaic systems suitable for use by homeowners of the future contemplating installing photovoltaic systems. Also technical information on photovoltaics suitable for use by utility managers and engineers is very sparse, e.g. no design handbooks exist showing engineers how to successfully design a large-scale photovoltaic system. The same can be said for small-scale decentralized systems.

University professors who are knowledgeable of photovoltaics and skilful writers could help overcome this barrier if properly funded.

It is clear that for the coming technical and business explosion in photovoltaics to succeed information in the right form and level will be a key ingredient.

- Cost Analysis Procedures for complete realistic photovoltaic systems appear to need much work aimed toward arriving at generalized realistic procedures covering the greatest number of cases. This facet of photovoltaics presently appears highly customized. Such procedures will eventually be needed by utility planners; one does not have to wait until the $500/KW cost goal is reached to then codify this general body of economic knowledge.

- Legal And Institutional Barriers undoubtedly will exist for photovoltaics in the coming few years. Legalists are working on the sticky and important 'sun-rights' issue. The building code laws will undoubtedly require modification, to allow photovoltaic systems, perhaps combined with thermal systems.[2] The issue of whether an owner of a photovoltaic system should be allowed income tax credits to spur this technology hasn't yet arisen; but it will. Utility opposition to residential photovoltaic systems possibly may be an institutional barrier of the future.

- Public Reaction to realistic photovoltaic systems, as on residences or large central stations, is a relatively unexplored area. No known serious studies have been made in this area. There is widespread belief within the photovoltaic community that the public will welcome photovoltaic systems; but the belief is largely unsubstantiated, and no well conceived statistically controlled survey has been run.

- General Lack of Utility Involvement in photovoltaic systems R&D is especially conspicuous. Currently this entire new industry is

---

[1] This researcher's senior level Solar-Electric Engineering course at the University of Florida.

[2] To this researcher's incomplete knowledge, there is no known U. S. building code in 1978 written around photovoltaic systems.

growing up outside the utility industry, with a few exceptions.

It is felt the utilities have much to offer photovoltaics, expecially in the systems area; conversely photovoltaics potentially have much to offer the utilities in potential fuel savings. To make such utility involvement a reality however will first require more positive attitudes toward advanced systems R&D than has heretofore been the case among some utility managements and their R&D organization; then utility funding of more and various important systems orientated photovoltaic studies and projects might become a reality. Such utility involvement could be mutually beneficial to both the photovoltaic community and the utility industry.

The above are the principal barriers seen in 1978. Others undoubtedly may see different barriers or express those elucidated above differently.

2-8 Future R&D Needed For Integration

After surveying an appreciable fractional part of the published literature on photovoltaic systems it is this researcher's opinion that the items listed below--not necessarily in order of priority--are in need of additional R&D. The viewpoint is constricted to that of the systems engineer interested in integrating photovoltaics into residences and utility grids.[1] These areas are listed without extensive arguments and justifications.

- The Cost of Photovoltaic Systems Nobody appears to have made a responsible statistical survey asking potential photovoltaic users, especially in the residential area, how many dollars/KW they'd be willing to pay for a photovoltaic system and then comparing this with responsible photovoltaic industry quotations for what it can produce systems. Is there a great disparity? None?

- Wear-Out Mechanisms within both cells and arrays need more specific identification, and the scientific base needs to be understood much better than at present. Such R&D necessarily would encompass:

  - Obtaining a baseline for the measured life of representative present cells and arrays under suitable defined terrestrial conditions.

  - Correlating accelerated and actual life testing.

  - Studying modes of failure and their sensitivity to environmental factors and array performance, e.g. how is number of thermal cycles an array is subjected to related to the array life? Will the life be modified if the depth of the thermal cycles is not so severe? How much? When failure occurs, what specifically failed? Why did it fail? What are the effects of moisture and dust on the array life?

[1] Naturally other research, not identified here, must be executed in the general area of photovoltaics devices.

The importance of this, or some similar, R&D cannot be underestimated for both residential and utility photovoltaic systems. The life expectancy needed for both are thought to be 20 years or more.

- Testing And Qualification Program for arrays bought for utility use. More thought needs to be devoted to such a program, for utility companies will surely eventually face the problem of "Here is the array we ordered from company x. How do we tell whether it is acceptable before we place it into service?" Should use be made of the national test facilities at LeRC Sandia and other places? Should all utilities use these? What kind of tests should a homeowner do, if any, on a newly purchased photovoltaic system?

- Photovoltaic Siting Considerations need a more in-depth examination than has been preliminarily assembled in this report. What considerations not mentioned in this report are encountered in an actual photovoltaic siting? Are the ones tabulated earlier really valid? Doesn't a body of knowledge on siting need to be assembled which would be applicable to the widest spectrum of potential photovoltaic applications?

- Photovoltaics and Lightning effects need in-depth study by some competent lightning researchers. The simple act of taking a small array and deliberately exposing it to a known high lightning area, as the St. Petersburg-Tampa, Fla. area,could have wide implications for the forthcoming photovoltaic power systems industry.

  Is lightning protection generally needed? What specific form is best?

- Standardization efforts, building on the work already started at NASA LeRC should continue and expand to eventually encompass all solar cell and array manufacturers. Standardization of some kind will be absolutely essential for the photovoltaic industry to grow to the wide acceptance state. It appears preferable for standards to be set by the industry rather than government imposed.

- A State-of-The-Art Study, beyond what has been assembled in this report, appears to be needed in concentrated high efficiency photovoltaics, as this area increasingly appears to be growing in importance. The R&D work done to date needs pulling together in more depth than could be done in this study.

- Array Design information needs to be accumulated and structured in useful ways so younger less experienced engineers and those interested in applications can build on the knowledge of this generation. Not a single book was uncovered showing how to design a photovoltaic array, including design alternatives, pros and cons, how to design support structures, whether to track and how to do so, design to permit easy fault location, isolation and repair of cells when they fail, and many other design-related topics.

- <u>Inverters</u> need further R&D to increase efficiency and decrease the cost. R&D also needs to be initiated studying the effects of the harmonics in inverter output on the utility system operation, especially when and if large numbers of such inverters feed appreciable power into the grid. This is a technically complex problem. Surprisingly, no one in the photovoltaic community seems to have studied it in-depth, though it is known that a similar problem occurs in industrial motor control. Perhaps the results from that application bear on the photovoltaic driven inverter.

  The Radio Frequency Interference (RFI) effects of inverters also needs examination, including recommended kinds of filtering and shielding to prevent RFI. No one appears to have addressed this problem within the photovoltaic community.

  The kind and type of lightning protection needed on inverters also needs an in-depth examination before we have thousands of inverters installed many of which could be damaged irreversibly by the first lightning on the power grid.

  Also more R&D would appear needed on megawatt level inverters designed specifically for photovoltaics if centralized photovoltaic stations are to come into existence by the utilities.

- <u>System Optimization</u> studies including factoring in load factor analysis need to be increased in number and depth as photovoltaic power systems evolve. While some beginning has been made, as cited earlier in this report, much more appears needed before a utility system management can be absolutely certain a given photovoltaic system has been optimally designed both from the customer and the utility viewpoints.

The successful execution of the above R&D over the next several years would constitute strong progress toward smoothing the integration of direct solar-electrics into the utility industry and residences.

REFERENCES

1. Wolf, Martin, "Photovoltaic Power," Chapter Seven in Solar Energy For Earth: An AIAA Assessment, 21 April 1975. Available from American Institute of Aeronautics and Astronautics, 1290 Ave. of the Americas, New York, N. Y. 10019.

2. Balcomb, Douglas, "The Department of Energy Solar Program: A Question of Balance," Solar Age, Vol. 3, No. 5, May 1978, p. 12ff.

3. National Photovoltaic Program Plan, March 1978. Available from Government Printing Office as DOE/ET-0035(78).

4. Photovoltaic Conversion Program: Summary Report, 1977. Available from Government Printing Office as DOE/ET-0019/1.

5. Photovoltaic Conversion Program: Summary Report, November 1976. Available from Government Printing Office as ERDA 76-161.

6. Herwig, Lloyd O., "ERDA Solar Energy Program," in Description of The Solar Energy R & D Programs In Many Nations, Final Report, February 1976. Available from NTIS as SAN/1122-76/1.

7. An Assessment of Solar Energy As A National Energy Resource. Prepared by the NSF/NASA Solar Energy Panel, December 1972. Available from NTIS as No. PB 221-659.

8. Cherry, W. R. and J. A. Zoutendyk, "State of the Art In Solar Cell Arrays for Space Electrical Power," Space Power Systems, [New York, N. Y.: Academic Press, Inc., 1966].

9. Energy, A weekly subscription abstract service of NTIS. Photovoltaic abstracts are mixed with other solar abstracts. Prices are cited for ordering ease.

10. Solar Energy Update. A monthly subscription service of DOE. Available through Technical Information Center, Oak Ridge, Tennessee.

11. Solar Cells and Solar Panels, April 1977. Available from NTIS as AD-A039100.

12. Solar Energy: A Bibliography - Citations, March 1976, U. S. Energy Research & Development Administration. Available through DOE, Technical Information Center, Oak Ridge, Tennessee, as TID-3351-R1P1.

13. Solar Energy: A Bibliography - Indexes, March 1976, U. S. Energy Research & Development Administration. Available through DOE, Technical Information Center, Oak Ridge, Tennessee, as TID-3351-R1P2.

14. Backus, Charles E., (Ed), Solar Cells, [New York, N. Y.: IEEE Press, 1976], 504 pages.

15. Rosenblatt, Alfred I., "Energy Crisis Spurs Development of Photovoltaic Power Sources," Electronics, Vol. 47, No. 7, April 4 1974, pp. 99-111.

16. Chang, Sheldon S. L., Energy Conversion, [Englewood Cliffs, N. J.: Prentice-Hall, Inc., 1963].

17. Daniels, Farrington, Direct Use of The Sun's Energy, [New York, N. Y.: Ballantine Books, Inc., 1974 (First printed 1964)].

18. Sutton, George W., Direct Energy Conversion, [New York, N. Y.: McGraw-Hill Book Co., 1966].

19. Walsh, Edward M., Energy Conversion, [New York, N. Y.: The Ronald Press Co., 1967].

20. Altman, Manfred, Elements of Solid-State Energy Conversion, [New York, N. Y.: Van Nostrand Reinhold Co., 1969].

21. Kettani, M. Ali, Direct Energy Conversion, [Reading, Mass.: Addison-Wesley Pub. Co., 1970].

22. Merrigan, Joseph A., Sunlight To Electricity [Cambridge, Mass.: The MIT Press, 1975].

23. Angrist, Stanley W., Direct Energy Conversion, Third Edition, [Boston, Mass.: Allyn and Bacon, Inc., 1976].

24. Krenz, Jerrold H., Energy: Conversion and Utilization, [Boston, Mass.: Allyn and Bacon, Inc., 1976].

25. Palz, Wolfgang, Solar Electricity, [Woburn, Mass.: Butterworth (Publishers), Inc., 1978].

26. Rappaport, Paul, "The Photovoltaic Effect And Its Utilization," RCA Review, Vol. 20, September 1959, pp. 373-397. Reprinted in Backus, Charles E., Solar Cells, [New York, N. Y.: IEEE Press, 1976].

27. Hayes, Denis, Energy: The Global Prospect, Worldwatch Paper 11, March 1977. Available from Worldwatch Institute, 1776 Massachusetts Ave., N. W., Washington, D. C. 20036.

28. Masden, G. W., "Increased Photovoltaic Conversion Efficiency Through Use of Spectrum Splitting and Multiple Cells," Special Short Abstracts, 13th IEEE Photovoltaics Specialist Conference, June 5-8 1978.

29. Moon, R. L., "Performance of Individual Solar Cells With Different Bandgaps, Applicable to Multigap Solar Cells," Special Short Abstracts, 13th IEEE Photovoltaics Specialist Conference, June 5-8 1978.

30. Bennett, A., "Analysis of Multiple-Cell Concentrator/Photovoltaic Systems," Special Short Abstracts, 13th IEEE Photovoltaics Specialist Conference, June 5-8 1978.

31. Lamorte, M. F., "Studies of Two-Junction Monolithic Cascade Solar Cells," Special Short Abstracts, 13th IEEE Photovoltaics Specialist Conference, June 5-8 1978.

32. Cape, J. A., "Spectrally Split Tandem Cell Converter Experiments," Special Short Abstracts, 13th IEEE Photovoltaics Specialist Conference, June 5-8 1978.

33. Fraas, L. M., "AlGaAs/GaAs/Ge Monolithic Dual Junction Solar Cell," Special Short Abstracts, 13th IEEE Photovoltaics Specialist Conference, June 5-8 1978.

34. Milnes, A. G., "Concepts For Rheotaxially-Grown Thin-Film Tandem III-V Solar Cells," Special Short Abstracts, 13th IEEE Photovoltaics Specialist Conference, June 5-8 1978.

35. Arienzo, M., "Investigation of Potentially High Efficiency Photovoltaic Cells Consisting of Two Heterojunctions On A Common Wide Band Gap Semiconductor Base," Special Short Abstracts, 13th IEEE Photovoltaics Specialist Conference, June 5-8 1978.

36. Electricity From The Sun, a DOE pamphlet, n. d. Available from DOE Technical Information Center, P. O. Box 62, Oak Ridge, Tennessee.

37. Solar Program Assessment: Environmental Factors--Photovoltaics, March 1977. Available from NTIS as ERDA 77-47/3 UC-11, 59, 62, 63A.

38. Allison, J. F., et al, "A Comparison of the Comsat Violet and Non-Reflective Cells," Record of the Tenth Intersociety Energy Conversion Engineering Conference, August 18-22 1975, pp. 1038-1040. Available from IEEE.

39. Solar Cell Array Design Handbook, Vol. I, October 1976. Available from NTIS as N 77-14193.

40. Kirpich, A., et al, "Performance and Cost Analysis of Photovoltaic Power Systems for On-site Residential Applications," Proceedings Eleventh Intersocity Energy Conversion Engineering Conference, Vol. II, September 12-17 1976, pp. 1300-1307. Available from American Institute of Chemical Engineers, 345 East 47th St., New York, N. Y. 10017.

41. Wolf, M., "Limitations and Possibilities for Improvement of Photovoltaic Solar Energy Converters," Proc. IRE, Vol. 48, pp. 1246-1263, July 1960. Reprinted in Backus' book, Solar Cells.

42. Loferski, J., J. Appl. Physics, 27, 777 (1956).

43. Wolf, Martin, and Hans Rauschenbach, "Series Resistance Effects on Solar Cell Measurements," Advanced Energy Conversion, Vol. 3, pp. 455-479, April-June 1963. Reprinted in Backus' book, Solar Cells.

44. Prince, M. B. and Wolf, M., J. Brit. IRE 18, 583 (1958).

45. Wolf, M., et al., Review and Evaluation of Past Solar-Cell Development Efforts, June 1968. Prepared under contract NASW-1427 for NASA.

46. Spratt, J. P., and R. F. Schwarz, "Concentrated Photovoltaic Power Generation," IECEC '75 Record, pp. 404-412. Available from IEEE.

47. Wysocki, Joseph J. and Paul Rappaport, "Effect of Temperature on Photovoltaic Solar Energy Conversion," J. Appl. Phys., Vol. 31, Mar. 1960, pp. 571-578. Reprinted in Backus' book, Solar Cells, p. 110.

48. Prince, M. B., "Silicon Solar Energy Converters," J. Appl. Phys., Vol. 26, May 1955, pp. 534-540. Reprinted in Backus' book, Solar Cells, p. 89.

49. Skolnik, Howard, "A System For The Evaluation of Solar Cell Samples," A presentation at the 1971 International Solar Energy Society Conference, Goddard Space Flight Center, May 10, 1971.

50. Slesser, M. and I. Hounam, "Solar Energy Breeders," Nature, July 22 1976. Cited in Hayes, Denis, Energy: The Global Prospect, Worldwatch Paper 11, March 1977.

51. Wihl, Manfred, and Alan Scheinine, "Simulation and Analysis of the Energy Source of the Future: THE SOLAR BREEDER," Special Short Abstracts, 13th IEEE Photovoltaics Specialist Conference, June 5-8, 1978.

52. Wolf, M., "The Present State-Of-The-Art of Photovoltaic Solar Energy Conversion," Solar Energy, 5, 3, July/September 1961, p. 83.

53. Rau, Hans, Solar Energy, tr. by Maxim Schur, [New York, N. Y.: The Macmillan Co., 1964].

54. Cherry, W. R., "Status of Photovoltaic Solar Energy Converters," IEEE Trans. on Aerospace and Electronic Systems, Vol. AES-1, No. 1, August 1965, p. 10-19.

55. Workshop Proceedings, Photovoltaic Conversion of Solar Energy for Terrestrial Applications, October 23-25 1973, Cherry Hill, N. J. Organized by JPL. Sponsored by NSF RANN program. In two volumes: Vol. 1 Working Group and Panel Reports, Vol. 2 Invited Papers.

56. The Photovoltaic Power And Its Applications In Space and On Earth, 2-6 July 1973. Available from Centre National D'Etudes Spatiales, B P No. 4, 91220-Bretigny-Sur-Orge (France).

57. Assessment of the Technology Required to Develop Photovoltaic Power Systems for Large-Scale National Energy Applications, October 15 1974. Available from NTIS as NSF-RA-N-74-072.

58. Li, Sheng S., "Thin Film Solar Photovoltaic Cells for Terrestrial Applications," presented at the 18th Annual Conference of Society of Vacuum Coaters, Miami, April 7-9 1975.

59. Essigman (Principal Investigator), Research In Energy Conversion, Final Report, AF 19 628-3836, October 1, 1963 - September 30, 1966. (From Northeastern University). He cites 7 references dealing with the Carnot efficiency limitation.

60. Mytton, R. J., Review Article, "The Present Potential of Cds Solar Cells as A Future Contender for Photovoltaic Space and Terrestrial Power Applications," Solar Energy, Vol. 16, No. 1, August 1974, pp. 33-44.

61. Barrett, M. J., Solar Cell Performance Mathematical Model, 15 March 1970. Available from NTIS as N70-25608.

62. Rauschenbach, H. S. and P. S. Gaylard, "Prediction of Fatigue Failures in Solar Arrays," Conference Proceedings, 7th Intersociety Energy Conversion Engineering Conference, 1972, pp. 665-675. Copy available from American Chemical Society, 1155 16th St., N. W., Washington, D. C. 20036.

63. Garner, N. R. and H. S. Rauschenbach, "Statistical Analysis Applied To Solar Cell Fatigue Failure." Conference Proceedings, 7th Intersociety Energy Conversion Engineering Conference, 1972, pp. 661-664. Copy available from American Chemical Society, 1155 16th St., N. W. Washington, D. C. 20036.

64. Eakins, T. C., "Results of Solar Cell Welded Interconnection Development," Conference Proceedings, 7th Intersociety Energy Conversion Engineering Conference, 1972, pp. 651-660. Copy available from American Chemical Society, 1155 16th St., N. W., Washington, D. C. 20036.

65. Curtin, D. J. and W. J. Billerbeck, "Advanced Interconnect Systems for Lightweight Solar Arrays," The Photovoltaic Power and Its Applications in Space and on Earth, 2-6 July 1973. Copy available from Centrė National D'Etudes Spatiales, B P No. 4, 91220-Bretigny-Sur-Onge (France).

66. Fayet, P., et al., "Prospects for the Use of Stable and Transparent Plastics As Coatings on Solar Arrays," The Photovoltaic Power and Its Applications in Space and on Earth, 2-6 July 1973, pp. 283-288. Copy available from Centre National D'Etudes Spatiales, B P No. 4, 91220-Bretigny Sur-Onge (France).

67. Carmichael, D. C., Studies of Encapsulant Materials for Terrestrial Solar-Cell Arrays, 1st Quarterly Report, December 22, 1975. Available from NTIS as ERDA/JPL-954328-75/1.

68. Bartels, F. T. C., Low Cost Silicon Solar Cell Array, September 1974. Available from NTIS as NASA CR-134743.

69. Taylor, William E., and Fred M. Schwartz, Topical Report-Demonstration of the Feasibility of Automated Silicon Solar Cell Fabrication, October 1975. Available from NTIS as NASA CR-134981.

70. Graham, C. D., et al., Research and Development of Low Cost Processes for Integrated Solar Arrays, January 1976. Available from NTIS as COO/2721-76/1.

71. Hilborn, R. B., Jr., and J. W. Faust, Jr., Web-Dendritic Ribbon Growth, December 12 1975. Available from NTIS as ERDA-JPL 954344 75-1.

72. Ray, W. E., Solar Panel Test Set, February 1970. Available from NTIS as AD 707345.

73. Terrestrial Photovoltaic Measurements, Workshop Proceedings, March 19-21 1975. Available from NTIS as N 76-71615/26.

74. Brandhorst, Henry, et al., Interim Solar Cell Testing Procedures for Terrestrial Applications, July 1975. Available as NASA TM X - 71771, 1975. Also available from NTIS as TID 26879.

75. Bifano, W. J., and A. F. Forestieri, "The NASA Program for Standardizing Silicon Solar Cells, 1974 Proceedings, 9th Intersociety Energy Conversion Engineering Conference, pp. 90-92. Copy available from the American Society of Mechanical Engineers, 345 East 47th St., New York, N. Y. 10017.

76. Kennerud, K. L., "A Technique for Identifying the Cause of Performance Degradation in Cadmium Sulfide Solar Cells," Proceedings, Fourth Intersociety Energy Conversion Engineering Conference, 1969, pp. 561-566. Copy available from American Institute of Chemical Engineers, 345 East 47th St., New York, N. Y. 10017.

77. Berman, Paul A., "Design of Solar Cells for Terrestrial Use," Solar Energy, Vol. 11, Nos. 3 and 4, 1967, pp. 180-185.

78. Preliminary Evaluation of Two-Element Optical Concentrators For Use in Solar Photovoltaic Systems, a report by Spectrolab, Inc. to Argonne National Laboratory, June 30, 1975. Available from NTIS as ANL K 75 3191 1.

79. Bell, R. O., et al., Photovoltaic Engineering Services Pertinent To Solar Energy Conversion, a report from Mobil Tyco Solar Energy Corporation to Argonne National Laboratory, June 1976. Available from NTIS as ANL-K75-3171-1.

80. Draft ERDA Photovoltaic Program Plan Sets Goal of 10-30¢/PWE, 50 PGWE in 2000, Solar Energy Intelligence Report, Vol. 3, No. 27, August 15, 1977.

81. Balzhiser, Richard E., Solar Photovoltaic Energy Conversion, EPRI Journal, June/July 1977, p. 49-51.

82. de Winter, F., and J. W. deWinter, Description of the Solar Energy R & D Programs in Many Nations, February 1976. Available from NTIS as SAN/1122-76/1.

83. Solar Energy, DFVLR Activities, 11 February 1976. Available from NTIS as ERDA-TR-143.

84. A Directory of Australian Solar Energy Research and Development, Dept. of National Development, Word Processing Centre, CSIRO, June 1978.

85. Hovel, Harold J., "Solar Cells for Terrestrial Applications," Solar Energy, Vol, 19, No. 6, 1977, pp. 605-615. This paper also appears in Sharing The Sun: Solar Technology In the Seventies, Vol. 6, Photovoltaics and Materials, Aug. 15-20 1976 (ISES Winnipeg meeting).

86. 13th IEEE Photovoltaics Specialist Conference, Special Short Abstracts, June 5-8 1978, Washington, D. C. (To be published by IEEE later as a Proceedings Cat. No. 78 CH 1319-3ED with complete papers).

87. Dunbar, P. M. and J. R. Hauser, A Theoretical Analysis of the Current-Voltage Characteristics of Solar Cells, August 1976. Available from NTIS as N76-32648.

88. Lindholm, Fredrik A., Studies of Silicon PN Junction Solar Cells, July 1976. Available from NTIS as N77-18557.

89. Lindholm, Fredrik A., "Fundamental Electronic Mechanisms Limiting the Performance of Solar Cells," IEEE Trans. On Elec. Dev., Vol. ED-24, No. 4, April 1977, pp. 299-304.

90. Fossum, Jerry G. "Physical Operation of Back-Surface-Field Silicon Solar Cells," IEEE Trans. on Elec. Dev., Vol. ED-24, No. 4, April 1977, pp. 322-325.

91. Brandhorst, H. W. (Ed.), Special Issue on Photovoltaic Devices, IEEE Trans. on Elec. Dev., Vol. ED-24, No. 4, April 1977.

91A. Boes, E. C., The Effects of Spectral Variations On Silicon Cell Output, c. 1976. Available from NTIS as SAND 76 9142.

92. Ertel, Danny, "The Application of Color Response Data of Silicon Cells For Improving Photovoltaic Efficiency," Proceedings of the 1977 Annual Meeting, Am. Sec. ISES, June 6-10 1977, pp. 23-20 through 23-24.

93. Ertel, D., "Photovoltaic Conversion-Can Its Efficiency Be Improved Through the Application of Color Response Data," Proceedings of Condensed Papers, Miami International Conference on Alternative Energy Sources, Dec. 5-7, 1977.

94. Brandhorst, Henry W., Jr., Invention Abstract, Improved Back Wall Cell, filed January 19, 1977. Available from NTIS as N 7717565.

95. Thornhill, J. W., Automated Fabrication of Back Surface Field Silicon Solar Cells With Screen Printed Wraparound Contacts, August 1977. Available from NTIS as NASA CR-135202.

96. Sodha, M. S., and A. K. Agarwal, "Performance of Difused Vertical Multi-junction Solar Cell," Solar Energy, Vol. 18, No. 3, 1976, pp. 265-268.

97. Fossum, Jerry G., Numerical Analysis of Back-Surface-Field Silicon Solar Cells, c. 1975. Available from NTIS as SAND 75 5730.

98. Wohlgemuth, J. and C. Wrigley, "Characterization of Vertical-Junction Silicon Solar Cells," Special Short Abstracts, 13th IEEE Photovoltaics Specialist Conference, June 5-8 1978.

99. Chu, T. L., "Polycrystalline Silicon Solar Cells On Low Cost Foreign Substrates," Solar Energy, Vol. 17, No. 4, September 1975, p. 229-235.

100. Feldman, C., et al., "Vacuum-Deposited Polycrystalline Silicon Solar Cells," a section in quarterly report, Energy Programs at the Johns Hopkins University APL, October-December 1976. Available from NTIS as ADA 038096.

101. Chu, T. L., Low Cost Polycrystalline Silicon Solar Cells, Proceedings of the National Workshop, May 18-19, 1976. Available from NTIS as PB-266 563.

102. Belouet, C., "Growth and Characteristics of Polysilicon Layers Achieved by the Ribbon Against Drop (RAD) Pulling Process," Special Short Abstracts, 13th IEEE Photovoltaics Specialist Conference, June 5-8 1978.

103. Hanoka, J. I. and P. S. Kotval, "Efficient Polycrystalline Solar Cells Made From Low-Cost Refined Metallurgical Silicon," Special Short Abstracts, 13th IEEE Photovoltaics Specialist Conference, June 5-8 1978.

104. Daud, T., "Effect of Copper Impurities On the Performance of Polycrystalline Silicon Solar Cells," Special Short Abstracts, 13th IEEE Photovoltaics Specialist Conference, June 5-8 1978.

105. Dapkus, P. D., et al, "The Properties of Polycrystalline GaAs Materials and Devices for Terrestrial Photovoltaic Energy Conversion," Special Short Abstracts, 13th IEEE Photovoltaics Specialist Conference, June 5-8 1978.

106. Yeh, Y. C. M., et al., "High Efficiency Polycrystalline GaAs AMOS Solar Cell," Special Short Abstracts, 13th IEEE Photovoltaics Specialist Conference, June 5-8 1978.

107. Kuper, Alan B., "Poly vs Single Crystal Silicon Solar Cells," Special Short Abstracts, 13th IEEE Photovoltaics Specialist Conference, June 5-8 1978.

108. Lindmayer, J., "Characteristics of Semicrystalline Silicon Solar Cells," Special Short Abstracts, 13th IEEE Photovoltaics Specialist Conference, June 5-8 1978.

109. Fabre, E., and Y. Baudet, "Solar Cells on R. A. D. Polycrystalline Silicon," Special Short Abstracts, 13th IEEE Photovoltaics Specialist Conference, June 5-8 1978.

110. Chu, T. L., "Thin Polycrystalline Silicon Solar Cells," Special Short Abstracts, 13th IEEE Photovoltaics Specialist Conference, June 5-8 1978.

111. Lindmayer, Joseph, Development of a High Efficiency Thin Silicon Solar Cell, December 1975. Available from NTIS as N 76-15593.

112. Chu, T. L., et al., Development of Low-Cost Thin-Film Polycrystalline Silicon Solar Cells for Terrestrial Applications, July 1976. Available from NTIS as PB-260 627.

113. Iles, P. A., Research, Development and Pilot Production of High Output Thin Silicon Solar Cells, November 1976. Available from NTIS as N77-19573.

114. Kirkpatrick, A. R., Silicon Thin Film Crystallization and Solar Cell Fabrication, June 1976. Available from NTIS as PB-258 949.

115. Chu, T. L., Development of Low Cost Thin Film Polycrystalline Silicon Solar Cells For Terrestrial Applications, January 1977. Available from NTIS as PB 266 057.

116. Shirland, F. A., Thin Film Solar Cells for Terrestrial Applications, February 1977. Available from NTIS as PB 265 983.

117. Chin, Barry Lee, An Investigation of Cu_x S/Si Photovoltaic Cell, December 1975. Available from NTIS as LBL 4553.

118. Böer, K. W., et. al., "CdS/Cu_2S Solar Cells, Their Potential and Limitations," IECEC '75 Record, pp. 387-391. Copy available from IEEE as Cat. No. 75 CHO 983-7 TAB.

119. Windawi, H. M., "Life Expectancy of Cd S/Cu_2S Solar Cells on Terrestrial Surfaces," IECEC '75 Record, pp. 392-395. Copy available from IEEE as Cat. No. 75 CHO 983-7 TAB.

120. Burton, L. C., and G. Haacke, "A New Type of Cu_2 S/CdS Backwall Solar Cell," IECEC '75 Record, pp. 396-399. Copy available from IEEE as Cat. No. 75 CHO 983-7 TAB.

121. Trivich, Dan, Cuprous Oxide Photovoltaic Cells, July 1976. Available from NTIS as PB-259 255.

122. DeMeo, Edgar A., Assessment of Cadmium Sulfide Photovoltaic Arrays for Large Scale Electric Utility Applications, Final report, February 1976. Available from EPRI as EPRI ER-188, Project WS 75-21.

123. Olsen, L. C., Investigation of Low Cost Solar Cells Based on Cu_2O, 1976. Available from NTIS as PB 252 301.

124. Olsen, Larry C., Investigation of Low Cost Solar Cells Based on Cu_2O, August 1976. Available from NTIS as PB 258-746.

125. Loferski, J. J., "Investigation of Thin Film Cadmium Sulfide/Copper Ternary Alloy Heterojunction Photovoltaic Cells," Special Short Abstracts, 13th IEEE Photovoltaics Specialist Conference, June 5-8 1978.

126. Shewchun, J., "The Properties of Vacuum Evaporated Cu_xS Films and Solar Cells on Si and CdS", Special Short Abstracts, 13th IEEE Photovoltaics Specialist Conference, June 5-8 1978.

127. Armantrout, Guy, et al., "Electronic Transport Properties of Sputtered Cu_xS-CdS Photovoltaic Heterojunctions," Special Short Abstracts, 13th IEEE Photovoltaics Specialist Conference, June 5-8 1978.

128. Shea, S. P. and L. D. Partain, "Heat Treatment Effects on the Minority Carrier Diffusion Lengths Surface Recombination Velocity and Junction Collection Factor of CuS/CdS Solar Cells," Special Short Abstracts, 13th IEEE Photovoltaic Specialist Conference, June 5-8 1978.

129. Rothwarf, A., et al., "Junction Field and Recombination Phenomena in the CdS/Cu_2S Solar Cell: Theory and Experiment," Special Short Abstracts, 13th IEEE Photovoltaic Specialist Conference, June 5-8 1978.

130. Baron, B., et al., "Formation and Properties of Cuprous Sulfide for Thin Film CdS/Cu_2S Photovoltaic Devices," Special Short Abstracts, 13th IEEE Photovoltaics Specialist Conference, June 5-8 1978.

131. Bragagnolo, J. A., "Photon Loss Analysis of Thin Film CdS/Cu_2S Photovoltaic Devices," Special Short Abstracts, 13th IEEE Photovoltaics Specialist Conference, June 5-8 1978.

132. Singh, V. P., "Characteristics of Cu_xS/CdS Cell On Sprayed Thin CdS Films," Special Short Abstracts, 13th IEEE Photovoltaics Specialist Conference, June 5-8 1978.

133. Jones, Kenneth A., "The Effects of Misfit Dislocations On The Photovoltaic Properties of CdS Based Solar Cells," Special Short Abstracts, 13th IEEE Photovoltaics Specialist Conference, June 5-8 1978.

134. Edenburn, M. W., et al., "Status of the DOE Photovoltaic Concentrator Development Project," Special Short Abstracts, 13th IEEE Photovoltaics Specialist Conference, June 5-8 1978.

135. Burgess, E. L., "Status of the Photovoltaic Concentrator Applications Experiments", Special Short Abstracts, 13th IEEE Photovoltaics Specialist Conference, June 5-8 1978.

136. Watkins, J. L., "Real-Time Environmental and Performance Testing of Concentrating Photovoltaic Arrays," Special Short Abstracts, 13th IEEE Photovoltaics Specialist Conference, June 5-8 1978.

137. Burgess, E. L., et al., "Performance of a one Kilowatt Concentrator Photovoltaic Array Utilizing Active Cooling," Special Short Abstracts, 13th IEEE Photovoltaics Specialist Conference, June 5-8 1978.

138. Lowe, J., "Preliminary Evaluation of A GaAlAs Solar Concentrator Space Power Unit," Special Short Abstracts, 13th IEEE Photovoltaics Specialist Conference, June 5-8 1978.

139. Wilkening, H. A., "Design of A 10 KW Photovoltaic Concentrator", Special Short Abstracts, 13th IEEE Photovoltaics Specialist Conference, June 5-8 1978.

140. James, L. W. and J. K. Williams, "Fresnel Optics for Solar Concentration on Photovoltaic Cells," Special Short Abstracts, 13th IEEE Photovoltaics Specialist Conference, June 5-8 1978.

141. Zimmerman, Donald K., "Evaluation of A Plastic Parabolic Concentrator for Terrestrial Photovoltaic Applications," Special Short Abstracts, 13th IEEE Photovoltaics Specialist Conference, June 5-8 1978.

142. Neuner, G. J., "A Novel Approach for A Low-Cost Photovoltaic Concentrator," Special Short Abstracts, 13th IEEE Photovoltaics Specialist Conference, June 5-8 1978.

143. Rapp, C. F., and N. L. Boling, "Luminescent Solar Concentrators," Special Short Abstracts, 13th IEEE Photovoltaics Specialist Conference, June 5-8 1978.

144. Williams, R. C. and B. D. Wood, "Passive Cooling For Linearly Concentrated Solar Cell Arrays," Special Short Abstracts, 13th IEEE Photovoltaics Specialist Conference, June 5-8 1978.

145. Swart, P. L. and J. D. van Wyk, "Source Tracking and Power Flow Control of Terrestrial Photovoltaic Panels For Concentrated Sunlight," Special Short Abstracts, 13th IEEE Photovoltaics Specialist Conference, June 5-8 1978.

146. Lindmayer, J. and C. Wrigley, "Photovoltage In Concentrator Cells," Special Short Abstracts, 13th IEEE Photovoltaics Specialist Conference, June 5-8 1978.

147. Hu, Chenming, and Clifford Drowley, "Open Circuit Voltage of High Intensity Silicon Solar Cells," Special Short Abstracts, 13th IEEE Photovoltaics Specialist Conference, June 5-8 1978.

148. Chappell, Terry I., "A Silicon Solar Cell With The Potential for 24% Conversion Efficiency in 50 x sunlight," Special Short Abstracts, 13th IEEE Photovoltaics Specialist Conference, June 5-8 1978.

149. Masden, Glenn W., "Optimization of the Geometrical Parameters For Arrays of Tracking Solar Collectors To Give Minimum Energy Cost," Special Short Abstracts, 13th IEEE Photovoltaics Specialist Conference, June 5-8 1978.

150. Robb, S. P., et al., "Performance of Silicon and Gallium Arsenide Concentration Cells," Special Short Abstracts, 13th IEEE Photovoltaics Specialist Conference, June 5-8 1978.

151. Castle, John, "Test and Evaluation of Silicon Cells Optimized For High Efficiency Under Concentrated Sunlight," Special Short Abstracts, 13th IEEE Photovoltaics Specialist Conference, June 5-8 1978.

152. Van der Plas, H. A., et al., "Performance of AlGaAs/GaAs Terrestrial Concentrator Solar Cells," Special Short Abstracts, 13th IEEE Photovoltaics Specialist Conference, June 5-8 1978.

153. Ewan, J., et al., "GaAs Solar Cells For High Solar Concentration Applications," Special Short Abstracts, 13th IEEE Photovoltaics Specialist Conference, June 5-8 1978.

154. Sahai, R., et al., "High Efficiency AlGaAs/GaAs Concentrator Solar Cell Development," Special Short Abstracts, 13th IEEE Photovoltaics Specialist Conference, June 5-8 1978.

155. Dalal, Vikram, "Design, Technology and Cost Considerations for High Intensity Silicon Solar Cells," Special Short Abstracts, 13th IEEE Photovoltaics Specialist Conference, June 5-8 1978.

156. Kemthong, S., et al., "Fabrication Experience with High Efficiency Silicon Concentrator Cells," Special Short Abstracts, 13th IEEE Photovoltaics Specialist Conference, June 5-8 1978.

157. Selikson, Bernard, and Reid Mickelsen, "Reliability Analysis of Contacts to Silicon Solar Cells for Use in Concentrated Sunlight," Special Short Abstracts, 13th IEEE Photovoltaics Specialist Conference, June 5-8 1978.

158. Donovan, R. L., et al., "Ten Kilowatt Photovoltaic Concentrating Array," Special Short Abstracts, 13th IEEE Photovoltaics Specialist Conference, June 5-8 1978.

159. Castle, John A., "Performance Evaluation of 10 KW Photovoltaic Concentrator System," Special Short Abstracts, 13th IEEE Photovoltaics Specialist Conference, June 5-8 1978.

160. Luque, A., et al., "Project of the 'Ramon A reces' Concentrated Power Station," Special Short Abstracts, 13th IEEE Photovoltaics Specialist Conference, June 5-8 1978.

161. Evans, D. L., and Florschuetz, "Terrestrial Concentrating Photovoltaic Power System Studies," Solar Energy, Vol. 20, No. 1, 1978, pp. 37-43. This same paper was earlier published in Sharing The Sun: Solar Technologies In The Seventies, Vol. 6. Photovoltaics and Materials, p. 93-105.

162. Terrestrial Photovoltaic Power Systems with Sunlight Concentration, Progress report for January 1, 1975-July 31, 1975. Available from NTIS as COO 2590 1.

163. Mollé, James E., Silicon Solar Cells With Fresnel Lens Concentrators Are Cost Competitive With Other Forms of Power Generation, 1977. A publication of Oregon Solar Institute.

164. Gorski, A. J., et al., "Photovltaic Power Generation By Use of Compound Parabolic Concentrators," Proceedings Of Condensed Papers, Miami International Conference On Alternative Energy Sources, 5-7 December 1977, pp. 787-789.

165. Proceedings of the ERDA Semiannual Solar Photovoltaic Program Review Meeting, Silicon Technology Programs Branch, San Diego, California, January 18-20 1977. Available from NTIS as CONF-770112.

166. Heinbockel, John H. and A. S. Roberts, Jr., Analysis of GaAs and Si Solar Energy Hybrid Systems, March 1977. Available from NTIS as N77-20564.

167. Backus, C. E., et al., "Considerations For Using Solar Concentrators In Photovoltaic Systems," Proceedings of the 12th IECEC, 1977, pp. 1147-1153.

168. Yekutieli, G., et al., "Solar Cell Array For Concentrated Sunlight," Proceedings of the 12th IECEC, 1977, p. 1154-1158.

169. O'Donnell, D. T., et al., "Characteristics of Solar Cells Designed for Concentrator Systems," Proceedings of the 1977 Annual Meeting, American Section of ISES, pp. 23-1 through 23-5.

170. IBM Diffusion Process Increases Solar-Cell Efficiency, Electronic Engineering Times, Monday, May 16 1977, p. 4.

171. McClintock, Mike, "MIT Breaks The Sun Barrier," Popular Mechanics, September 1976, p. 6.

172. Burgess, E. L., "Photovoltaic Energy Conversion Using Concentrated Sunlight," Circa 1976. Available from NTIS as CONF-760832-5.

173. Fossum, Jerry G., "Design Optimization of Silicon Solar Cells for Concentrated-Sunlight, High Temperature Applications," circa 1976. Available from NTIS as CONF-761207-1.

174. Davis, R., and J. R. Knight, "Operation of GaAs Solar Cells at High Solar Flux Density" Solar Energy, Vol. 17, No. 2, May 1975, p. 145.

175. Horne, W. E., "Solar Thermal Photovoltaic Electric Power Generator," preprint. For an abstract see Proceedings of Condensed Papers, Miami International Conference on Alternative Energy Sources, 5-7 December 1977, pp. 793-795.

176. Swanson, R. M., and R. N. Bracewell, Silicon Photovoltaic Cells In Thermophotovoltaic Conversion, February 1977. Available from EPRI as EPRI ER-478.

177. Interconnect and Bonding Technologies For Large Flexible Solar Arrays, July 26, 1976. Available from NTIS as LMSC-D492654.

178. Anagnostou, E., et al., Encapsulated Solar Cell Modules, patent disclosure letter to NASA, November 30, 1976. Available from NTIS as N77 15490.

179. Sierawski, D. A. and C. G. Currin, "The Use of Silicone Gel for Potting Photovoltaic Arrays," Proceedings of the 1977 Annual Meeting, American Section ISES, June 10, 1977. Available from ISES headquarters, Kileen, Texas.

180. Investigation of Test Methods, Material Properties, And Processes For Solar Cell Encapsulants, November 1976. Available from NTIS as ERDA/JPL/954527-76-2.

181. Investigation of Test Methods, Material Properties and Processes For Solar Cell Encapsulants, August 27, 1976. Available from NTIS as ERDA/JPL/954527-76/1.

182. Kolyer, J. M. and N. R. Mann, Second Quarterly Progress Report on Accelerated/Abbreviated Test Methods, Study 4 of Task 3 (Encapsulation) of the Low-Cost Silicon Solar Array Project, October 1976. Available from NTIS as ERDA/JPL/954458-76/2.

183. Carmichael, D. C., Studies of Encapsulation Materials For Terrestrial Photovoltaic Arrays, September 28, 1976. Available from NTIS as ERDA/JPL/954328-76/3.

184. Carmichael, D. C., et al., Review of World Experience and Properties of Materials For Encapsulation of Terrestrial Photovoltaic Arrays, July 21, 1976. Available from NTIS as ERDA/JPL/954328-76/4.

185. Carmichael, D. C., Studies of Encapsulation Materials For Terrestrial Photovoltaic Arrays, June 28, 1976. Available from NTIS as ERDA/JPL/954328-76/2.

186. Carmichael, D. C., Studies of Encapsulant Materials For Terrestrial Photovoltaic Arrays, March 26, 1976. Available from NTIS as ERDA/JPL/954328-76/1.

187. Broder, Jacob D., Comparison of Types A and C Fluorinated Ethylene Propylene (FEP) as Cover Materials For Silicon Solar Cells, March 1976. Available from NTIS as NASA-TM X-3375.

188. Bereny, Justin A., and Francis deWinter, Survey of The Emerging Solar Energy Industry, 1977 Edition. Available from Solar Energy Information Services, P. O. Box 204, San Mateo, Calif. 94401.

189. Bickler, Donald B., et al., "A Candidate Low-Cost Processing Sequence For Terrestrial Silicon Solar Cell Panel," Special Short Abstracts, 13th IEEE Photovoltaics Specialist Conference, June 5-8 1978.

190. Grenon, L. A. and M. G. Coleman, "Silicon Solar Cells, A Manufacturing Cost Analysis," Special Short Abstracts, 13th IEEE Photovoltaics Specialist Conference, June 5-8 1978.

191. Coleman, M., Phase I Of The Automated Array Assembly Task of the Low Cost Silicon Solar Array Project, July 1976. Available from NTIS as ERDA/JPL/954363-76/2.

192. Coleman, M., Phase I Of The Automated Array Assembly Task of the Low Cost Silicon Solar Array Project, April 1976. Available from NTIS as ERDA/JPL/954363-76/1.

193. Carbajal, Bernard G., "A 1982 Low Cost Photovoltaic Module Factory Study," Special Short Abstracts, 13th IEEE Photovoltaics Specialist Conference, June 5-8 1978.

194. Carbajal, Bernard G., Automated Array Assembly Task, Phase 1, July 1976. Available from NTIS as ERDA/JPL/954405-76/2.

195. Kran, Alexander, "Silicon Ribbon Technology Assessment 1978-1986. A Computer-Assisted Analysis Using Pecan," Special Short Abstracts, 13th IEEE Photovoltaics Specialist Conference, June 5-8 1978.

196. Breneman, W. C., et al., "Preliminary Process Design & Economics of Low-Cost Solar-Grade Silicon Production," Special Short Abstracts, 13th IEEE Photovoltaics Specialist Conference, June 5-8 1978.

197. Buhs, Rolf H. A., "Silicon Solar Cell Arrays: Technical Status, Terrestrial Applications, and Future Developments for Low Cost Productions," Special Short Abstracts, 13th IEEE Photovoltaics Specialist Conference, June 5-8 1978.

198. Thornhill, J. and W. E. Taylor, Demonstration of the Feasibility of Automated Silicon Solar Cell Fabrication, August 1976. Available from NTIS as N77-15492.

199. Williams, B. F., Automated Array Assembly, September 1976. Available from NTIS as ERDA/JPL/954352-76/3.

200. European Space Agency: Specification for Silicon Solar Cells, Issue No. 1, c. 1976. Available from NTIS as 77-7586 2/E.

201. Cull, Ronald C., and Americo F. Forestieri, "The DOE/LeRC Photovoltaic Systems Test Facility," Special Short Abstracts, 13th IEEE Photovoltaics Specialist Conference, June 5-8 1978.

202. Forestieri, A. F., et al., Photovoltaic Test and Demonstration Project, March 1976. Available from NTIS as N76-18673.

203. Kolecki, Joseph C. and Suzanne T. Gooder, Laboratory 15 KV High Voltage Solar Array Facility, January 1976. Available from NTIS as NASA TM X-71860.

204. Shepard, N. F., Jr. and L. E. Sanchez, "Development of A Shingle-Type Solar Cell Module," Special Short Abstracts, 13th IEEE Photovoltaics Specialist Conference, June 5-8 1978.

205. Forestieri, Americo F., et al., Solar Cell Shingle, Invention Abstract, August 24, 1976. Available from NTIS as N77-10645.

206. Romaine, W. R., et al., "The Mead 25 Kilowatt Photovoltaic System," Special Short Abstracts, 13th IEEE Photovoltaics Specialist Conference, June 5-8 1978.

207. Nation's First Solar Photovoltaic Cell Powered Irrigation System Under Operation In Nebraska, ERDA News, Vol. 2, No. 16, August 8 1977, p. 6.

208. Imamura, M. S., et al., Definition Study for Photovoltaic Residential Prototype System, September 1976. Available from NTIS as N77-13533.

209. Bartels, F. T. C., "Design, Analysis and Specification of Photovoltaic Solar Terrestrial Power Systems," Conference Record, Tenth IEEE Photovoltaic Specialists Conference, Nov. 13-15, 1973, pp. 258-263. Reprinted in Backus' book, Solar Cells, p. 387.

210. Partain, Larry, "Effects of Shading On Solar Arrays and Optimal Array Design With Cu_2 S/CdS Solar Cells," 1974 Proceedings, 9th Intersociety Energy Conversion Engineering Conference, pp. 362-369.

211. Böer, K. W., "Direct Solar Energy Conversion For Terrestrial Use," Conference Record, Ninth Photovoltaic Specialists Conf., May 2-4 1972, pp. 351-358. Reprinted in Backus' book, Solar Cells, p. 379.

212. Pope, M. D. and R. W. Matlin, "Photovoltaic Power System Field Tests," Special Short Abstracts, 13th IEEE Photovoltaics Specialist Conference, June 5-8 1978.

213. Thomas, R. E., and G. B. Gaines, "Procedure For Developing Experimental Designs of Accelerated Tests For Service-Life Prediction," Special Short Abstracts, 13th IEEE Photovoltaic Specialist Conference, June 5-8 1978.

214. Ravin, Jerry W., Study Terrestrial Applications of Solar Cell Powered Systems, September 1973. Available from NTIS as CR-134512.

215. Vincze, Stephen A., "Should Solar Electricity be Centralized?," Solar Energy, Vol. 15, No. 4, April 1977, p. 317.

216. Executive Report of Workshop Conference on Photovoltaic Conversion of Solar Energy for Terrestrial Applications, held at Cherry Hill, N. J., October 23-25 1973. Available from NTIS as NSF-RA-N-74-073.

217. Proceedings of the ERDA Semiannual Solar Photovoltaic Program Review Meeting, Silicon Technology Programs Branch, January 18-20, 1977. Available from NTIS as CONF-770112.

218. Schueler, D. G., Status of the DOE Photovoltaic Systems Engineering and Analysis Project, Special Short Abstracts, 13th IEEE Photovoltaics Specialist Conference, June 5-8 1978.

219. Schueler, Donald G., The ERDA Photovoltaic Systems Definition Project, Proceedings of the 11th IECEC, September 1976, p. 1297-1299. Copy available from AIChE.

220. Electric Utility Solar Energy Activities; 1976 Survey, January 1977. Available from EPRI as EPRI ER-321-SR.

221. Proceedings of Semiannual EPRI Solar Program Review Meeting and Workshop, February 1977. Available from EPRI as EPRI ER-371-SR.

222. Proceedings of First Semiannual EPRI Solar Program Review Meeting and Workshop Held In San Diego, California, on March 8-12, 1975, Vol. II: Solar Electric Power. Available from NTIS as PB-260595.

223. Pittman, P. F., and C. R. Chowaniec, "Solar Photovoltaic Intermediate Power System Operating Strategies and Economic Considerations", preprint. Special Short Abstracts, 13th IEEE Photovoltaics Specialist Conference, June 5-8 1978.

224. Biran, D. and M. S. Erlicki, "Model of Solar-Cell Array For Terrestrial Use," Solar Energy, Vol. 17, No. 6, 1975, pp. 325-329.

225. Bartels, F. T. C., and D. W. Moffett, "Design Analysis and Specifications of Photovoltaic Terrestrial Power Systems," Conf. Rec., Tenth IEEE Photovoltaic Specials Conf., Nov. 13-15, 1973, pp. 258-263. Reprinted in Backus' book, Solar Cells, p. 387-391.

226. Landsman, E. E., "Maximum Power Trackers For Photovoltaic Arrays," preprint. Special Short Abstracts, 13th IEEE Photovoltaics Specialist Conference, June 5-8 1978.

227. Mosher, D. M., et al., "The Advantages of Sun Tracking for Planar Silicon Solar Cells," Solar Energy, Vol. 19, No. 1, 1977, pp. 91-97.

228. Blake, F. A., and K. L. Hanson, "The 'Hotspot' Failure Mode for Solar Arrays," Proceedings Fourth IECEC, 1969, pp. 575-581.

229. Sullivan, Ralph M., Shadow Effects On A Series-Parallel Array of Solar Cells, Circa 1968. NASA Goddard Space Flight Center report No. X-636-65-207.

230. Gooder, S. T., Series-Parallel Method of Direct Solar Array Regulation, September 1976. Available from NTIS as N77-10641.

231. Lahri, R., et al., "An Experimental Study of Series/Parallel Configurations of Solar Cells for Power Output Optimization," Special Short Abstracts, 13th IEEE Photovoltaics Specialist Conference, June 5-8 1978.

232. Solar Cell Array Design Handbook, Volume 2, October 1976. Available from NTIS as N77-14194.

233. LSSA (Low-Cost Silicon Solar Array) Project Quarterly Report, April 1976-June 1976. Available from NTIS as N77-10637.

234. LSSA Low Cost Silicon Solar Array Project, Project Quarterly Report 1, April 1976-June 1976. Available from NTIS as ERDA/JPL 1012-76/6.

235. LSSA (Low-Cost Silicon Solar Array) Project First Annual Report For The Period January 1975-March 1976. Available from NTIS as N76-33627.

236. Carmichael, D. C., et al., "Evaluations of Candidate Low-Cost, Long-Life Encapsulation Designs and Materials For Silicon Photovoltaic Arrays," Special Short Abstracts, 13th IEEE Photovoltaics Specialist Conference, June 5-8 1978.

237. Keeling, M. C., et al., "Fixed Angle And Seasonably Adjustable Structural Support Concepts For Solar Collectors," Special Short Abstracts, 13th IEEE Photovoltaics Specialist Conference, June 5-8 1978.

238. Forman, Stephen E., "Field Testing And Evaluation of PV Module Performance," Preprint. Special Short Abstracts, 13th IEEE Photovoltaics Specialist Conference, June 5-8 1978.

239. Watkins, J. L. and D. A. Pritchard, "Real-Time Environmental And Performance Testing of Concentrating Photovoltaic Arrays," Special Short Abstracts, 13th IEEE Photovoltaics Specialist Conference, June 5-8 1978.

240. Anagnostou, E. and A. F. Forestieri, "Endurance Testing of First Generation (Block I) Commercial Solar Cell Modules," Special Short Abstracts, 13th IEEE Photovoltaics Specialist Conference, June 5-8 1978.

241. Pope, M. D. and R. W. Matlin, "Photovoltaic Power System Field Tests," Special Short Abstracts, 13th IEEE Photovoltaics Specialists Conference, June 5-8 1978.

242. Hoffman, Alan R., and Ronald G. Ross, Jr., "Environmental Qualification Testing of Terrestrial Solar Cell Modules," Special Short Abstracts, 13th IEEE Photovoltaics Specialist Conference, June 5-8 1978.

243. Treble, F. C., "Progress In Europe Towards Terrestrial Photovoltaic Measurement Standardization," Special Short Abstracts, 13th IEEE Photovoltaics Specialist Conference, June 5-8 1978.

244. Bond, John W., "Lightning Field Effects On Large Solar Electric Power Systems," Special Short Abstracts, 13th IEEE Photovoltaics Specialist Conference, June 5-8 1978.

245. Kroeger, Frederick R., "Status of Power Conditioning Technology," Proceedings of the ERDA Semiannual Solar Photovoltaic Program Review Meeting, Silicon Technology Programs Branch, San Diego, California, January 18-20, 1977. Available from NTIS as CONF-770112.

246. Corry, T., Frequency Converter: Portable, Alternating Current, Multi-frequency, 10 KW, Final Technical Report, Vol. I (May 1974), Vol. II (May 1974), Vol. III (January 1975). Available from NTIS as ADA 035043, ADA 035044, ADA 035045.

247. Corry, T., Generator Set, 100 KW Frequency Converter, Final Technical Report, July 1975. Available from NTIS as ADA 035046.

248. Bedford, B. D., and R. G. Hoft, Principles Of Inverter Circuits, [New York, N. Y.: John Wiley and Sons, Inc., 1964].

249. Gyugyi, L. and B. R. Pelly, Static Power Frequency Changers, [New York, N. Y.: John Wiley and Sons, Inc., 1976].

250. Pelly, B. R., Thyristor Phase-Controlled Converters And Cycloconverters, [New York, N. Y.: John Wiley and Sons, Inc., 1971].

251. Pickrell, Roy L., "System Test Facility Power Conditioning Hardware Testing," Proceedings of the ERDA Semiannual Solar Photovoltaic Program Review Meeting, Silicon Technology Programs Branch, San Diego, California, January 18-20, 1977. Available from NTIS as CONF-770112.

252. Pickrell, Roy L., et al., "An Inverter/Controller Subsystem Optimized For Photovoltaic Applications," preprint. Special Short Abstracts, 13th IEEE Photovoltaics Specialist Conference, June 5-8 1978.

253. Roesler, Dietrich J., "A 60 KW Solar Cell Power System With Peak Power Tracking and Utility Interface," Special Short Abstracts, 13th IEEE Photovoltaics Specialtist Conference, June 5-8 1978.

254. Urish, J. M., and J. L. Watkins, "Design and Construction of a 50 KVA Photovoltaic Power Conditioning Unit," Special Short Abstracts, 13th IEEE Photovoltaics Specialist Conference, June 5-8 1978.

255. Wood, P., and B. R. Pelly, AC/DC Power Conditioning And Control Equipment For Advanced Conversion And Storage Technology, August 1975. Available from NTIS as PB-247 217. Their much shorter report of the same title is available from EPRI as EPRI EM-271, November 1977.

256. Chowaniec, C. R., et al., "Energy Storage Operation of Combined Photovoltaic/Battery Plants In Utility Networks," Special Short Abstracts, 13th IEEE Photovoltaics Specialist Conference, June 5-8 1978.

257. Eldridge, Frank R., Mitre Photovoltaic Energy System Study, October 1973. Available from The Mitre Corp., McLean, Virginia, as report M73-238.

258. Haas, Gregory M., and Steven Bloom, "Experience To Date With The MITRE Photovoltaic Energy System," paper presented at the ISES 1975 Conference, July 29, 1975. Publication source uncertain.

259. Haas, Gregory M., and Steven Bloom, MITRE Terrestrial Photovoltaic Energy System, preprint, circa 1975. Publication source uncertain.

260. Linn, James K., "Optimization of Terrestrial Photovoltaic Power Systems," preprint. Special Short Abstracts, 13th IEEE Photovoltaics Specialist Conference, June 5-8 1978.

261. Masden, Glenn W., "Optimization of the Geometrical Parameters for Arrays of Tracking Solar Collectors To Give Minimum Energy Cost," Special Short Abstracts, 13th IEEE Photovoltaics Specialist Conference, June 5-8 1978.

262. Ross, R. G., "Photovoltaic Design Optimization For Terrestrial Applications," Special Short Abstracts, 13th IEEE Photovoltaics Specialist Conference, June 5-8 1978.

263. Watkins, J. L. and E. L. Burgess, "The Effect Of Solar Cell Parameter Variation On Array Power Output," preprint. Special Short Abstracts, 13th IEEE Photovoltaics Specialist Conference, June 5-8 1978.

264. Bauer, Paul, "Computer Simulation of Satellite Electric Power Systems," Proceedings IECEC, 1969, pp. 33-41.

265. Schwartzburg, Mel, "Performance Analysis of Satellite Electric Power Systems by Computer Simulation," Proceedings, IECEC, 1969, pp. 42-50.

266. Cherry, William R., "The Generation of Pollution Free Electrical Power From Solar Energy," Transactions of the ASME: Journal of Engineering for Power, Vol. 94, No. 2, April 1972, pp. 78-82. Originally published March 1971 by Goddard Space Flight Center as X-760-71-135.

267. Ralph, E. L., "Large Scale Solar Electric Power Generation," paper presented at ISES Conference, Greenbelt, Md., May 10 1971. Publication source uncertain.

268. Leonard, S. L., "Central Station Power Plant Applications For Photovoltaic Energy Conversion," preprint. Special Short Abstracts, 13th IEEE Photovoltaics Specialist Conference, June 5-8 1978.

269. Matlin, R. W. and W. R. Romaine, The 25-Kilowatt, Photovoltaic-Powered Agricultural Experiment At Mead, Nebraska, 21 April 1978. Available from NTIS as COO/4094-4.

270. Roesler, Dietrich J., "A 60 KW Solar Cell Power System With Peak Power Tracking and Utility Interface," preprint. Special Short Abstracts, 13th IEEE Photovoltaics Specialist Conference, June 5-8 1978.

271. Lyon, E. F., "Design of the Natural Bridges National Monument 100KW Photovoltaic Load Center," Special Short Abstracts, 13th IEEE Photovoltaics Specialist Conference, June 5-8 1978.

272. Solar Cells To Power New Arkansas College Building. ERDA News, Vol. 2, No. 18, September 5 1977.

273. Maycock, Paul, "National Photovoltaic Program Overview," August 30, presentation at 1978 Annual Meeting of the American Section of ISES, Denver, Colorado.

273A. Backus, Charles E., "A Solar-Electric Residential Power System," Conference Proceedings, 7th IECEC, 1972, pp. 704-711. Available from American Chemical Society, 1155 16th St., N. W., Washington, D. C. 20036.

274. Imamura, M. S., and J. A. Sanders, "Residential Photovoltaic Prototype System Definition Study," Proceedings, 1977 Annual Meeting of ISES, June 6-10 1977, pp. 23-6 through 23-10.

275. Federman, E. F., et al., "New Concepts In Solar Photovoltaic Electric Power Systems," Proceedings, 11th IECEC, 1976, pp. 1308-1315.

276. Sater, Bernard L., and Chandra Goradia, "An Integrated Photovoltaic/Thermal High Intensity Solar Energy System (HISES) Concept for Residential Applications," Proceedings, 11th IECEC, 1976, pp. 1316-1323.

277. Eltimsahy, Adel H. and Richard DeLombard, "Experimental Investigation of A Solar Cell/Inverter System," Proceedings, 1977 Annual Meeting of ISES, June 6-10 1977, pp. 23-11 through 23-15.

278. White, J. R., J. F. Marshall, and R. D. Offenhauer, "Perspectives On Solar Electric Power For Small Homes," Proceedings of Condensed Papers, Miami International Conference on Alternative Energy Sources, 5-7 December 1977, p. 791-92.

279. Federman, E. F., et al., "Potential for Stand-Alone Photovoltaic On-Site Total-Energy Residential Systems," Special Short Abstracts, 13th IEEE Photovoltaics Specialist Conference, June 5-8 1978.

280. Kirpich, A., and G. Tully, "Regional Analysis of Residential Photovoltaic System Concepts," Special Short Abstracts, 13th IEEE Photovoltaics Specialist Conference, June 5-8 1978.

281. Böer, K. W., "A Combined Solar Thermal and Electrical House System," The Sun In The Service of Mankind, International Congress, Paris, 2-6 July 1973, page EH 108-2 through EH 108-11.

282. Samara, George A., Integrated Photovoltaic-Thermal Solar Energy Conversion Systems, July 1975. Available from NTIS as SAND 75-5717.

283. Chobotov, V. and B. Siegel, "Analysis of Photovoltaic Total Energy System Concepts For Single-Family Residential Applications," Pre-print. Special Short Abstracts, 13th IEEE Photovoltaics Specialists Conference, June 5-8 1978.

284. Smith, D. R., et al., "Combined Photovoltaic/Thermal Collector Testing," Special Short Abstracts, 13th IEEE Photovoltaics Specialist Conference, June 5-8 1978.

285. Kern, E. C., et al., "Combined Photovoltaic and Thermal Hybrid Collector Systems," Special Short Abstracts, 13th IEEE Photovoltaics Specialist Conference, June 5-8 1978.

286. Landsman, E. E., "Motor Starting With PV Systems," Preprint. Special Short Abstracts, 13th IEEE Photovoltaics Specialist Conference, June 5-8 1978.

287. Bifano, William J., et al., "Design, Fabrication and Installation of A Photovoltaic Power System For The Papago Indian Village Schuchuli (Gunsight), AZ," Special Short Abstracts, 13th IEEE Photovoltaics Specialist Conference, June 5-8 1978.

288. Ratajczak, Anthony F., "Description and Status of NASA-LeRc/DOE Photovoltaic Applications Systems Experiments," Special Short Abstracts, 13th IEEE Photovoltaics Specialist Conference, June 5-8 1978.

289. Weiner, Howard, et al., "Computer Program For Design And Performance Analysis of Navigation-Aid Power Systems," Proceedings, 11th IECEC, 1976, pp. 1330-1337. Available from AIChE.

290. Medler, Leon, "A Lightweight Photovoltaic Power Unit For Military Applications," Special Short Abstracts, 13th IEEE Photovoltaics Specialist Conference, June 5-8 1978.

291. DeMeo, Edgar A., Assessment of Cadmium Sulphide Photovoltaic Arrays For Large Scale Electric Utility Applications, Final Report, February 1976. Available from EPRI as EPRI ER-188.

292. Bradley, Jerry O., and Dennis R. Costello, "Technoeconomic Aspects of Central Photovoltaic Power Plants," Midwest Research Institute Report. Later presented at the 1976 ISES Meeting and Abstracted in Photovoltaics and Materials, Vol. 6 from that conference. The complete paper was published in Solar Energy, Vol. 19, No. 6, 1977, pp. 701-709.

293. Chowaniec, C. R., et al., "Solar Photovoltaic Power Stations," Proceedings, 1977 ISES, Vol. one, pp. 22-9 through 22-13.

294. Cole, R., et al., "Estimated Cost of Electricity Produced By Four Types Of Compound Parabolic Concentrators," Proceedings, 1977 ISES, Vol. one, pp. 23-25 through 23-30.

295. Evans, D. L., and L. W. Florschuetz, "Cost Studies On Terrestrial Photovoltaic Power Systems With Sunlight Concentration," Solar Energy, Vol. 19, No. 3, 1977, pp. 255-262.

296. Böer, K. W., "Direct Energy Conversion For Terrestrial Use," Conf. Rec., Ninth IEEE Photovoltaic Specialists Conf., May 2-4, 1972, pp. 351-358. Reproduced in Backus' Solar Cells book.

297. Wolf, Martin, "Cost Goals For Silicon Solar Arrays For Large Scale Terrestrial Applications," Conf. Rec. Ninth IEEE Photovoltaic Specialists Conf., May 2-4, 1972, pp. 342-350. Reproduced in Backus' Solar Cells book.

298. Tsou, Peter, and Walter Stolte, "Effects of Design on Cost of Flat-Plate Solar Photovoltaic Arrays For Terrestrial Central Station Power Applications," preprint. Special Short Abstracts, 13th IEEE Photovoltaics Specialist Conference, June 5-8 1978.

299. Ross, R. G., "Photovoltaic Design Optimization For Terrestrial Applications," preprint. Special Short Abstracts, 13th IEEE Photovoltaic Specialist Conference, June 508 1978.

300. Hayes, Denis, The Solar Energy Timetable, Worldwatch Paper 19, April 1978. Available from Worldwatch Institute, 1776 Massachusetts Ave., N. W., Washington, D. C. 20036.

301. Williams, B. F., Automated Array Assembly, Quarterly Report No. 2, June 1976. Available from NTIS as ERDA/JPL/954 352-76/2.

302. Carbajal, B. G. Automated Array Assembly, Quarterly Report No. 2, July 1976. Available from NTIS as ERDA/JPL/954405-76/2.

303. Carbajal, B. G. and S. N. Rea, Automated Array Assembly, Quarterly Report No. 1, March 1976. Available from NTIS as ERDA/JPL/954405-76/1.

304. Iles, P. A., and H. McLennan, Low Cost Solar Cell Arrays, May 1975. Available from NTIS as NASA CR-134815.

305. Medler, Leon, "Potential Near Term DOD Applications of Photovoltaic Power Systems And Barriers To Market Development," Preprint. Special Short Abstracts, 13th IEEE Photovoltaics Specialist Conference, June 5-8 1978.

CHAPTER 3

WIND-ELECTRICS

"All things considered, a cleaner, safer, less disruptive source of energy is hard to imagine."
Denis Hayes [1, p34]

3-1 Introduction

Chapter 2 dealt with the direct conversion of solar energy to electricity. We now deal with an indirect solar-electric technology, windpower.

The inexhaustible energy resource of the wind arises from nonuniform solar heating of the earth and its atmosphere. Thus when we work with windpower we are dealing with indirect solar energy. As is well-known, the wind may blow day or night, and it is therefore not subject to the daytime-only collection of other solar-electrics.

In the mid 1970's it was believed by many that windpower could play a significant future role in the nation's energy structure. This chapter summarizes some of the principal prior and present art of wind energy conversion systems (WECS) and emphasizes some of the technical aspects of integrating WECS with the power grid.

3-2 Wind And The Atmosphere

Wind results from air in motion. Air in motion arises from a pressure gradient [2, Ch. 4]. On a global basis one primary forcing function causing surface winds from the poles toward the equator is convective circulation. Solar radiation heats the air near the equator, and this low density heated air is buoyed up. At the surface it is displaced by cooler more dense higher pressure air flowing from the poles. In the upper atmosphere near the equator the air thus tends to flow back toward the poles and away from the equator. The net result is a global convective circulation with surface winds from north to south in the northern hemisphere, as Eldridge [3, p3] has shown.

It is clear from the above oversimplified model that the wind is basically caused by solar energy irradiating the earth. This is why wind utilization is considered a part of solar technology. The correlation between solar and wind energies at a single site has been investigated by Arnold [4].

In actuality the wind is much more complex. The above model ignores the earth's rotation which causes a coriolis force resulting in an easterly wind velocity component in the northern hemisphere.

There is the further complication of boundary layer frictional effects between the moving air and the earth's rough surface. Mountains, trees, buildings, and similar obstructions impair streamline airflow. Turbulence results, and the wind velocity in a horizontal direction markedly increases with altitude near the surface [5].

Then there is the obvious fact of land and water with their unequal solar absorptivities and thermal time constants. During daylight the land heats up rapidly compared to nearby sea or water bodies, and there tends to be a surface wind flow from the water to the land. At night the wind reverses because the land surface cools faster than the water.

Superimposed on each of the above identifiable phenomena is the added effect at the poles where a large high pressure air mass breaks off from the polar air. Such a cold air mass may be a thousand miles across. This 'bubble' of cold air then tends to move--perhaps up to 48 Km/Hr. (30 MPH)--in a south-easterly direction across the United States. The well-known cold fronts with their violent gusty winds, thunderstorms, and sometimes cyclones and tornadoes result along the frontal edge of the air mass as it wedges its way south, interacting with the warmer air at the front. As the front passes a surface point there usually will be rapid changes of surface wind direction and magnitude.

Nor does the above model account for the movements, generally northeast-ward across the United States, of the well-known low pressure cells. Substantial surface winds may also be associated with these. Additionally there may be brief winds caused by the well-known summer-time build up of the awesome cumulonimbus storms particuarly severe in Florida and to a lesser extent and frequency in other parts of the nation. Finally the initial model does not account for the extremely violent winds associated with the sporadic hurricanes which are prone to meander across the southern and eastern states.

Some summary statements about wind may now be made:

- Wind is an indirect manifestation of solar energy; solar energy is the basic forcing function causing the wind.
- Wind is a locally measurable phenomenon which is part of a larger much more complex meteorological science beyond our scope here.
- Wind modeling is both complex and imprecise. Our knowledge of how to model the wind is still growing; but the present science largely rests on empirical wind data taken at various points in space and time by many organizations. In the United States the National Climatic Center in Asheville, N. C., a part of the National Oceanic and Atmospheric Administration (NOAA), archives meteorological and climatic data. Justus [6, Ch. 2] and Changery [7, p326] nicely summarize the services available.

- Surface wind is inherently linked to global and localized phenomena and terrain, e.g. latitude, altitude, mountains, plains, valleys, buildings and trees.
- Winds suitable for electric generation vary widely across the earth's surface and within the United States. The midwestern plains, New England, Alaska, and Hawaii, are prime sites for small and large-scale wind-electric utilization. It is evident that a wind-electric site must be selected with care.
- Wind suitable for wind-electrics may be available day or night or both. The possible availability of the wind at night is in sharp contrast to other solar-electric methods which have a strong diurnal character tied to the direct solar insolation. The night-time availability characteristic is a major plus for windpower utilization.
- Winds, in general, can be expected to have some diurnal and seasonal variabilities, in accord with general and local weather.
- The energy in the wind at any given site may be subject to wide variability from calms to hurricanes; but the long-term average wind velocity at a given site with a previous wind history is quite predictable.

  Because of structural limitations virtually all present art wind-electric apparatus cannot convert the wind energy in violent storms into useful electricity.
- Wind velocities and directions on an area basis are known for most of the earth's surface as a result of the various international and national weather services.

Thus it is easy to see that the answer to "What causes the wind?" is not simple. The engineering approach is to simply accept the wind as is where it is and attempt to convert it to useful energy without probing the question deeper than we have here.

3-3 Maximum Energy Via Wind-Electrics

From the large-scale systems view it is important, before proceeding with developing windpower, to establish that the potential exists in windpower to supply a significant fraction of electric energy for a nation. There are two perspectives: (1) maximum energy and power in the winds worldwide and (2) maximum energy and power in the winds within the United States. We should understand a priori that the maxima are only estimates made by knowledgeable people, and they vary widely, depending on the assumptions made. An exact maximum energy and power for either case may never be known. We should also recognize there is considerable confusion in the literature; energy gets called power and vice versa.

The Worldwide Case Earlier in Table 1-2 we showed that Thring estimated that of the solar radiation impinging on the earth about 26,000 x 10^{12} KWH is converted into wind. He [8, p279] makes the important point that "By far the greatest part of this energy is a priori unattainable for use by men because it is stored high up at the border of the troposphere and stratosphere at altitudes between 30,000 ft. and 40,000 ft. above sea level." Golding [2, p4] makes essentially the same point, though his energy estimate cited from his reference by Parker [9] is considerably lower, only 13 x 10^{12} KWH. Heronemus [10, p30] calculates that about 16,100 x 10^{12} KWH "are available as kinetic energy in moving air particles across the globe at any time, which compares with about 298 x 10^{12} KWH total energy consumed by the world in 1970," i.e. there is, according to Heronemus' figures, about 54 times the wind energy available to that used by the world for all purposes.

Hardy [11] in addressing the wind energy resource says, "The mean motion of the atmosphere represents, however, a large 'flywheel' in which the equivalent of approximately one day's total incident solar flux is stored. Von Arx [12] noted that wind power, in large-scale global collection, cannot be extracted at a rate exceeding the replenishment rate due to the solar flux. Thus the natural rate of kinetic energy dissipation and replenishment (which are in balance) places an upper limit on windpower collection."

Heronemus, in an Appendix to one of Sen. Gravel's speeches [13], finds that the total windpower of the atmosphere is estimated to be about 3 x 10^{17} kilowatts[1] while the windpower available for turbines at favorable sites throughout the world is of the order of 2 x 10^{10} KW, according to the World Meteorological Organization's estimates. The present world power demand is close to 10^{10} kilowatts, according to Kornreich [14, p857].

The United States Case Heronemus [13], in a 1971 estimate, indicates that "if aeroturbines to the total of 2 x 10^{10} KW were installed at all favorable sites, if the annual production were 5000 KWH per annum per installed KW, total energy extracted would be 10^{14} KWH per year (100 x 10^{12} KWH)." Krenz [15, p284], cites a later estimate made for the National Science Foundation. That estimate was for "a maximum yearly energy of 1.29 x 10^{12} KWH"[2] and "an average power of 1.4 x 10^{11} watts is anticipated for the Continental United States. Since the present electrical energy consumption rate is on the order of 2 x 10^{11} W,[3] wind power could, if the estimate is realized, supply 75% of the electrical power."

Virtually all persons involved in WECS recognize that a vast number of aeroturbines will be required, somewhere between 780,000 and 3 million large-scale units, depending on the sites chosen and the assumptions made

---

[1] Apparently this figure originated with Brunt [16].

[2] This compares with 1.536 x 10^{12} KWH estimated by the Solar Energy Panel [17, p69] possible by year 2000.

[3] His data must have been from the mid 1960's, from our curve in Fig. 1-26. The figure is higher now by a factor of almost 2. The percentage of electrical power would correspondingly be reduced.

regarding wind speed and its frequency distribution.

"Thus, the wind energy resource of the U. S., though imprecisely known at present, represents a significant energy source," according to Hardy [11, p2]. Our problem then becomes one of economically tapping this large inexhaustible energy resource and integrating these WECS with our existing electric energy systems, whether small-scale or large-scale.

We close this section by noting that the question of whether WECS can be net producers of energy is an important one related to the foregoing. Devine [18] has investigated this issue on the basis of lifetime energy expenditure for an aeroturbine using five indices one of which was:

$$\frac{\text{Delivered Electrical Energy (BTU/yr.)}}{\text{Total Primary Energy Subsidy (BTU/r.)}} = 40$$

He concludes that "the wind-electric generating station used in the fuel displacement mode is clearly a net producer of energy."

3-4 Site Selection Considerations

WECS Siting And Systems Viewpoint The site choice for a single or a spatial array of WECS is an important matter when wind-electrics is looked at from the systems point of view of aeroturbine generators feeding power into a conventional electric grid. If the WECS sites are wrongly or poorly chosen the net wind-electric generated energy/yr. may be suboptimal with resulting high capital costs for the WECS apparatus, high costs for the wind generated electric energy, and low returns on investment. Virtually all researchers to date who have constructed successful WECS have recognized that the saleable electric energy generated/yr. from windpower is strongly linked to the site chosen. Even if the WECS is to be a small generator not tied to the electric grid the siting must be carefully chosen if inordinately long break-even times are to be avoided.

In reviewing the wind literature, one is struck by the natural systems point of view taken by some of the early pioneers in windpower, particularly Putnam [5] and Golding [2]. They saw a particular wind-electric generator as a component in a larger electric energy system whose useful component of the output was, in a major way, determined by the site chosen for the WECS. Thus they saw one of the keys to successful economic payback of WECS with the utility grid lies in the siting, and they therefore paid careful and meticulous attention to the siting as a means of later insuring successful integration of the particular aeroturbine into the electric system. Many more recent papers on various aspects of siting underscore the importance of this subject: Gordon [19], Sittler [20], Wentick [21], Corotis [22], Knox [23], Smith [24], Reed [25], Hardy [26], Coty [27], Hurlebaus [28], Meroney [29], and Justus [30].

The siting of WECS generators is analogically kin to the siting of a conventional fossil fuel or nuclear plant. There a systems view is taken, and literally every facet of a proposed site is examined in detail by electric systems engineers and others. The viewpoint is multidisciplinary. Technical, economic, environmental, social, and other factors are examined before a decision is made to erect a generating plant on a specific site. The process may require several years and cost many hundreds of thousands of dollars before a site is chosen. Trade-offs on many factors are made, and the site finally chosen is never ideal in all respects. The dominant thinking throughout is that of the systems view.

Yet windpower siting is different, and in some ways more critical, than siting a fossil or nuclear plant. When considering several alternative sites for a fossil plant the electric energy each generates/yr. at the various sites may be only weakly site dependent. In contrast, for several WECS sites being considered--even in the same general area--there may be a marked difference in yearly electric energy outputs.

So the WECS systems engineer trying to integrate one or more aero-generators into an electric energy system has to be initially concerned with siting. It is worth mentioning that even with all the care[1] taken in the 1940 siting of the pioneering Smith-Putnam aero-generator on Grandpa's Knob, Vermont, that, as Merriam [31, p56] has pointed out, "The predicted average wind speed was 24 MPH. The actual average wind speed turned out to be 17 MPH. If the 1250 KW Smith-Putnam machine had been intended as a commercial power producer the error in siting would have been an economic disaster. Since it was only intended as a test unit no harm was done."

In the following paragraphs we briefly highlight site selection. Details will be found in the references.

A Fundamental Requirement to the successful use of WECS, obviously, is an adequate supply of wind. The wind velocity is the critical parameter. The power in the wind, P_W, through a given cross-sectional area for a uniform wind velocity, V, is[2]

$$P_W = K\,V^3 \qquad \text{where K is a constant.} \tag{3-1}$$

It is evident, because of the cubic dependence on wind velocity that small increases in V markedly affect the power in the wind, e.g. doubling V increases P_W by a factor of 8.

It is obviously desirable to select a site for a WECS with high wind velocity. Thus a high average wind velocity is the principal fundamental parameter of concern in initially appraising a WECS site. For a more detailed appraisal, one would like to have the average of the velocity cubed.

Eq. 3-1 implies that the velocity is that which exists at the aeroturbine--not in the upper atmosphere--for the windpower near the surface is what we have to work with practically, as all aeroturbines are mounted on the ground.[3]

---

[1] After considerable study over several years and consulting with the most knowledgeable people of that time.

[2] We shall show its derivation in Sec. 3-5 and be more explicit there.

[3] Perhaps some visionary may eventually figure out how to convert the 200 MPH jet streams into useful electric power. No known way of achieving this exists in 1978. This approach to wind-electrics does not appear promising.

The average wind velocity at the hub height[1] of the proposed aeroturbine is of chief interest.

Unfortunately, due to frictional effects at the air-surface boundary the horizontal component of wind velocity increases rather radically as one leaves the surface. Putnam [5] investigated rather completely wind velocities at various heights over New England sites. Golding [2 , Ch. 7] summarizes the state of knowledge as of about 1955 and some of the difficulties, particularly with hills. More recently Reed [32] [33] has shown that over relatively <u>flat</u> terrain useful wind increases in proportion to the one-seventh power of height above ground. He claims a maximum error of 4 percent in this approximation. He investigated heights up to about 100 meters. Finally he developed in reference [32] a combined extrapolative equation for speed and height:

$$V/V_a = \left(Z/Z_a\right)^{0.27960 + 0.03265 \ln Z_a} \cdot V_a^{0.10528 \ln Z_a - \left(0.09831 + 0.05502 \ln Z + .00642 \ln Z \ln Z_a\right)} \quad (3\text{-}2)$$

where:

V = wind speed at height Z

V_a = wind speed known from anemometer at height Z_a.

He claims, based on Sandia and Argonne tower data, the above relation showed excellent agreement with upper level records. Further research is necessary to find height relationships over rougher terrain.

Anemometer data is normally based on wind speed measurements from a height of 10 m; but there are many anemometers mounted at nonstandard heights reporting data. The anemometer height must be known for the most accurate windpower predictions. Also for the most accurate assessment of windpower potential it is absolutely essential that anemometer data be obtained at the precise site and hub height for any proposed WECS.

<u>Strategy For Siting</u> is generally recognized, according to Freeman [34, p356], to consist of:

- Survey historical wind data.
- Contour maps of terrain and wind are consulted.
- Potential sites are visited.
- Best sites are instrumented for approximately 1 year.
- Choose optimal site.

This is the approximate siting strategy which was used by Putnam [5] on the Grandpa's Knob machine and similar to that later outlined by Golding [2 , p66].

The strategy appears still valid in 1978.

---

[1] Hub height, for a horizontal axis aerogenerator, is the vertical distance from the hub of the aeroturbine to the ground surface.

Systems Factors Entering Into Site Choice It is helpful to summarize the principal factors which past WECS investigators have found to be important in choosing an optimal site for one or more aerogenerators. Only the barest information can be presented here. Researchers wishing to pursue the siting matter further will find much more in the references.

The factors tabulated below are not necessarily grouped in their order of importance nor do we attempt to weight them.

- Whether One or Many Aerogenerators are proposed. The case of a single aerogenerator to be sited is the simplest case; but, as will be seen, responsibly siting it is a major undertaking in its own right, the Smith-Putnam case [5] being a good example. In this case one is concerned only with a single set of wind statistics at the proposed site. Most of the large-scale wind-electric generators actually constructed to date anywhere in the world have been single generators.

  If many generators are proposed for an area, two sub-cases arise:

  - Clustered The generators are to be grouped in a cluster relatively near to each other, perhaps on the same hill or prominence. Golding [2] has considered this case, pointing out some of the possible pitfalls such as not spacing the machines too close together to avoid air turbulence interactions between machines.

    So far as can be determined from the literature, it appears such clusters of aerogenerators have been proposed by various individuals but not actually built anywhere in the world.[1] The success of these 'wind farms' obviously hinges on single generators, and the state-of-the art of wind-electrics has not yet actually advanced to such systems of WECS; But clearly clustered systems have been envisioned by previous researchers [3 5].
  - Spatially Dispersed aerogenerators over an area. Here one must be vitally concerned with wind statistics at multiple locations at simultaneous times. By space diversifying, one can improve the available power and hence the reliability of wind generated electric power, for the chances are that the wind may be blowing at one site while another nearby site in the same area may experience a calm.

    This is a much more complex case than the clustered one above; only now is research underway to study it via computer simulation, the work of Justus [36] [37] for New England and Central U. S. being a notable example. His preliminary results indicate that by so dispersing the aerogenerators, significant peak load displacement--from the utility viewpoint--can be achieved without the use of storage.

---

[1] Clusters of conventional windmills have, however, existed for years on various Mediterranean islands. But these are all nonelectric and of relatively small-scale.

● National And Regional Wind Maps In the earliest stages of site planning one should have some perspective on the general regions of the nation where, based on reliably measured wind records, wind is generally available. Perhaps equally important is to know where it is not available in useful amounts. Numerical procedures for regional studies on Oahu have been generated by Hardy [26].

The National Climatic Center in Asheville, N.C. is the principal data source for the government's network of 686 wind reporting stations in the continental United States, and many investigators have drawn from this source.

Surface wind data on a national or regional basis is usually presented in the forms of:

▲ Isovents or contours of constant average wind velocity (M/sec.). The averaging period seen in the literature varies widely, but monthly, quarterly, and yearly averages are commonly seen. It is important to know what the data averaging period is when examining a given isovent contour map, for the winds change seasonally.

▲ Isodynes are contours of constant wind power (watts/m^2 of area perpendicular to the wind flow). Again, it is important to know the averaging period. An example of this kind of map is in Fig. 3-1 taken from Radice [38]. This is a computerized plot of the National Climatic Center data where the computer was programmed to plot the isodynes for intervals of 100 W/m^2 over the range of 100 to 1,600 W/m^2. Maps showing Standard Metropolitan Statistical Areas (SMSA) were superimposed over the isodynic plots. More than half of the U. S. population lives within these irregular areas shown in Fig. 3-1.

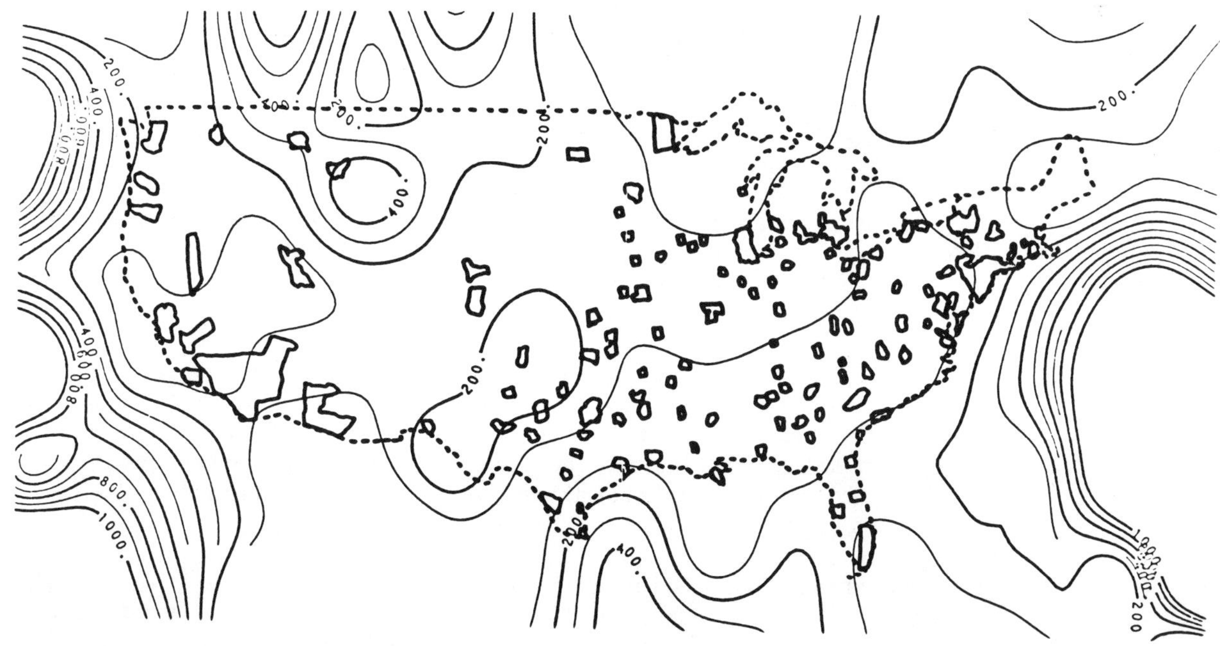

FIG. 3-1 Isodyn plot of available windpower (watts/M^2) within the United States. Yearly average. From Radice [38].

From the systems viewpoint Radice's procedure could help the wind site investigator by not losing sight of the fact that desirable windpower sites should be reasonably close to a sizeable user demand zone.

Radice's plot also nicely shows the much higher windpower density available over the oceans than over land; but with a few exceptions--Heronemus' proposal [39] for wind generators off New England's shores--such power is so far removed from the population centers as to be at present impractical to tap.

▲ Wind Roses are a polar means of plotting, for a given site, the annual average wind speed from different directions. Thus prevailing winds, if they exist, can be quickly spotted by this means. Putnam [5, Ch. 2] found these useful in siting as did Golding [2 , Ch. 5]. A Wind rose also shows the percentage of calms at that site. The systems engineer could immediately use this number as a preliminary indicator of maximum WECS reliability at that site.

● Is Anemometry Data Available At Or Near Proposed Site? is another important siting factor. The principal object is to measure the wind speed which basically determines the WECS output power; but there are many practical difficulties with the instrumentation and measurement methods, as Putnam [5 , Ch. 3] and Golding [2 , Ch. 8] both found as have later investigators. The anemometer height above ground, accuracy, linearity, location on the support tower, shadowing and inaccurate readings therefrom, icing, inertia of rotor, whether it measures the horizontal velocity component or vertical, and temperature effects are a few of the many difficulties mentioned in the literature.

As indicated under strategy for siting, virtually all WECS researchers underscore the wisdom of having anemometry data avilable over some time period at the precise spot where any proposed WECS is to be built and that this should be accomplished before a siting decision is made.

In the absence of anemometry data at a proposed site, Putnam [5 , p74] was the first to discover a useful short-cut if there is a nearby meteorological station. "Then the long-term mean velocity at the survey station may be determined as follows:

> Determine the ratio between the velocity recorded at the survey station and that recorded at the control station in the same period of time. Multiply the long-term mean velocity at the control station by this ratio to obtain the long-term mean velocity at the survey station."

Later systems engineers have found Putnam's algorithm useful in an initial site investigation before an anemometer is set up. Using it, approximate long-term average wind speed can be found at the proposed site. Then Reed's eq. 3-2 can be used to estimate the speed at hub height of the proposed WECS machine.

- Is Wind V(t) Curve Available at the proposed site? This important curve determines the maximum energy in the wind and hence is the principal initial controlling factor in predicting the electrical output and hence revenue return of the WECS machine. It is desirable to have average wind speed, $\overline{V}$, such that $\overline{V} \geq$ 8-10 MPH which is about the lower limit at which present large-scale WECS generators 'cut-in', i.e. start turning. The V(t) curve also determines the reliability of the delivered WECS generated power, for if the V(t) curve goes to zero there will be no generated power during that time. If there are long periods of calm the WECS reliability will be lower than if the calm periods are short. In making such reliability estimates it is desirable to have measured V(t) curves over about a 5 year period for the highest confidence level in the reliability estimate.

  It is also evident the V(t) curve influences the system energy storage required, if any. For periods of calm, energy storage may later be needed, and as storage is a separate subsystem from the WECS itself, the busbar energy costs will be increased.

  From the V(t) curve for the wind at a specific site one can also compute the velocity-duration curve ($\overline{V}$ vs hours/yr. See Putnam 5, Ch. 4.) which in turn determines the energy output of that WECS. Thus the 'plant load factor' defined as:

  $$C_p = \frac{\text{Electric Energy/yr. actually generated (KWH/yr.)}}{\text{Installed rated energy capacity (KWH/yr.)}} \tag{3-3}$$

  as shown by Golding [2, Ch. 10] and graphically portrayed in Sec. 3-5 can be computed. Systems engineers immediately recognize this as identical to the plant load factor of a conventional electric generating plant [Hill 40, Ch. 7], also sometimes called the capacity factor.[1] Golding finds it useful to multiply C_p by 8760 hrs./yr. to obtain the specific output in KWH/yr./KW which he uses as a criterion for ranking potential WECS sites.

  It is clear that the wind V(t) furnishes vital information to a systems engineer attempting to integrate WECS with the utility.

- What is the Wind Structure at the proposed site? The ideal case for a WECS would be a site such that the V(t) curve was flat, i.e. a smooth steady wind that blows all the time; but a typical site is always less than ideal. "Wind especially near the ground is turbulent and gusty, and changes rapidly in direction and in velocity. This departure from homogenous flow is collectively referred to as 'the structure of the wind'" [5, p22]. Justus [6, Ch. 3] shows how gusts are vectorially handled. He also shows gust definitions based on the World Meteorological Organization's standards permitting practical interpretations from a measured V(t) curve.

[1] See Fig. 1-34 for its economic importance.

In the wind structure classification it is also of considerable importance for the systems planning engineer to determine whether a proposed WECS site has a prior history of tornadoes, cyclones, or hurricanes. If it has, then the WECS structure must be designed to withstand such extremes, and this will increase the plant cost.

- What Is The Terrain like, and what are the aerodynamics of it? If the WECS is to be placed near the top--but not on the top--of a not too blunt hill facing the prevailing wind, then, as Putnam found [5, Ch. 2], it may be possible to obtain a 'speed-up' of the wind velocity over what it would otherwise be. Also the wind here may not flow horizontal making it necessary to tip the axis of the rotor so that the aeroturbine is always perpendicular to the actual wind flow.

  It may be possible to make use of hills or mountains which channel the prevailing winds into a pass region, thereby obtaining higher wind power.

  The complex subject of wind flow over hills is beyond our scope here; the reader will find an excellent treatment of this subject in Golding [2, Ch. 7]. Prediction of wind velocities over irregular surfaces is still more art than science, and more basic research in the area appears needed to improve the accuracy of WECS siting predictions.

  Finally, there are the geomorphologists [41] [42] [43] who are currently trying to relate the terrain structure in windy areas of the West as seen from satellites to the average wind velocity and hence to site selection.

- What Is the Altitude of the proposed site? It affects the air density and thus the power in the wind and hence the useful WECS electric output. Also, as is well-known, the winds tend to have higher velocities at higher altitudes.

  One must be careful to distinguish altitude from height above ground. They are not the same except for a sea level WECS site.

- In Region Where Icing, particularly of the aeroturbine blades, is a potential problem? Putnam [5, p34], had a great fear of icing and its potential adverse effects on their machine. They took the view that the test unit must not fail structurally, and that if it did, the project would die. Because of this siting restriction they accordingly sought to design their machine with much excess strength. They found that "ice was no bar to turbine operation at 4000 feet in the Green Mountains." These results could be entirely different at other sites, however.

- What Is The Local Ecology Like at the proposed site? If the surface is bare rock it may mean lower hub heights hence lower structure cost. If trees or grass or vegetation are present, all of which tend to destructure the wind, then higher hub heights will be needed resulting in larger system costs than the bare ground case.

Putnam [5, p53, p82] indicates that the ecology in New England gives valuable clues to the average wind velocity at that site and that these clues can be used by a trained observer in a preliminary site visit. He says that balsam tree deformation starts to occur when the average wind velocity is 17 MPH. At 27 MPH the balsams grow like a carpet to only 1 foot height as a result of wind deformation.

- Salt Spray or Blowing Dust present at the site at any time of the year? If so the systems engineer would want to know it early, as such an environment is generally adverse to machinery and electrical apparatus.

- Distance To Roads or Railways? This is another factor the systems engineer must consider, for heavy machinery, structures, materials, blades and other apparatus will have to be moved into any chosen WECS site.

- Accessibility of Summit? Does a road have to be built? Can it carry heavy loads? Is the road safe?

- Nature of Ground? Will foundations for a WECS be secure thereon? Is the ground surface stable? Is erosion a problem which could possibly later wash out the foundations of a WECS, destroying the whole system?

- Favorable Land Cost needs some consideration, as this along with other siting costs, enters into the total WECS system cost.

- Nearness Of Site To Load Center/Users? This obvious criterion minimizes transmission line length and hence losses and costs. After applying all the previous siting criteria, hopefully as one narrows the proposed WECS sites to one or two they would be relatively near to the users of the generated electric energy.

It is evident from the above factors that site selection for one or more WECS is an important systems aspect of the overall problem of integrating WECS systems into existing electric energy systems. The total process clearly involves mature judgment with trade-offs between many factors, just as is involved in siting a conventional fossil power plant. Finally, the WECS siting process is important whether trying to locate large or small-scale aerogenerators, as siting determines the maximum useful energy output for either case.

It would appear highly probable that siting predictive techniques will grow in the future as more WECS are constructed--both small and large-scale. The general siting area appears particularly fertile for further advancement by future multidisciplinary systems engineers.

3-5 Basics of Wind Energy Conversion

In this Sec. we present a few basic topics of interest to the systems engineer attempting to integrate wind energy conversion systems into electric systems. It is not our intent here to go into the many interesting design details of an aerogenerator. We deal in a general way only with those basics having the widest possible applicability. The steady-state case[1] is emphasized, i.e. constant nongusty wind from a fixed direction turning the aeroturbine at a constant rate. Emphasis is on the horizontal axis WECS.

Power In The Wind An air mass, M, moving with a speed, V (m/s), has a kinetic energy, E, of

(3-4) $$E = \frac{1}{2} M V^2 \quad \text{Joules}$$

Since the volume of air passes through an area, A, normal to the air velocity, and since the air mass per unit volume, ρ, equals the air density, then the kinetic energy in the volume of air passing through the area of interest during the observation time t is

(3-5) $$E = \frac{1}{2} (\rho A V t) V^2 = \frac{1}{2} A V^3 t \quad \text{Joules}$$

Since power is defined as the time rate of energy use, we immediately see from (3-5) that the wind power available, P_a, is

(3-6) $$P_a = \frac{1}{2} \rho A V^3 \quad \text{Watts}$$

Eq. 3-6 tells us that the maximum windpower available--the actual amount will be somewhat less because all the available energy is not extractable--is proportional to the cube of the wind speed. It is thus evident that small increases in wind speed can have a marked effect on the power in the wind. Equation 3-6 gives us insight into why such great emphasis is placed on wind speed throughout the wind literature.

Eq. 3-6 also tells us that the power available is proportional to air density (1.225 Kg./m^3 at sea level). It may vary 10-15 percent during the year because of pressure and temperature changes according to Golding [2, p23]. It changes negligibly with water content.

Equation (3-6) also tells us that the wind power is proportional to the interception area. Thus an aeroturbine with a large swept area is higher power than a smaller area machine; but there are added implications. Since the area is normally circular of Diameter D in horizontal axis aeroturbines, then $A = \pi \frac{D^2}{4}$, which, when put into (3-6) gives

(3-7) $$P_a = \frac{1}{2} \rho \pi \frac{D^2}{4} V^3 \quad \text{Watts}$$

This equation tells us that the maximum power available from the wind varies according to the square of the diameter of the interception area, normally taken to be the swept area of the aeroturbine. Thus doubling the diameter of the interception area results in a factor of 4 increase in the

[1] See Preuss [44] for an entrée to the much more complex unsteady state case.

available power. Eq. (3-7) gives us insight into why the designer of an aeroturbine for wind-electric use would place such great emphasis on the turbine diameter. Justus [6, p11] plots eq. (3-7) for various diameters.

Returning to eq. (3-6), but now recognizing that V, in actuality, is not constant but is represented by a statistically 'noisy' wind speed-time curve, V(t), then the instantaneous power in the wind would be

$$P_a(t) = \frac{1}{2}\rho A\,[V(t)]^3 \quad \text{Watts.} \tag{3-8}$$

Since we are normally more interested in average power, we must time average both sides of (3-8), signified by the bar below

$$\overline{P_a(t)} = \frac{1}{2}\rho A\,\overline{[V(t)^3]} \quad \text{Watts} \tag{3-9}$$

Equation (3-9) tells us that for a nonsteady state wind it is necessary to cube the measured wind speeds and then take the average to find the average windpower available.[1] It is immediately obvious that this non steady state case is more complex than the simple steady state case, and it is why for the former case such great emphasis is placed on anemometry data at a proposed WECS site.

Transposing eq. (3-9) results in

$$\frac{\overline{P_a(t)}}{A} = \frac{1}{2}\rho\,\overline{[V(t)]^3} \quad \text{Watts/m}^2 \tag{3-10}$$

which says that the average available windpower per unit area is directly related to the average of the wind speed cubed. This is one useful method of characterizing the potential specific power in the wind over a geographic area. Fig. 3-1 was an example of such an isodyn plot.

Basic Components of a WECS are shown in Fig. 3-2 in block diagram form. Ramakumar [45] nicely summarizes the system operation. "Aerotrubines convert energy in moving air to rotary mechanical energy. In general, they require pitch control and yaw control (only in the case of horizontal or wind-axis machines) for proper operation. A mechanical interface consisting of a step-up gear and a suitable coupling transmits the rotary mechanical energy to an electrical generator. The output of this generator is connected to the load or power grid as the application warrants.

The purpose of the controller is to sense wind speed, wind direction, shaft speeds and torques at one or more points, output power and generator

---

[1] In general for a noisy V(t) wind curve, this is not the same thing as first finding $\overline{V(t)}$ and then cubing that average. Following this procedure is only a rough preliminary indication of windpower. The technically correct way to do the averaging is as shown in Eq. (3-9).

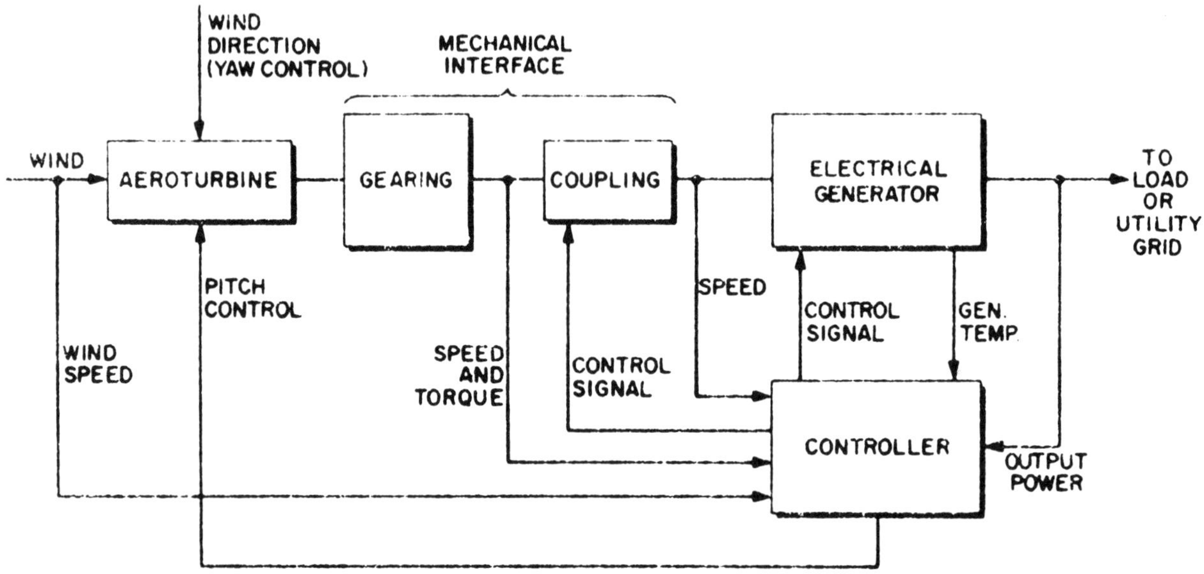

FIG. 3-2 Basic components of a Wind-Electric system. From Ramakumar [45].

temperature as necessary and appropriate control signals for matching the electrical output to the wind energy input and protect the system from extreme conditions brought upon by strong winds, electrical faults, and the like."

The physical embodiment for such an aerogenerator is shown in a generalized form in Fig. 3-3. Such a machine typically is a large impressive structure. Note that the aeroturbine is downwind from the nacelle containing the transmission and generator. Smaller aerogenerators for individual houses typically have many of these basic elements except for omission of the pitch changing apparatus. They also typically are made to face into the wind.

Classes Of WECS Systems First, there are two broad classifications:

- Horizontal Axis Machines The axis of rotation is horizontal and the aeroturbine plane is vertical facing the wind.

- Vertical Axis Machines The axis of rotation is vertical. The sails or blades may also be vertical, as on the ancient Persian windmills, or nearly so, as on the modern Darrieus rotor machine.

Then they may be classed according to size, as determined by their useful

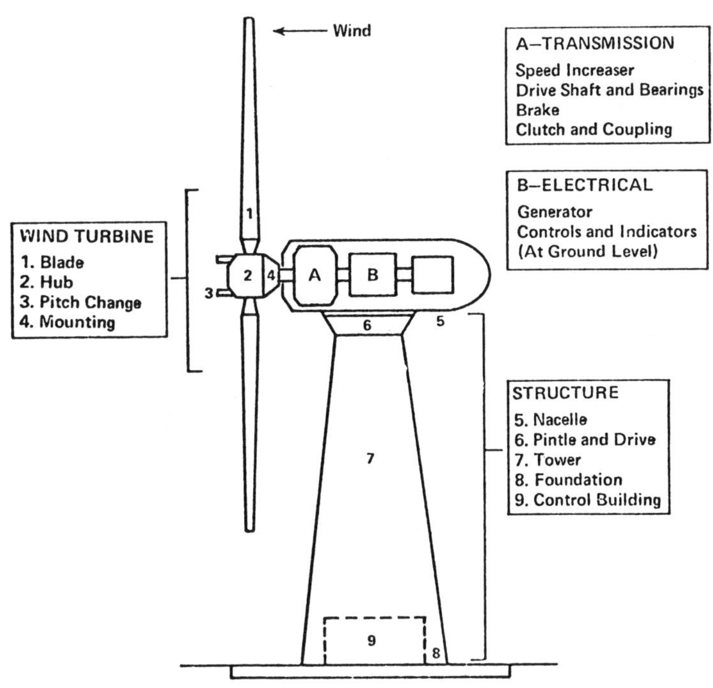

FIG. 3-3 Physical embodiment of wind-electric generating station. From Devine [18]

electrical power output. There appears to be little agreement in the literature on the ranges, and we here adopt Merriam's [31, p10] arbitrary classification:

- Small-Scale 0-2 KW. These might be used on farms, remote applications, and other places requiring relatively low power.
- Medium Scale 2-100 KW. Might be used to supply several residences or small businesses.
- Large-Scale 100 KW and up. Might be used to feed the power grid in a fuel saver mode supplying electric energy in a localized area. There are two subclasses:
  - Single Generator at a single site.
  - Mulitple Generators sited at several places over an area.

Wind-Electrics can also be classified by type of output, after Ramakumar [45]:

- DC Output
  - DC generator.
  - Alternator-rectifier.
- AC Output
  - Variable Frequency, variable or constant voltage AC.
  - Constant Frequency, variable or constant voltage AC.

They may also be classified by rotational speed of the aeroturbine, again after Ramakumar [45]:

- Constant Speed with variable pitch blades. This mode implies use of a synchronous generator with its constant frequency output.

- Nearly Constant Speed with fixed pitch blades. This mode implies an induction generator.

- Variable Speed with fixed pitch blades.
  This mode could imply, for constant frequency output:
  - Field Modulated generator system.
  - AC-DC-AC link.
  - Double output induction generators.
  - AC commutator generators.
  - Other variable speed constant frequency generating systems.

He nicely summarizes the merits of each generation scheme in his paper which is highly recommended reading for systems engineers.

A final classification can be made according to how the WECS output is to be utilized [Ramakumar, 45]:

- Battery Storage.

- Direct Connection to an electromechanical energy converter.

- Other Forms (thermal, potential, etc.) of storage.

- Interconnection with conventional electric utility grids.

The systems engineer seeking to integrate WECS will, naturally, be most interested in the latter case but should be aware that WECS offer other options as well.

Aerodynamic Forces acting on a blade element tending to make it rotate are another basic to a systems engineer asking the fundamental question, "Why does the aeroturbine rotate?"

First, we recognize that the whole general problem of forces on a wind generator blade are quite involved and beyond our scope here. Putnam, in his pioneering work [5, p67] dealt with this problem in its most general sense. Surprisingly, he showed no vector diagram of the aerodynamic forces acting on a blade element. Golding [2, p199] however, treats the case of the forces on a plane blade element.

There are several basic types of blades an aeroturbine may have, e.g. sails, planes, and aerodynamic surfaces based on the aircraft wing cross section for which there are many kinds. The early history of windmills is based on the first two; modern higher efficiency wind-electric generators are based on use of blades with aerodynamic surfaces.

Consider the aerodynamic blade shown in Fig. 3-4. The blade can be thought of as a typical cross-sectional element of a two-bladed aeroturbine. The element shown is at some radius r from the axis of rotation. It is moving to the left. Because the blade is moving in the plane of rotation it sees a tangential wind velocity component, V_T, in the plane of rotation. This velocity component added vectorially to the impinging wind velocity gives the resulting wind velocity, V_R, seen by the rotating blade element. At right angles to V_R is the lift force, F_L, caused by the aerodynamic shape of the blade. The drag force, F_D, is parallel to V_R. The vector sum of F_L and F_D is F_R which has a torque producing component, F_T and a thrust producing component. The former is what drives the aeroturbine rotationally, and the latter tends to flex the blade and also overturn the aerogenerator.

The vector diagram is centered on the center of lift of the aerodynamic blade. As is well-known from aircraft wing theory, one of the critical parameters is α, the angle of attack of the aerodynamic element. It determines lift and drag forces and hence speed and torque output of the aeroturbine. These quantities can be varied by changing the blade pitch angle, β, and this is the basic torque control method used on large variable pitch wind-electric generators. The torque would determine the AC output power if a synchronous generator was used.

Since V_T increases linearly as we go out radially, r, on an actual aeroturbine blade, it is necessary to adjust β with r so as to always have a positive angle of attack and to maintain reasonable stress levels within the blade. This means that at large r, β is made small while at small r, β

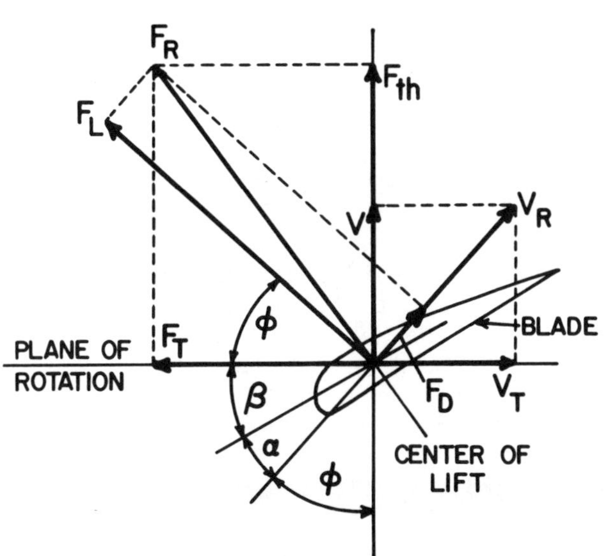

V = IMPINGING WIND VELOCITY.
V_T = WIND VELOCITY IN PLANE OF ROTATION DUE TO BLADE TURNING.
V_R = RESULTANT WIND VELOCITY SEEN BY AERO TURBINE BLADE.
F_L = LIFT FORCE (NORMAL TO V_R).
F_D = DRAG FORCE (PARALLEL TO V_R).
F_R = RESULTANT FORCE ON BLADE.
F_T = TORQUE PRODUCING COMPONENT OF F_R MAKING AEROTURBINE ROTATE.
F_{Th} = THRUST FORCE COMPONENT OF F_R.
α = ANGLE OF ATTACK OF BLADE.
β = BLADE PITCH ANGLE.

FIG. 3-4 Vector diagram of forces on a elemental blade section of an aeroturbine. The blade is rotating to the left.

is large. Thus the blade 'bites' the air more in close than near the tips. These considerations result in an aeroturbine blade with an apparent twist in it. The need for twisting windmill sails was recognized hundreds of years ago and widely used on Dutch windmills, as Stokhuyzen's [49] photographs show.

Having now achieved an elementary understanding of the basics of what turns the aeroturbine, we close by observing that the airfoil orientation for an aeroturbine driven by the wind is exactly opposite to the orientation of a classical airplane propeller which is mechanically driven and whose function is to give a lift force in the axial direction. Thus, though aircraft wing theory is applicable to the operation of an aeroturbine, direct use of a classical aircraft propeller on an aeroturbine would not produce the most efficient aeroturbine because the aerodynamic surface is oriented backwards to what it should be.

WECS Efficiency is of interest to both aerogenerator designers and systems engineers. As WECS is a capital intensive technology, it is desirable for the overall wind-electric plant to have the highest efficiency possible, thus optimally utilizing capital resources and minimizing the busbar electric energy cost.

The overall conversion efficiency, η_{OV}, of an aerogenerator of the general type shown in Fig. 3-2 is:

$$\eta_{OV} = \frac{\text{Useful Output Power}}{\text{Wind Power Input}} = \eta_A \, \eta_G \, \eta_C \, \eta_{Gen} \tag{3-11}$$

where

η_A = Efficiency of the aeroturbine

η_G = Efficiency of the gearing

η_C = Efficiency of the mechanical coupling

η_{Gen} = Efficiency of the generator

We immediately recognize eq. (3-11) as an application of cascaded energy conversion treated earlier in Sec. 1-7. There we said that the overall efficiency will be strongly determined by the lowest efficiency converter in the cascade. For the aerogenerator this is the aeroturbine; the efficiency of the remaining three elements can be made quite high but less than 100 percent. It is now evident why so much emphasis is placed on the efficiency of the aeroturbine in wind literature.

Consider an arbitrary aeroturbine[1] of cross-sectional area A driven by the wind. Its efficiency would be:

$$\eta_A = \frac{\text{Useful Shaft Power Output}}{\text{Wind Power Input}} \equiv C_p = \text{Coefficient of Performance} \tag{3-12}$$

[1] Note that aeroturbine $\neq$ aerogenerator.

Thus the coefficient of performance of an aeroturbine is the fraction of power in the wind through the swept area which is converted into useful mechanical shaft power. The coefficient of performance is widely utilized throughout recent wind literature. Golding [2, Ch. 12] indicates that Betz, in his 1927 studies of the windmill applied simple momentum theory for the ship's propeller to the windmill. He summarizes Betz's assumptions and analysis which led to the conclusion that the theoretical maximum value C_p could have is 59.3 percent. He also indicates studies other investigators made trying to determine whether, in fact, this is the theoretical upper limit of efficiency. The consensus appears to be that it is.

Thus the systems engineer should recognize a priori: that the theoretical maximum upper efficiency for a wind-electric generator is 59.3 percent brought on by the aeroturbine. This theoretical efficiency limitation on a wind energy conversion system is loosely analogically similar to the thermodynamic Carnot efficiency limitation on a conventional thermal power plant. In actuality, real aeroturbines have lower efficiencies, as we shall now see.

Golding [2, p199], starting with a blade element model similar to our Fig. 3-4 except that it was planar, derived the aeroturbine efficiency to be:

$$\eta_A = \frac{1 - k\dfrac{V_T}{V}}{1 + k\dfrac{V}{V_T}} = C_p \tag{3-13}$$

where

$$k = \frac{F_D}{F_L} = \text{drag/lift ratio.}$$

V_T = wind velocity of blade element in plane of rotation due to blade turning.

V = Impinging wind velocity.

Clearly if there were no drag, i.e. $k = 0$, then the efficiency would be unity. In actuality k can be made very small, depending on the airfoil chosen and the angle of attack. Also eq. (3-13) tells us the efficiency would be low if V_T/V were very large or again if it were small. One suspicions that there exists an optimum ratio of V_T/V. We shall soon see that this is so.

Since V_T will be largest at the end of a given aeroturbine blade, then it is easy to see that V_T/V becomes the so-called 'tip speed ratio' which is widely recognized throughout the wind literature to be a basic in determining the coefficient of performance of an aeroturbine, and eq. (3-13) links tip speed ratio to C_p.

If one assembles models of various types of aeroturbine blades and puts them in a wind tunnel and runs carefully controlled experiments[1] of their efficiencies as functions of their tip speed ratios, then one obtains a family of curves similar to that shown in Fig. 3-5. We first note that there are rather large differences between the various kinds of aeroturbine blades. The Savonius rotor and the American Multi-Blade (farm windmill) are obviously intended for low speed operation whereas the modern two blade type and Darrieus rotor type are intended for high speed operation more compatible with generating electric energy. The classical Dutch blade is intermediate. We also observe that all the blade performances fall below

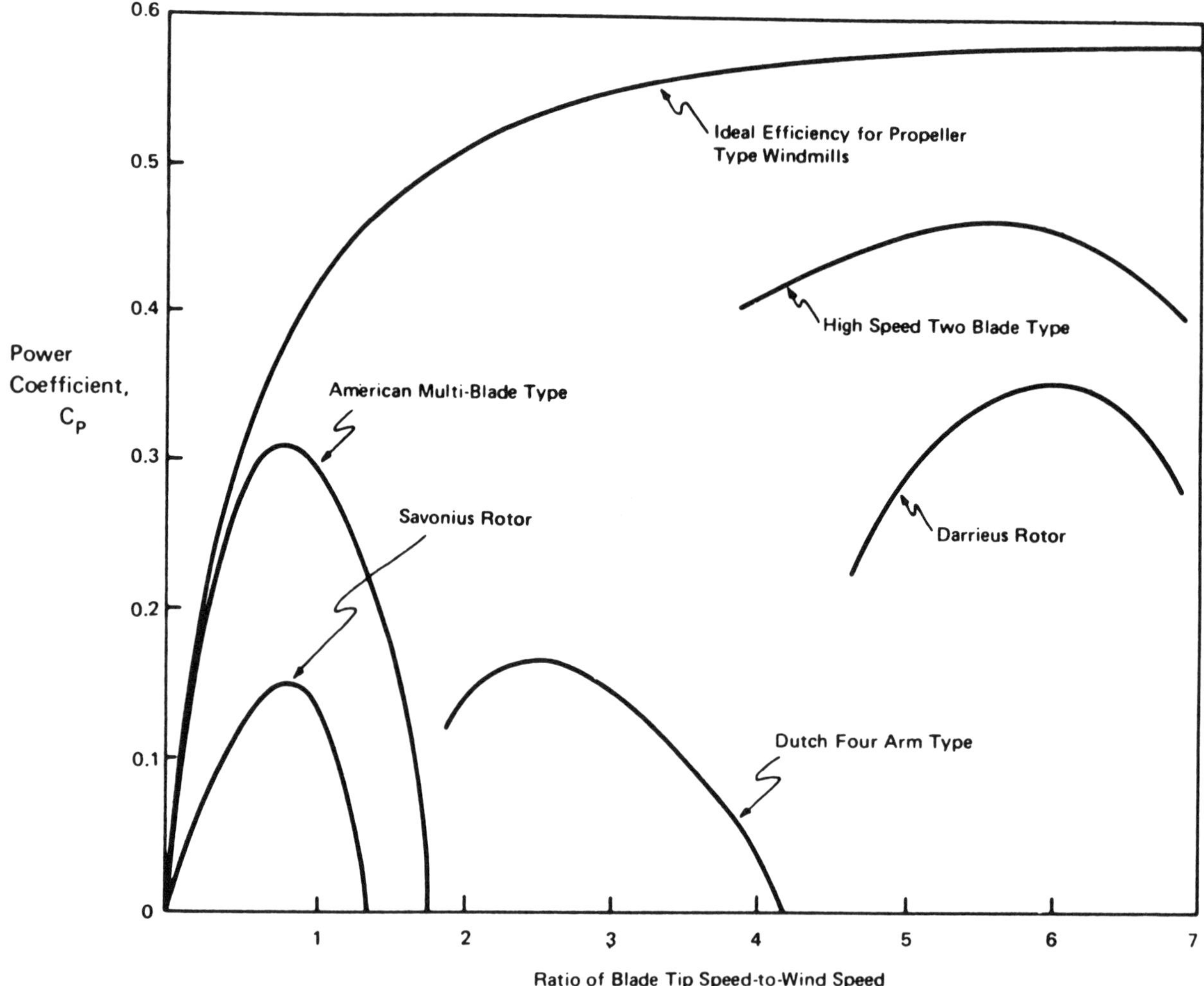

FIG. 3-5 Typical power coefficients of various kinds of wind machines. From Eldridge [3].

---

[1]See Rohrbach [46] for an example of this type research. There are other similar efforts in the wind literature.

the ideal efficiency curve which has a maximum of 59.3 percent. It is interesting to observe that the modern two blade aeroturbine has a peak power coefficient which is a large fraction of the theoretical maximum, i.e. they are well designed.

The aeroturbine designer obviously would be concerned with the optimal airfoil geometry and angle of attack and how these variables affect the performance curves. These details are beyond our scope here.

Ramakumar [35] discusses the use of power coefficient curves in connection with constant vs variable speed operation[1] of an aerogenerator. He concludes that variable speed constant frequency (VSCF) operation has the potential for higher annual energy yields per rated KW, i.e. the plant factor will be highest for this mode of operation. This thinking led him to develop his field modulated conversion system to give constant frequency output.

We close this brief treatment with the observation that the aerodynamic design of the aeroturbine is clearly critical to WECS success. It points up, once again, the multidisciplinary nature of wind energy conversion--a viewpoint familiar to systems engineers.

<u>A Typical WECS Transfer Curve</u> It is helpful to the systems engineer to have some kind of curve relating the wind speed input to the useful power output of a wind-electric machine. This is the transfer curve shown in Fig. 3-6. Putnam [5, p89] made use of such a curve as have many later researchers.

We first note the dotted available wind power curve which we showed in eq. 3-9 to increase with the cube of the wind speed averaged. In light of our treatment in the previous paragraphs about coefficient of performance it is evident that the transfer curve for an actual aerogenerator must lie below the dotted curve as shown. Note that this curve doesn't start until a certain 'cut-in speed', V_{in}, is reached. Below this speed the aeroturbine blades are stalled, and there is no lift force available to turn them. The cut-in speed is one of the critical parameters in determining the electrical energy output into the power grid.

As wind speed increases beyond V_{in}, electrical power output continues to rise rapidly. When the wind reaches V_r the generator output reaches the rated value. For wind speeds beyond this the aerogenerator controls become operative to limit the output power to a constant value. Without this control the generator output would continue to rise with wind speeds beyond V_r and the generator temperature would increase dangerously until possibly it burned up. Thus the transfer curve is flat up to wind speed V_{out}. Here the aeroturbine blades are furled, rotation stops, and the power output drops to zero. It is necessary to furl the machine at some chosen high wind speed to avoid potential irreversible structural damage to the aerogenerator.

The measured wind speed vs time curve, shown in the lower right permits

---

[1] Martinez-Sanchez also addresses this issue [47].

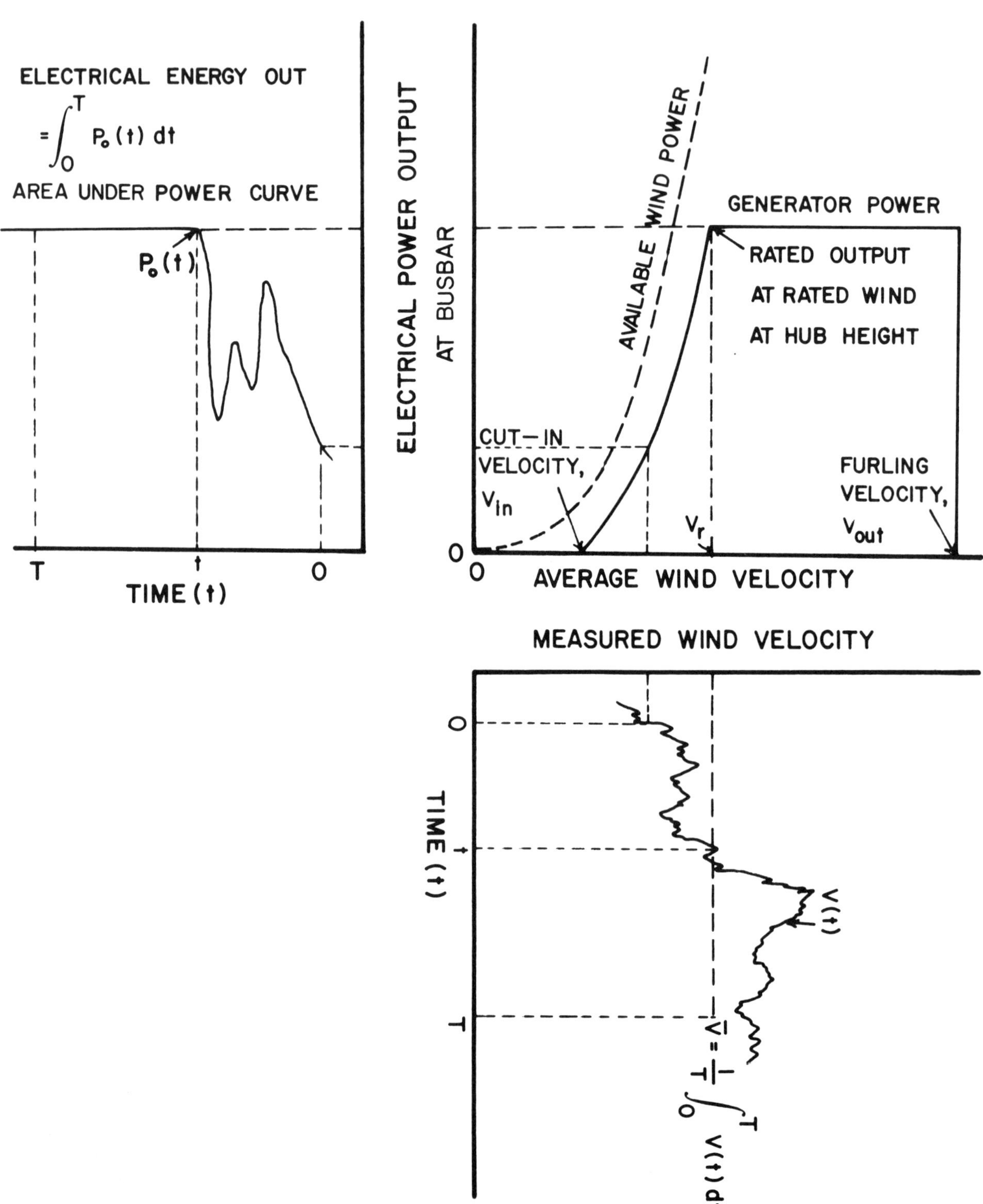

FIG. 3-6 Transfer curve (upper right) for a generalized WECS. It relates the wind velocity curve to the useful busbar power output.

us to project through the transfer curve and hence make a plot of the output power vs time from the aerogenerator.[1] It is immediately evident that when the wind average speed is less than V_r, then, because of the slightly less than cubic shape of the transfer curve in the range $V_{in} \leq V \leq V_r$, then there will be jagged peaks in the output power curve, i.e. the delivered power will be 'noisy' instead of being constant as for a classical electric generator driving a power grid. The effects of feeding this 'noisy' power into the electric grid constitutes an obvious concern for a systems engineer.

Once the average wind speed exceeds V_r, then the output electrical power of a WECS is constant as shown. During such power delivery times the aerogenerator would then be functioning like any classical generator feeding power into the electric grid.

The electrical energy out of the WECS is clearly the area under the P(t) curve, as shown in Fig. 3-6.

Thus curves of the general type shown in Fig. 3-6 permit the systems engineer to draw together the WECS characteristics and relate input to output. It should now be obvious _why_ the measured V(t) wind curve discussed in Sec. 3-4 so strongly influences the useful output power of a wind-electric machine.

We now treat the related topic of predicting the energy output of a WECS generator.

Generalized Energy Output of A WECS is a final matter here of basic concern to the systems engineer, for it is the saleable product from wind-electric conversion.

An overview of how the electric energy output calculation for a WECS can be made is shown in Fig. 3-7. The basic method appears to have been first created by Golding [2, Ch. 10], but his integration of it with the wind speed duration curve was not so clear. Ramakumar [45] combined the wind speed duration curve with the other plot we shall need, giving a clearer picture of the relationships; we use his curve in Fig. 3-7 in the following brief explanation.

First there is the actual wind speed curve V(t) shown in Fig. 3-6. It is obtained from anemometry data, say over a 1 yr. time span at the proposed site, and ideally at the proposed hub height. From it the statistical wind speed--duration curve is prepared as shown in the upper portion of Fig. 3-7. On it the various speeds are indicated, some (V_p, V_R, and V_m) depend on the

---

[1] For this process to be valid, the V(t) curve must not change too fast, as, strictly speaking, the transfer curve itself is based on the _average_ wind speed input and not the instantaneous wind speed. The aerogenerator cannot respond instantaneously to rapid changes of wind-speed, rotor inertia being one limiting factor.

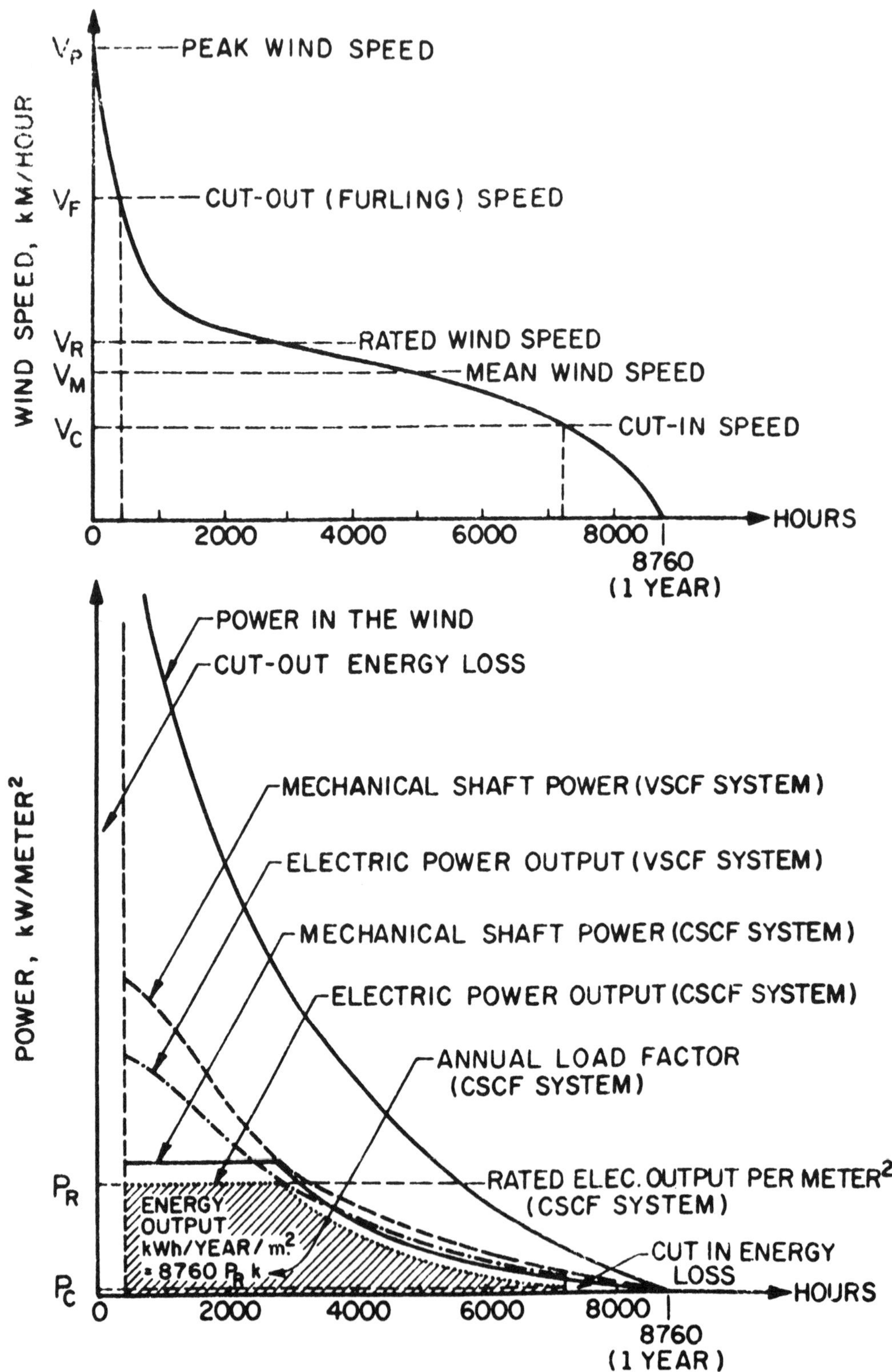

FIG. 3-7 Key generalized curves used in estimating the energy output/yr. from a wind-electric generator. From Ramakumar [45].

site's wind statistics while others (V_F and V_c) depend on the aerogenerator design.

Beneath the wind-duration curve is plotted the specific power (KW/M^2) vs time for the power in the wind. This curve might be obtained from a combination of eq. (3-10) and the wind duration curve. Then the WECS transfer curve in Fig. 3-6 and the wind-duration curve provide enough information to plot the locus of points on the upper shaded area for the aerogenerator. Note that in doing this we have implicitly included aeroturbine coefficient of performance and gearing, coupling and generator efficiencies, all of which were implied in the transfer curve of Fig. 3-6. The cut-in energy loss area is then sketched in on the right side of Fig. 3-7. Finally an area to the left is blocked out to account for cut-out loss.

The resulting shaded area will be the calculated energy output/year on a normalized basis, i.e. $KWH/yr./M^2$. To find the gross energy output/yr. of the aerogenerator we multiply this figure by the swept area of the aeroturbine.

Were the rated power, P_R, supplied constantly during the year the energy output would be 8760 P_R $KWH/yr./m^2$. Thus the WECS capacity factor is from a normalized form of eq. (3-1):

$$\text{(3-14)} \quad \text{Plant Capacity Factor} = \frac{\text{Shaded Area}}{\text{Area of } 8760 \times P_R}.$$

Golding [2, p161-166] has an excellent discussion of possible discrepancies between energy calculated in the above general way and what one would actually achieve when the WECS plant is built.

It is clear from the above procedure that the annual electrical energy output from a WECS generator is strongly dependent on both site wind statistics as well as the machine design, principally the coefficient of performance of the aeroturbine.

3-6 Prior Wind-Electric Methods

In this Section we mention in overview fashion some of the more significant prior art of possible interest to the systems engineer. The treatment is both limited and non-exhaustive.

The idea of using the wind for performing man's tasks is very old. One example is the sail whose origins are lost in antiquity;[1] yet it was largely the sail which, until comparatively recent times, permitted adventuresome men to explore the earth on great sea voyages, those of Columbus and Magellan

[1] Eldridge [3, p8] shows a 5000 year old Egyptian drawing of a Nile river craft which, he claims, is the oldest recorded depiction of the use of a sail to produce translational motion.

being classics.

The idea of a windmill appears to have originated in ancient Persia (Modern Iran). Such machines were used for lifting water and grinding grain. The machine [48] [3, p9] had a vertical axis. The sails were made of reeds, and half of them were blocked off so that the wind force was exerted on the other half yielding a net torque. A typical stone windmill structure was about 32 feet high and 20 feet square with a wood rotor diameter of about 14 feet. The windmill was fixed orientated in the direction of the prevailing wind. Merriam [31, p1] observes that many machines of this same type are still used in Iran today.

The Babylonian emperor Hammurabi, in the seventeenth century B.C. planned to use windmills for his ambitious irrigation schemes, according to Golding [2, p7].

The Chinese may have used windmills--also of the vertical axis type [2, p79]--perhaps as early as 2000 B.C. Some historians think possibly the Chinese windmill designs originated in Persia.

In the Western world the windmill came much later. Hero of Alexandria appears to have invented the horizontal axis windmill which was used to supply air to a pipe organ [2, p8]. How and when the various ideas of the horizontal axis windmill arrived in Western Europe, principally Holland, and England is obscure.

Windmills are said to have existed in Holland from about 1200 A.D. [49, p13] where they were used as water pumps for reclaiming vast areas of land from the sea. Windmills were also used for irrigation, sawing lumber, grinding corn and wheat, mechanical power to run small factories, and salt extraction and desalination of seawater. In addition to serving utilitarian purposes, the Dutch windmills have been immortalized by various painters.

The decline of Dutch windmills[1] took place in the 19th century when the steam engine came into widespread use. At that time (about 1850), according to Stokhuyzen [49, p14], there were about 9,000 windmills in Holland which, in toto, imparted a special character to the country. Windmills of this era appear to have had an individual power output of 55 Hp. (for an 86 ft. diameter rotor of four improved sails) at most. Most operated at considerably lower output power--about 2 Hp. according to Merriam [31, p3]. He also points out that the great virtue of the English windmillmill-- a variation of the Dutch design--is that they could be built even up to rather large sizes without factory technology or high performance materials.

Only in the last century have windmills[2] been used to generate electrical

---

[1]To arrest the further decline, in 1923 there was founded an Association for the Preservation of Windmills in the Netherlands. Stokhuyzen is a former chairman. His delightful and authoritative book [49] records and summarizes Dutch windmill technology and lore.

[2]The term 'windmill' which originally implied a mill for grinding grain becomes an obvious misnomer when applied to electric power generation. The term is still widely used, however. Aerogenerator avoids the difficulty.

power. Moses G. Farmer, in the late 1860's designed a magneto rotated by fans to produce electricity to light an incandescent bulb[1] [50, p97]. In 1894 the artic explorer, Fridtjof Nansen, powered an electric light bulb by windpower during his search for the North Pole. Surprisingly, the Dutch appear to have not investigated electric power from windmills until about 1948, according to Stokhuyzen's account.

Major scientific progress in windpower was made in Denmark in the 1890's when P. La Cour experimented with wind generation of electricity [51] [2, p15]. La Cour probably built the first wind tunnel in the world about 1895 at his experimental windmill. His mill produced electricity for his experiments as well as supplied electric lights and power to the township and the local high school at Askov where he was a professor.

In the following decade several hundred wind-electric generators of 5-25 KW capacity were installed according to his system. The output voltage was usually 110 or 220 volts D.C. with batteries for storage. During World War I when Denmark's crude oil supply was practically cut off Denmark relied on its wind power plants. The same situation was repeated in World War II.

In the United States in the 1920's and 1930's windpower lighted many farms. The technology grew out of the estimated 6.5 million windmills built between 1880 and 1930 to pump water for cattle and irrigation in the developing Western states.[2] The wind-electric farm unit used a simple two bladed propeller of 4-5 ft. diameter instead of a wheel. The propeller spun a generator which charged batteries to light the house, barn, and supply the radio which required several hundreds of watts in pre-transistor days. These decentralized wind-electric systems started to disappear in the late 1930's when the Roosevelt administration implemented the Rural Electrification Act (REA) which eventually brought cheap electric power to most farms [52, p13].

In 1931 the Russians built a large 100 KW wind-electric generator near Balaclava on a bluff overlooking the Black Sea [5, p105]. The propeller had two blades and was 100 ft. in diameter with a hub height[3] of 100 ft. above ground. It drove a 220 volt induction generator which was tied into a 20 MW steam plant at Sevastopol. The power output was varied by regulating the propeller pitch. Yaw control, the aiming of the rotor into the wind, was done by an electric motor mounted on a circular track. The yaw was directed by a vane on the windmill, a technique commonly used subsequently.

---

[1] This apparatus is displayed in the Smithsonian Institute.

[2] These multivane low speed machines had an enormous social and practical impact. The design still survives and is widely used in the West, Australia, and other countries. The multi-blade is not particularly suitable for electricity generation, however, because of its relatively low speed.

[3] Hub height is an important parameter in wind-electrics. The reason is that the wind velocity varies so markedly with height above ground.

Interesting features of this machine included main gears made of wood and blade skins of roofing metal. The efficiency was 24 percent at 30 RPM and a tip speed ratio[1] of 4.75. The power output was 100 KW in a 24.6 MPH wind. The machine provided 279,000 KWH of energy/yr. into the utility grid.

In 1942 a 200 KW generator was built at Gedser, Denmark,[2] but this project was overshadowed by an earlier pre World War II American project--a 1250 KW giant wind-electric generator built on Grandpa's Knob in central Vermont [5]. It had a 175 ft. diameter stainless steel two bladed propeller which was 11 feet wide and weighed over 15 tons! The hub height was 120 feet. It was by far the largest aeroturbine of any kind built anywhere up to that time. Power output was regulated by pitch control, and the turbine ran at a constant 28.7 RPM. Power output was obtained at a cut-in wind speed of 20 MPH; maximum output began at 30 MPH and was constant up to a cut-off speed of 60 MPH. The generator was a General Electric synchronous machine rated at 1250 KVA at 2400 V and 600 RPM. It supplied electric power at various times between 1941 and 1945 to the Central Vermont Public Service Co., a subsidiary of New England Public Service Co.

The project ended on March 26, 1945 when one of its 7 1/2 ton blades broke off in a 25 MPH wind. Field welding in a previous repair to the blade at a high stress point near the hub had contributed to its fatigue and failure. The major reason the project was not pursued further was economic. It was estimated to cost about $190/KW to instal such large-scale wind-electric systems in the future, but it was estimated to be worth only $125/KW to the power company. To have bridged this economic gap would have meant testing more ideas; but the project had already cost over $1.25 million dollars, so the S. Morgan Smith Co., a private firm which had entirely supported the research without government aid, decided to drop the project. Putnam, designer of the project, summed it up: "For the first time, wind had been harnessed to drive a synchronous generator feeding directly into the high-line of a utility network." Putnam and his group did an impressive piece or engineering, and those seriously interested in developing large-scale wind-electrics are well advised to study Putnam's book [5] carefully.

In the immediate post World War II years R&D on wind-electrics was at a minimum. The age of cheap fossil fuel was upon the U. S. along with the expectation of developing even cheaper limitless nuclear power. Few advocated further windpower development. Gradually it became apparent the nation was going to have to seriously examine a broad range of alternate energy sources, and in 1971 Professor Heronemus, in an Appendix to

---

[1] The importance of tip speed ratio on performance was treated in Sec. 3-5.

[2] There is a conflict in the literature about when this windmill was built. One report [121, p271] says the "Gedsermill is the last existing large windmill of the many turbines tested in Europe during the 1940's and 1950's." In 1977 a DOE-Danish 1 year program was under way to refurbish and test this wind-electric Gedersmill.

a speech by Senator Gravel [13], proposed a frontal attack on "the concept of expanded generation of electricity by windpower in Heartland America." Subsequently the 1972 NSF/NASA Solar Energy Panel [53] identified and recommended windpower as part of a worthy overall national solar energy program; later the Federal Wind Energy Program[1] became an active part of ERDA and more recently DOE [54]. Reference [54] summarizes the major federally supported wind R&D projects as of 1975, including project names, principal investigators, contract duration, and amounts.

In the mid 1970's a 100 KW wind-electric major project was built at the NASA-Lewis Plum Brook test area near Sandusky, Ohio, at a site where the yearly mean wind speed was 4.5 m/sec. (10 MPH) [55], [56], [57], [58] [59] [60] [61] [62] [63] [64] [65]. This machine was designated the Mod 0 and is shown in Fig. 3-8. According to Puthoff [66], it was largely based on Professor Hutter's design. It is clear, however, that the machine has many basic elements from the earlier Smith-Putnam machine and the even earlier Russian Balaclava unit. This important wind-electric machine manifested the nation's renewed interest in wind-electrics.

The aerogenerator had a 37.5 m (125 ft.) diameter twin propeller type blade of aluminum. It was horizontal axis and mounted on a 30 m (100 ft.) tower. The aeroturbine had a 7 degree coning angle and a coefficient of performance of about 0.4 at its peak. The cut-in speed was 8 MPH--much lower than previous large units, and full output of 100 KW was obtained from 18 MPH to 40 MPH wind speed with the rotor turning at a constant 40 RPM. Blade pitch control regulated the output, permitting spilling the excess wind in the 18-40 MPH range. The low cut-in speed allowed generation of an estimated 180 MWH/yr. which was as much energy as the much larger Smith-Putnam Grandpa's Knob machine generated. The NASA unit was also about 30 percent smaller. Unlike the earlier Smith-Putnam machine using stainless steel, the NASA design utilized aluminum blades. It also feathered the blades in high winds.

The synchronous alternator was an 1800 RPM self cooled type with brushless exciter. It was rated at 125 KVA, 0.8 power factor, 480 volts. It was a three-phase, 60 hertz, Y connected machine [66].

Leaving the NASA machine now, another signficant development in the prior art of interest to electrical systems engineers has been the creation of field modulated generator systems by Professor Ramakumar [67] [45] [35].

The National Research Council (NRC) of Canada was responsible for reviving the Darrieus vertical axis wind-electric machine [68]. This interesting 'eggbeater' looking machine has many distinct advantages, one being it produces power from wind of any direction. It does not need yaw control. We shall discuss this machine in Sec. 3-7.

---

[1] Federal R&D funding for wind energy went from $300,000 in fiscal year 1973 to about $12,000,000 in FY 76 [54, p2].

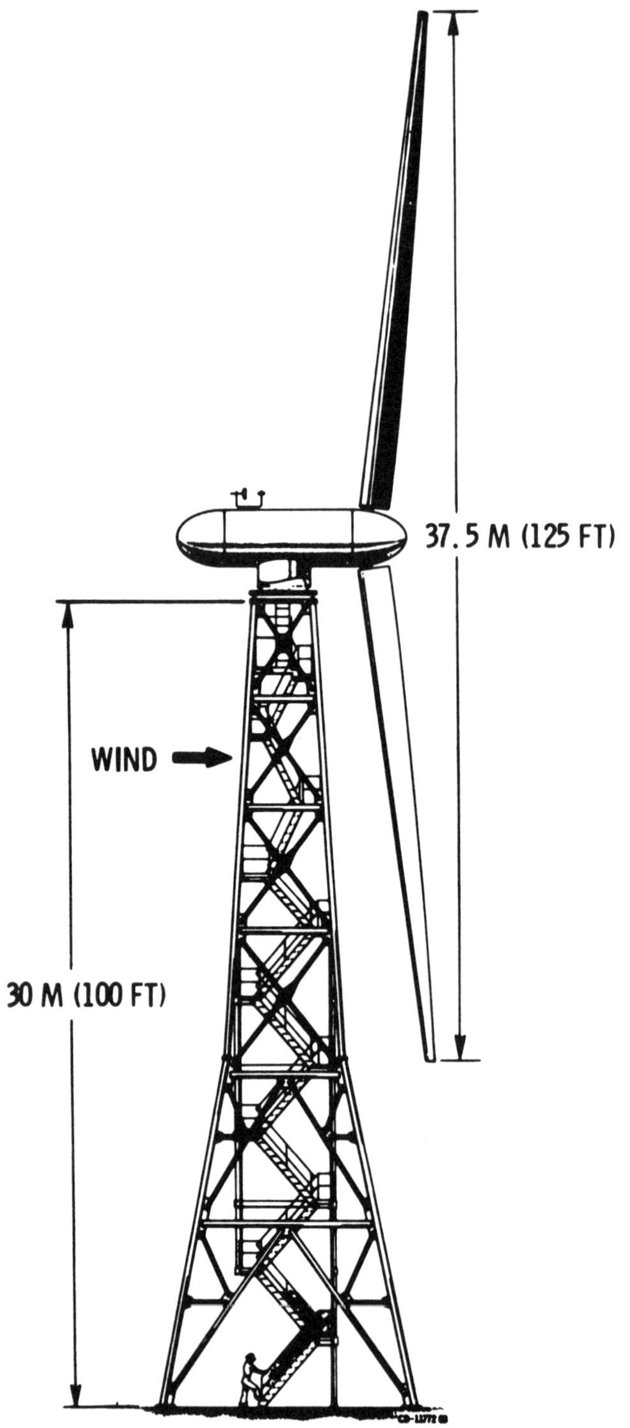

FIG. 3-8 100 KW Mod 0 experimental wind turbine generator built near Sandusky, Ohio in the mid 1970's. A later modification eliminated the stairway to minimize rotor and tower vibrations and stresses. From Puthoff [66].

In the late 1970's wind-electric R&D spans small, intermediate, and large-scale machines. One analysis [69, p32], however, shows the principal application to be in large machines, generally in the range of 1 MWe, for use by electric utilities in a fuel-saver mode. In spite of this, but with the exception of Southern California Edison which is reported [70, p60] about to purchase a 2,700 kilowatt wind-electric generator, there appears in the mid 1970's to be relatively little R&D activity by utilities in wind-electrics. Non-utility interest in wind-electric conversion systems is both widespread and growing in 1978; but we should recognize this interest has waxed and waned over a long time period.

Other large wind-electric generators have been built and tested in England, France, Germany, Italy, Russia, Denmark and other countries [3].

Unfortunately this all too brief overview of wind-electric prior art has had to omit the pioneering contributions of Smeaton, Betz, Savonius, and many others--some in recent times--whose work will be found in the literature. The prior art simply contains far more than can even be mentioned here. Golding [2, Ch. 2] has an excellent more detailed treatment of the early history of windmills. A detailed bibliography is in reference [119].

Numerous contributions, on a smaller scale, have also been made by "the non-credentialed, highly motivated and talented individuals and groups operating outside the mainstream" [31, p9], according to Merriam. They [71] are tangible reminders that, as Hayes [1, p3] says, "the 'think big' approach does not necessarily make sense."

3-7 Present Art

Here the term 'present art' implies that which existed roughly as of the 1975-1977 period. It cannot be precisely defined; rather it is more like a continuum from the prior art. Naturally this assessment of what constitutes the present art was based on the 1978 information available to this researcher which is nonexhaustive.

An examination of the rather large published wind energy literature clearly reveals there is much current R&D activity in the wind-electric area; further this total activity appears to be growing. Collectively it manifests a significant national interest in seriously investigating the viableness of wind-electric energy systems.

We can only highlight the present art here, touching those WECS machines believed to have the highest likelihood of impacting the systems engineer thinking about integrating WECS technology into the utility grid.

The systems engineer seeking perspectives on the present art of the major advanced and innovative WECS design concepts presently being studied would do well to study Weisbrick's paper [72]. He made an excellent summary of each type of WECS currently being researched, showed a sketch, the principal investigator and firm, principle of operation, the advantages claimed or anticipated for each concept, and the areas of concern. He treated present activity on:

- Darrieus Vertical Axis WECS
- Diffuser Augmented Wind Turbine
- Tornado Type Wind Energy System
- Windmill With Tip Vanes
- Wind Tip Concentrator
- Vortex Augmentor Concept
- Circulation Controlled Panemone
- Cylindrical Darrieus rotor
- Variable Geometry Vertical Axis Windmill
- Giromill
- Precessor System
- Tracked Vertical Airfoil
- Cylindrical Obstruction Type Concentrator
- Torroidal Accelerator for Rotor Platform

All these systems are potentially applicable to the generation of electricity, though it is much too early to identify the best candidates. It is very clear that the wind field does not suffer from a lack of creative ideas.

Large-Scale Horizontal Axis WECS In 1972 Heronemus [73] proposed the erection of large towers in the range of 340-700 ft. high off the New England coast, with each tower containing multiple aerogenerators arranged in array fashion. He argued that such wind-electric generators could make significant electric energy contributions to the historically fuel-short New England. Recent information on this project was not available.

An extension of the NASA Plum Brook work has been the recent erection and test of a 200 KW Mod OA aerogenerator at Clayton, N.M. where the average wind speed is about 17 MPH [74] [75]. This machine dedicated January 28, 1978 has a 125 ft. diameter rotor and generates 200 KW when operating at or above its rated wind speed of 19 MPH. The cut-in wind speed is 18 MPH, and the furling speed is 40 MPH. The rotor speed is constant at 40 RPM which is geared up to 1800 RPM for the generator. This aerogenerator is integrated with Clayton's electric energy system and supplies enough power for about 60 homes--about 3 percent of Clayton's total power needs.

Two additional 200 KW aerogenerators are being built under DOE sponsorship to be installed at Culebra Island, Puerto Rico and Block Island in Rhode Island [76].

A much larger machine, evidently building on the earlier NASA/DOE R&D, will be installed in late 1978 near Boone, North Carolina where the average wind speed is about 20 MPH. Its 200 foot fiberglass composite rotor is expected to supply 2 MW of power in a 24 MPH wind--enough for about 500 homes [75] [77] [78]. Its output will be integrated with the Blue Ridge Electrical Membership Corp., a northwest North Carolina cooperative. The rotor will turn at 40 RPM which will drive the alternator at 1800 RPM. It will be mounted on a 150 foot tower. This machine is designated Mod-1 by Thomas [79].

Planning is also presently underway by NASA and DOE for a 300 ft. diameter aerogenerator with an output of 1-2 MW for sites with a median wind speed near 12 MPH. It is designated Mod-2 [79]. The site for this machine is unclear from the literature surveyed.

An integral part of the present art is the ongoing research on the aerodynamics, modeling and design of the critical rotor blades for wind turbines.

The works of Wilson [80], Weber [81], Jordan [82] are samples of this kind of research. Other aerodynamic research is undoubtedly under way.

While our principal interest here is within the U. S., it is of interest to note that, according to one recent report [83], the Dutch are "spending $3.5 million on a five-year research program to decide whether to give windmills another whirl. Five thousand new windmills are needed to supply 10 to 20 percent of the nation's electricity, experts estimate. They plan mills made of metal and heavy plastic with sails 150 feet in diameter."

The Danes are also constructing a 2 MW wind-electric machine in Tvind [84] [85] claimed to be the largest windmill standing in the world in 1978.[1]

Vertical Axis WECS: The Darrieus Machine Early work in the United States on this interesting machine[2] was done by NASA Langley Research Center where experiments were made on small machines--about 15 ft. high--having an output of 1.3 (Hp) in a 15 MPH wind [86] [87]. Work on the Darrieus machine has also been done in Germany [88] and Canada [89].

Substantial R&D work on the Darrieus machine has been done by Sandia Laboratories since about 1973. This work is still in progress there. We cite a few highlights here from that experimental test work.

Fig. 3-9 shows Sandia's seven story high machine composed of three 'egg-beater' blades, the center sections of which have aerodynamic cross sections. The blades have an approximate troposkien[3] shape--the shape a rope would take if held at top and bottom and rotated about the vertical axis. It is approximately, though not exactly, a parabola [90]. The turbine swept area is about 180 m^2 (1937 $ft.^2$) [91]. The blades drive into the wind and accept wind from any direction. Earlier models built by both Langley Research Center [87] and Sandia Laboratories [92] had only two blades. The three blade version gives larger output power and appears to perform best. The turbine attains its rated output of 60 kilowatts at a wind speed of 32 MPH. At cut-out wind speeds over 26.8 m/s (60 MPH), the turbine is parked.

The transmission, starter, and generator are at the bottom. Operation of the turbine begins when the starter motor is activated at its minimum cut-in wind speed of 4.5 m/s (10 MPH). The starter motor then disengages and the turbine is under wind power. The turbine speed is about 30 to 50 RPM. A fixed-ratio transmission translates the turbine output to an 1800 RPM (for 42 RPM of turbine) generator speed. The generator is of the brushless synchronous type, 60 KW, 3-phase, 60 Hz, and 277/480 volts [91].

---

[1] This magnificent machine cost only $650,000 dollars and was provided privately by school teachers at Tvind. It was "conceived, organized, and mainly built by amateurs" [85]. It is the first large aerogenerator to operate at variable speed.

[2] It is based on the 1920's work of the French inventor, Darrieus.

[3] Troposkien is Greek for 'turning rope'.

Fig. 3-9 60 kilowatt Darrieus vertical axis wind turbine (VAWT) developed by Sandia Laboratories in mid 1970's. It is the largest of its type in the United States. Photo courtesy Sandia Laboratories.

Output power may be applied to a commercial power grid or directed to a test load.

The principal advantages of the vertical axis wind turbine (VAWT) over the conventional horizontal-axis wind turbines are cited in Sandia literature as:

- Cost which may be significantly lower than for standard wind turbines.
- Simplicity of the structure, hence ease of manufacture [93].
- Yaw And Pitch Controls not needed to bring it into the wind or operate in high winds. A constant speed VAWT automatically stalls at high wind speeds.
- Ground Level Mounting for the generator and gearing permits easy access and maintenance, and reduces tower costs.
- Overall Weight of the turbine may be substantially less than that of conventional systems. This is because of the small amount of material [94] involved in relation to the swept area which translates to lower cost.

The aerodynamic modelling of this machine has been dealt with by Strickland [95] and Templin [89]. Wind tunnel tests have been reported by Blackwell [96].

An early performance evaluation of the Darrieus machine, of possible interest to systems engineers, has been done by Banas [97].

It will be interesting to watch the future progress of the Darrieus machine which at this time appears to offer much promise for low cost small and intermediate power output wind-electric generation.

3-8 System Integration of Wind-Electrics

The Meaning of Integration We now leave the broader aspects of WECS outlined in the previous sections and focus on the issue of integration, as we did in Ch. 2. There are two facets of integration:

- Electrical Integration of the WECS with a nearby electrical power grid. The emphasis here is principally technical in nature.
- Technology Integration wherein the problems would be dealt with of introducing WECS generators, on a growing scale to the electrical energy system. This facet of integration would be predicated on the a priori assumption that overall technical and economic realisibility of WECS technology had been shown and that the principal problem would be to construct more such WECS generators. This facet of integration would deal with sites, schedules, money, socio-economic, legal, organizational, and institutional issues, construction and related topics--all aimed at arriving at the state of an electric

system driven by both conventional fuels and wind inputs with an appreciable fraction of the total power supplied being from wind energy. Thus the system would be physically 'integrated' with WECS.

The principal emphasis in this Sec. will be on electrical integration. The overall WECS technology appears a few years away from the second kind of integration, though hopefully eventually it will arrive at that state.

We further limit our consideration to that of integrating a single WECS with the electric grid. The problems of multiple WECS, i.e. systems of aerogenerators, feeding the power grid will be treated in Chapter 5.

We also do not include here the possible use of WECS being used as mechanical water pumps for filling a pumped storage reservoir where the water would later be used to generate electricity via conventional hydroelectric means. Obviously WECS used in this form would be a type of integration with the larger generating system as shown by Eldridge [3]. Such WECS applications would, on the surface, appear to have merit. Rather we here adopt the view of Smith [98] and his utility colleagues who advance the philosophy that "the preferred method from all points of veiw is that the wind plants would produce electric power into the network, and the storage facilities would draw electric power from the network as needed. Thus there would be no direct connection between the storage units and the wind plant; such a storage facility would stand as an independent operating entity, and the wind plants, if implemented, would be physically placed as determined by other factors." Mulcahy [99] and his utility working group succinctly sums up the matter:

Q. "Will energy storage be required?"

A. For utility applications:

No! a. if used to save fuel, or
b. if have hydroelectric with adequate pondage, or
c. if have adequate diversity of wind energy available, i.e. utility service area is such that minimum planned wind energy is always available at some locations in same.

Yes! a. in general

No! a. if an interruptible supply is acceptable to the user, e.g. irrigation.

Storage adds flexibility to use of WECS in that it permits peak shaving and capacity saving as well as fuel saving. However, it can be expensive, and each utility will have to evaluate whether or not to instal it."

Sorensen [100] has analyzed the need for storage in combination with WECS as a means of smoothing and finds that its addition as short-term storage (10-20 hrs.) makes the wind energy system as dependable as a large nuclear power plant.

Thus while some may see the topic of energy storage for WECS as an integral part of the meaning of 'integration,' in this volume we treat it as a separate topic, independent of WECS, in Chap. 4.

Having largely said what 'integration' is not for our present purposes, we now, in a general way indicate what its meaning is here. 'Integration', in its broadest sense, implies an interface between WECS and the utilities. There are, as Mulcahy and his utility working group found [99] "many aspects indeed to the interface area between WECS and utilities and other users. Also, this is an important area---."

Swanson [101] and his utility colleagues summed up the systems view implicit in the term 'integration': "We believe the best opportunity for large-scale, significant use of windpower, is in conjunction with existing electric utility networks. They have the capital resources, the diversity, and the existing capability for operation and maintenance of large wind energy conversion systems, and it is only through, large systems that significant impact on the energy economy can be made."

At no place is the term 'integration' so apparent as when a circuit breaker is closed connecting a wind-electric generator to the utility power grid. There are unique system technical problems involved in this copper-copper interconnection or 'integration', and we now deal with some of these after having limited the scope of 'integration' to this area.

Some Fundamental System Requirements On Integrating WECS The system engineering requirements for integrating with the power grid have been succinctly laid out by Mulcahy [99]. "WECS must be designed and engineered:

- To provide acceptable power to the utility grid.
- With associated equipment which is capable of providing full protection for both the WECS and the grid.

The voltage, frequency, and phase of the WECS output should meet the utility's requirements and, in addition, the necessary control equipment should be provided to ensure cut-in and cut-out of the WECS at the appropriate wind speeds. Included under protection would be lightning protection, isolating switches to provide isolation of the WECS or the line, as required, and reverse power relays to prevent the feeding of power from the line to the WECS."

We now amplify Mulcahy's concerns.

- The Constant Frequency requirement of the electrical output of WECS is an obvious one to integrate successfully with the 60 Hertz power grid in the United States. For large WECS this requirement translates to the necessity of turning the synchronous generator[1] at exactly a constant speed, usually 1800 RPM. This is a difficult system constraint in the design of a large-scale WECS because of the inherent non-constant nature of the wind speed. Thus the WECS systems designer is faced with designing an energy conversion system

[1] Virtually all the large-scale U.S. designed WECS to date have synchronous generators.

having a statistically variable input energy, the wind, which he must in some way translate to an absolutely constant shaft speed for an alternator.

The above difficult system requirements have led to the use of blade pitch control systems on operable WECS along with constant torque coupling between gearing and generator.

It is possible to somewhat circumvent the constant speed requirement by utilizing a variable speed rotor, and hence variable generator speed. These were classified in Sec. 3-5. Ramakumar [45] has shown (see our Fig. 3-7) that the variable speed system results in the largest annual generated electrical energy. The Tvind (Denmark) [85] aerogenerator also uses a variable speed generator, converting the variable frequency AC first to DC and then using modern solid state devices to invert to the constant frequency output required by the grid; but there are added integration complications of harmonics and concerns over output waveforms invariably associated with inverters. Thus the large U.S. designed and built WEC machines, to date have not gone the variable speed route.

- Importance Of The Alternator in successful large-scale WECS systems to-date is underscored as the principal reliable efficient means of converting shaft power to the desired electrical power output.

  The synchronous alternator has a long successful history of electric energy generation by the utilities. The machine has been extensively studied for both steady state and transient cases, and it is expected much of this knowledge from conventional generator design and systems integration would be applicable to WECS. The point is that this important component of a WECS is a highly developed well understood technology in 1978, and it is directly applicable, with some modifications, to WECS technology.

  It appears highly probable that alternators will be used for a long time--in some form or other--on future large-scale WECS.

  Electrical systems engineers with utility companies are well aware of the many system complexities of both practical and theoretical natures one encounters in using alternators in a dynamic power system.

- The Basic Circuit Model And Phasor diagram of a WECS synchronous generator feeding power into a grid is shown in Fig. 3-10. Systems engineers immediately recognize the model and phasor analysis method as being nearly the same as for a classical electrical power system.

  It is clear from Fig. 3-10 that if Eg and δ statistically vary in the time domain, as they do under dynamic conditions in a WECS with variable wind, then P_o, the power delivered into the

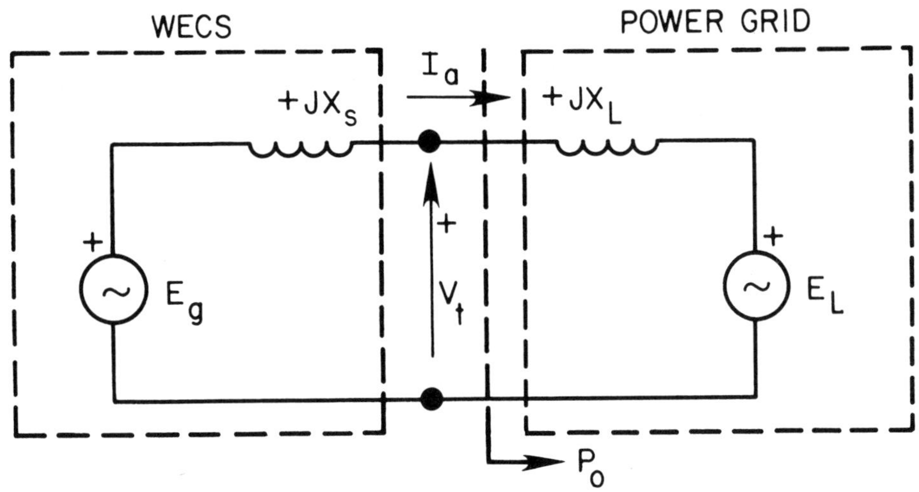

$$P_o = \frac{E_g V_t}{X_s} \sin \delta$$

δ = Power angle

ϕ = Power factor angle

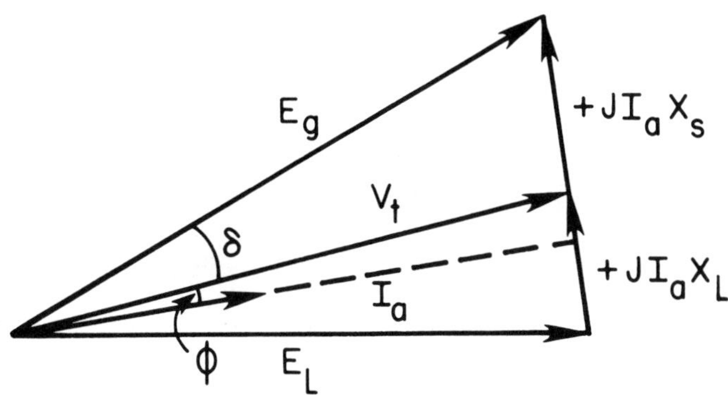

FIG. 3-10 Electric equivalent circuit model for one phase of an alternator, assuming negligible armature resistance. The phasor model relates the key parameters. The power factor is slightly lagging for the construction shown. Eg and δ are variables for a WECS.

grid will also statistically vary; this is the reason why the measured $P_o(t)$ curves on wind-electric generators tested to date at below rated wind and power yield 'noisy' power output curves as indicated in Fig. 3-6. The theoretical analysis, from the systems standpoint, of WECS in this area appears hardly explored. The use of perturbation techniques is an obvious starting point for such further systems work.

Steps To Integration of large-scale WECS are of concern to the systems engineer. We assume here that a single WECS has been designed and successfully built and that it is suitably instrumented. It is desired to synchronously integrate it with the power grid. The wind literature examined, except for the older works of Putnam [5, Ch. 8] and Golding [2, Ch. 11], is nearly silent on this subject. We draw heavily from their wisdom here which is founded on their successful experiences with early large WECS.

The integration of WECS into the electric grid involves two major steps:

- Preliminary Checkout of the machine. Its output is not connected to the grid.
- Synchronous Power Generation with the machine connected to the grid, delivering electric power.

We summarize and integrate a few of the kinds of things involved in both the above steps, according to Putnam and Golding. Additional details will be found in their respective books.

Preliminary Checkout

1. Gradually bring the WECS rotor up to speed in small increments. No load is on the generator output.
2. Observe the general operation of the machine. Is it smooth? Is there excessive vibration[1] in the tower or flutter in the blades? Are any resonance effects noted? At what speeds? Are the blade stresses excessive, particularly near the hub?
3. Does the rotor cut-in at the correct wind speed?
4. Does the rotor pitch control work properly and smoothly?
5. Ditto for the yaw control and mechanisms.
6. Do the oil seals leak?
7. Do the bearings[2] heat up? Make noise?

---

[1] Putnam's machine experienced a great deal of vibration in both horizontal and vertical planes, in addition to pitching, and yawing. Substantial design modifications to the machine were made to minimize these.

[2] Putnam also had much bearing trouble--hot and cracked.

8. Do hydraulic systems work properly, especially on the rotor pitch control?
9. Is the gearing and coupling working smoothly? Any overheating? Additional cooling needed?
10. Do all safety devices work properly, particularly the rotor overspeed protective device? Does the parking brake to stop the rotor work correctly?
11. Does the furling mechanism and control operate correctly?
12. Run it for awhile with no load. Observe astutely. Shut the machine down and carefully examine for any possible flaws or malfunctions.
13. Install an isolated resistive load and power measuring instruments to the alternator output, preparatory to making loaded tests.
14. Restart the aerogenerator and over a suitable time period to obtain wind of many speeds measure and plot the power output vs wind speed transfer curve similar to that shown in Fig. 3-6. Golding has much helpful advice on how to make this measurement.
15. Simultaneously with (14) and particularly near or at rated power output check for any bearings, gears, couplings[1] or generator overheating.
16. Determine and plot the overall WECS efficiency as a function of wind speed. It should peak up near the rated power for a well designed WECS, falling to either side.
17. Perhaps a suitable strip chart recording may be made to determine the noiseness of the load power as a function of time. It will give some indication of localized wind conditions, and give a preliminary assessment of what the machine would probably deliver into the power grid.
18. Check the generator output voltage, waveform and frequency. Verify that it is of a suitable quality to feed into the power grid.
19. Perform any other tests or modifications to the WECS to insure its probable capability of performing in an integrated fashion with the grid.

Synchronous Power Generation Assuming all the previous steps and tests have been performed with favorable results, the WECS is now ready to be integrated with the power grid.

1. Be sure a suitable wind is blowing, ideally steady, for the first integrated tests.
2. Bring the rotor up to rated speed. The generator RPM should be within a few RPM of the rated speed and above it for synchronous output.
3. With a suitable synchronizer between generator and grid determine relative phasing between WECS alternator output and grid voltage.
4. Adjust blade pitch control or torque control until phasing is within desired range of synchronization.

[1] Putnam experienced hydraulic coupling overheating after a few hours operation.

5. Check generator and grid voltages for compatibility. Adjust generator excitation as necessary.
6. Double check instruments and general WECS operation before closing circuit breaker.
7. Close output circuit breaker. The WECS should now be operating synchronously with the utility grid and delivering power to it.
8. Observe and listen to the WECS total machine for any sign of malfunction.
9. Continue observations and electrical tests, particularly measuring P(t) curve under different wind conditions.

How far the loaded tests should go, will of course be a function of the particular WECS and utility system into which one is integrating along with what one's objectives are. Since relatively few large-scale WECS to date have been fed into the power grid, caution and over-instrumenting appear in order until more operating experience is built up with WEC-utility systems operating as an integral energy system. Golding suggests that one additional test may be needed to determine how closely the actual electrical energy output corresponds to that for which the machine was designed. This is not a simple measurement to make, however. He gives some suggestions for how to do it.

Integration: Transient and Steady State Preliminary Analysis The previous paragraphs have more or less implied that the WECS output would be in an approximate steady state condition when integrated with the grid. It presumes everything is set up exactly correct to ensure synchronization and integration; but real-world WECS systems are seldom so ideal, and one quickly gets into questions of how close does the generator speed have to be to rated to insure synchronization? And suppose one closes the breaker at not quite the optimal time; will there be a serious transient? How long will it last? Will the resulting system be stable or unstable?

Several hundreds of papers on WECS were diligently searched for answers to the above kinds of integration questions. The fundamental electromechanical system of a WECS feeding an infinite power grid under dynamic conditons appears to have only[1] been recently examined in a significant paper by Hwang and Gilbert [102] who claim to be the first to have studied this important engineering problem. Their claim is confirmed by this investigator. With a digital computer they obtained complete solutions for rotor speed, generator power angle, electromagnetic torque, wind turbine torque, wind turbine blade pitch angle and armature current--all for the NASA Mod 0 wind turbine. Their model permitted simulating the effect of closing the output breaker when WECS conditions were near to but not ideal for synchronization; they could observe the resulting transients and determine whether the system was stable and the time required to reach stability.

The WECS-bus model Hwang and Gilbert used is reproduced in Fig. 3-11.

---

[1] A paper by Lotker [120] indicates that a group of Northeast utilities performed a computer study to analyze the dynamic interaction between the generator and network, but no details were given and none could be found.

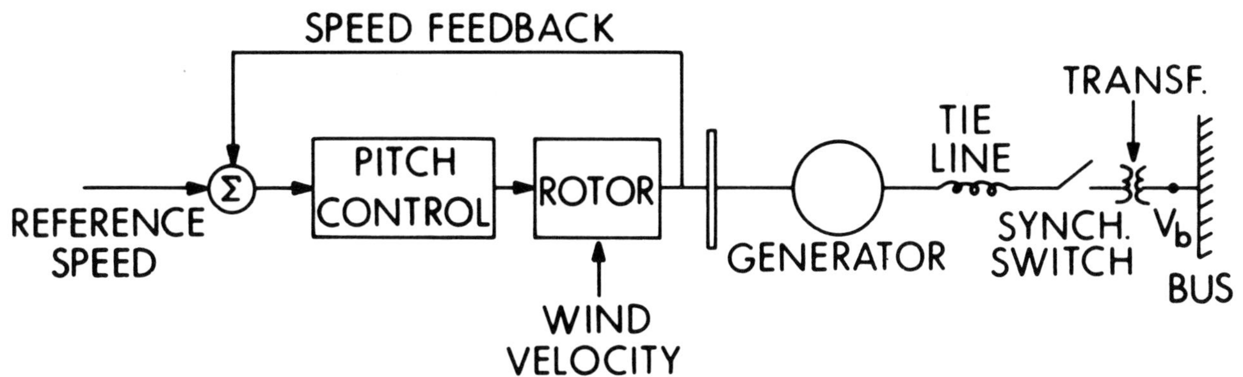

FIG. 3-11 Schematic of the WECS-bus model analyzed by Hwang and Gilbert [102] for the NASA Mod-0 100 KW aerogenerator.

Their principal results are in graphical form and are reproduced in Fig. 3-12. They considered the two cases of (1) where the synchronizing switch is closed when the generator is 7.5 RPM above the rated 1800 RPM and (2) when only 2.5 RPM above the rated.

It is clear from their results that all the curves are damped oscillations and that the critical rotor speed and generator power angle do eventually reach steady-states. There are some rather large transients, however--a not unexpected result in such a system. One therefore concludes the system will be stable, and they state that their "experimental results matched the computer study very closely and confirmed that the synchronization can be accomplished by means of the existing speed control system and an automatic synchronizer."

They recognized that their simulation model could be expanded to include the effects of strong wind gusts. We can clearly recognize that much theoretical work of a systems nature needs to yet be applied on this general integration problem. Hwang and Gilbert appear to have taken the first important step. It is somewhat surprising that this problem so fundamental to integration of WECS has not been addressed by others before now.

Integration: Nonsynchronous AC/DC/AC Case The idea of letting the alternator in a WECS operate at a variable speed instead of a constant speed has already been mentioned, its principal advantage being the production of a larger annual electrical energy. The variable speed alternator naturally would supply a variable frequency output--unsuitable for directly feeding into the constant frequency power grid. This frequency mismatch between WECS alternator output and utility grid can be overcome, with some power loss penalty, by rectifying the alternator's nonsynchronous AC output to DC and then using appropriate

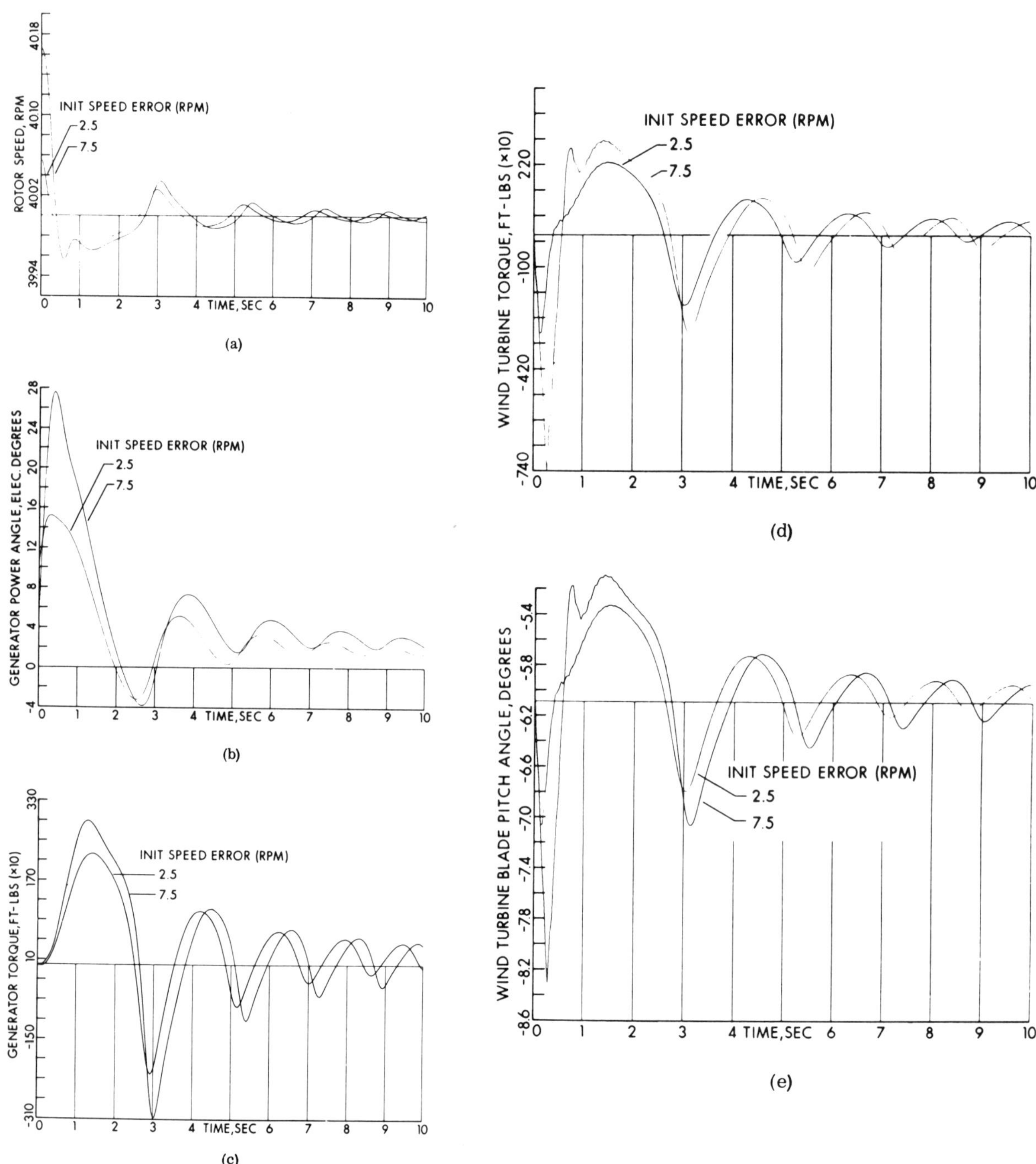

FIG. 3-12 Results of computer studies by Hwang and Gilbert [102] on the WECS-bus model of Fig. 3-11.

solid state switching of the DC to convert it into constant frequency AC.

The above general approach has been taken by Reitan [103] who in 1975 was investigating the possibilities of creating a 20 KW AC/DC/AC three-phase to single-phase link. The schematic diagram of the system he was proposing to test is shown in Fig. 3-13. He was hopeful of showing this general type of converter might be applicable up to several thousands of megawatts. He points out that one of the advantages of this system is that the AC/DC/AC link has over 10 years of development up to the 2000 MW, 800 KV level. He anticipated that the overall efficiency from alternator to utility grid would be 70 percent or better. He was estimating a cost of approximately $300/KWe for just the electric link in 20-100 KW sizes. Recent progress on this development is unknown.

More recently and at a much larger power, the Danes have utilized the variable frequency idea on their 2 MW wind-electric generator now built at Tvind. It is the world's largest WECS in 1978 [85]. Their 3300 volt variable frequency alternator output is transformed to 440 volts, rectified to DC, and inverted to 50 Hertz AC where it is synchronously phased with the local power grid. Their experience would seem to prove that the variable frequency AC/DC/AC method of integration can be made to work with the largest size WECS known to date, though technical details are lacking.

The principal disadvantages to this method of integration appear to be:

- Losses in the conversion circuits. Two energy loss conversions have to be made. Whether these energy losses exceed the energy gain by going to the variable speed case in the first place appears unanswered in the literature.
- Harmonics in the current and voltage delivered to the line appear inherent in this method, brought on chiefly by the last conversion from DC to AC. Reitan indicates that for the 3-phase case 5th, 7th, 11th, 13th, etc. harmonics are present and that "large, costly, but appropriate harmonic filters on the utility 3ϕ bus are a part of the normal link."

Neither of these disadvantages are encountered with the constant speed alternator approach which may account for its use in the large power U.S.-built WECS to date.

Small-Scale Integration Most of the small-scale wind-electric converters are constant blade pitch variable frequency output machines if they are AC. Many of these machines are simply DC generators. The principal integration methods for the small-scale class of WECS appear to be:

- Inverters to convert the DC WECS generator output to constant frequency AC output suitable for the utility grid. Meyer [104] and Lindsley [105] have described such a commercially available unit suitable for up to 8 KW and being, they claim, 98 percent efficient. They give no circuit details, however.

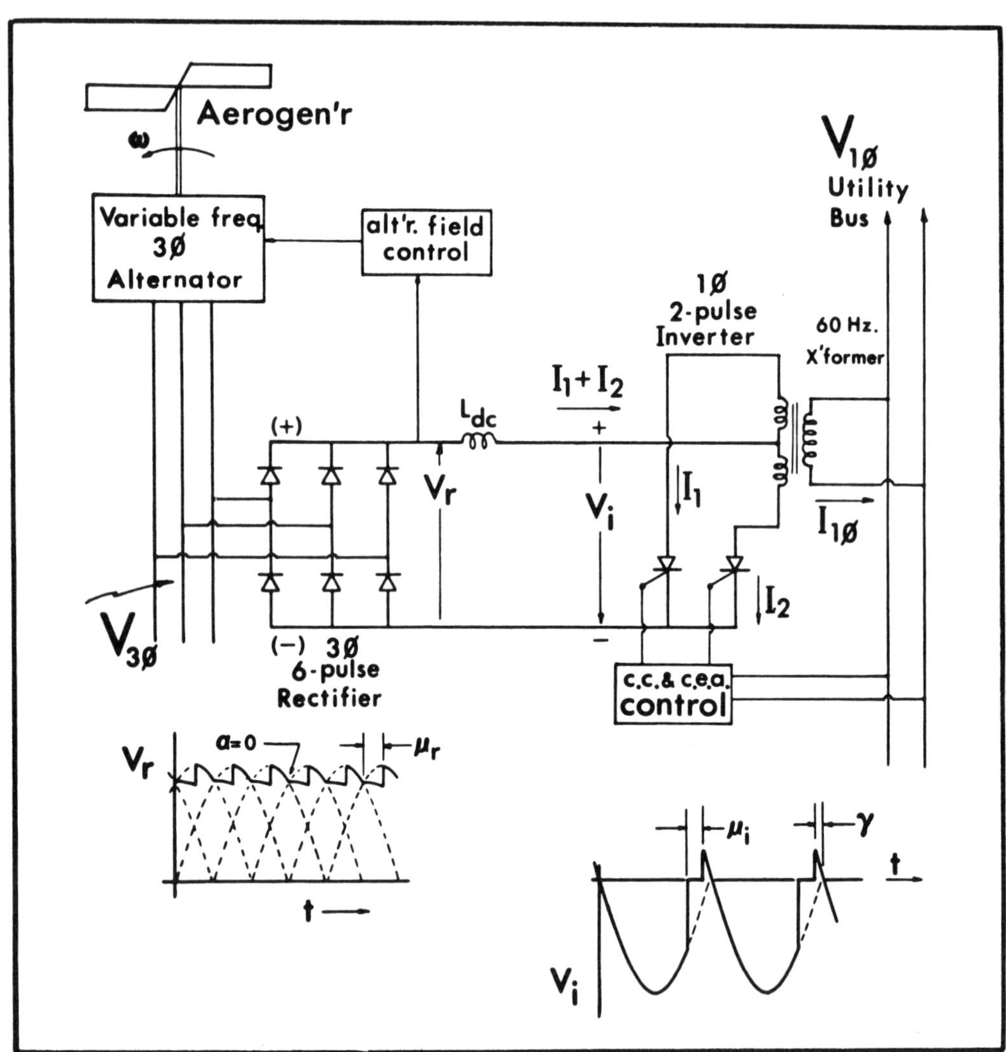

FIG. 3-13 Basic diagram of Reitan's [103] proposed three-phase to single-phase AC/DC/AC electric link for wind-electric generator use.

Inverters in the low power range are known to be a well developed technology in modern electronics, and this area of integration was not explored further in this research.

- AC-DC-AC link along the lines of the previous paragraphs.
- Field Modulated generators after Ramakumar's ideas [67] [35].

Undoubtedly the technical details of the above methods of achieving integration will have to be explored further if small-scale wind-electrics technology grows. This field presently appears dominated by the nation's resourceful 'do-it-yourselfers' with inputs from an all too few electrical engineers. Unfortunately, because of time limitations the potential impact of a few large-scale WECS or many small-scale WECS on a local utility's load vs time curve could not be explored in this research. There appears relatively little on this issue in the literature examined, however.

3-9 Economic Aspects of WECS

The subject of economics of WECS is an important topic, for, in the final analysis, it is here that WECS will be tried as a competing energy source with conventional electric plants.

Unfortunately, the scope and timing of this research did not permit an in-depth examination of the economics of wind-electrics; but during the course of this research, literature dealing with the cost issue was uncovered. Therefore what is presented briefly here should be viewed as only roughly mapping out this subject so that those interested in pursuing the economic aspects of WECS in more depth can quickly get a start.

First, it is clear after examining the wind literature that the subject of costs is a very complex one and far too detailed to include in any depth in this state-of-the-art survey. Costing necessarily involves both the design and design trade-offs of the WEC generating plant proper. It also impinges on the whole complex rate making process of utility companies [106].

It is clear that there are two cost parameters of fundamental importance in wind-electrics:

- The Busbar Energy Cost to the user of the WECS output, be it a local home owner or a utility firm purchasing the WECS generator power.
- The Capital Cost/KWe of the WECS plant.

Some would add a third factor: the busbar cost/dollar invested. In arriving at the costs it is important to adopt the view of life-cycle costing and not place undue emphasis on the initial capital cost. A WECS capital cost of $568/KWe may be a bargain over a fossil fuel plant capital cost of $400/KWe when figured over the life cycle, for there is no fuel involved with the WECS unit. The deciding cost factor then becomes the busbar energy cost. Bereny [107] says that, if one sums up WECS technology as of about 1977, "present estimates indicate that WECS, if located at sites with reasonably high average wind velocities (14-20) would be expected to produce electricity at busbar prices of 20 to 30 mills per kilowatt hour.[1] This compares favorably with conventional electrical generation plants using oil priced at $11 to $12 per barrel."

[1] Compare this with our Fig. 1-28 for perspective.

Putnam [5, Chs. 9, 11 and 13] dealt extensively and in great depth with the cost issue, as his firm had initially set out to explore the possibilities of delivering cost-effective electric power to a utility operating in conjunction with hydroelectric generation, and both the initial capital cost and the busbar cost was of great concern to him and his colleagues. He deals with the best size for a large turbine, shows tables of detailed cost calculations, shows a nomograph [p183] relating installed cost/KW, annual charges, annual energy output/KW and energy cost in mills/KWH, and finally offers an entire chapter [13] of suggestions on ways to reduce cost of windpower generators. Based on their experience in the early 1940's, he found that the capital cost (1945 dollars) was about $205/KW at the busbar whereas the worth of windpower to the Central Vermont Public Service Corp. was only $125/KW for a block of 9000 KW. As was cited earlier, the project was terminated by the blade failure before the cost gap was ever bridged.

Golding [2, Chs. 15 and 16] treats the economics of wind-electrics, though much less thoroughly than did Putnam.

It may be well to remember that neither Putnam or Golding had access to modern digital computers. Nevertheless they studied an enormous number of cost sensitive variables in WECS.

More recently, Meier [108] reveals some of the cost figures, graphs, and thinking involved in 100 KW and 1000 KW WECS. He shows that a plot of electric energy cost vs mean wind speed decreases, the energy cost being about 3 cents/KWH at a 20 MPH mean wind speed. At that same speed the direct capital cost is about $450/KW and the plant factor is about 33 percent. There <u>are</u> economies of scale in WECS, as he shows a 100 KW system costing $1980/KW while a 1000 KW unit only costs $568/KW.[1]

A serious investigator of WECS costs will also want to examine Quinn's paper [109] who studied the various cost factors in-depth and derives useful new cost formulas which had not previously appeared.

It is highly likely others have also addressed the cost issue of WECS during the past year, but as was stated earlier, no attempt was made here to be exhaustive in this area. Those interested in pursuing this topic further will find an entrée in [110].

3-10 Environmental Aspects of WECS

No comprehensive treatment on wind-electrics would be complete without a few statements regarding the known and potential environmental impacts of adding large numbers of WECS throughout the nation.

First, we should capture the perspective set by Merriam [31, p13] who said, "The environmental impact of wind energy use is so small as to be

---

[1] It may be interesting to compare this with the approximately $325/KW for the Danish 2 MW Tvind machine built largely with volunteer labor.

virtually nonexistent. It is hard to imagine any other primary source with less environmental impact." He goes on to identify some of the minor potential environmental impacts the addition of WECS may encounter and opines that they will be mostly negligible.

Rogers, et al [111] did an NSF study on the potential environmental effects of wind energy systems and concluded: "The operation of wind energy conversion systems does not appear to involve any unusual environmental effects. As presently conceived, wind energy systems will produce no significant thermal or chemical effluents. Land requirements are minimal and compatible with many other land uses." Their study was focussed on the NASA Plum Brook 100 KW machine.

Another study of wind energy conversion environmental factors was made in 1977 for ERDA [112]. This study looked at a wide spectrum of potential environmental effects of WECS--radio wave interference, fabrication and operation of the WECS plant, asthetic effects, microclimate effects, floral effects, faunal effects, insects and birds killed by flying into the blades, accidents and safety aspects, including throwing a blade, and other potential effects. It proposed an environmental research program on:

- <u>RF Interference Problem</u>
- <u>Wake Effects</u>
- <u>Structural Failure</u>

which is evidently now under way.

It is somewhat surprising that none of these in-depth environmental studies turned up the potential problems of:

- <u>Ice Throwing</u> by the blades in cold weather and the potential dangers to those nearby the machine.
- <u>Aircraft Hazard</u> potentiality in some locations. As WECS get larger--we are now with 200 ft. diameter rotors mounted on 150 ft. towers--care should be exercised to insure that no such machines are located near airports. Also such machines should be identified on the various aeronautical sectional charts used by pilots.

Of all the potential environmental problems with WECS, it appears the most serious is the radio frequency interference problem. Radio, and especially television signals, will be reflected from the large moving blades of a WECS and cause dark bars to drift across nearby television screens. This has already been a problem for one WECS installed on Block Island [76] where the government had to arrange for the installation of a TV cable system to alleviate the problem. The extent the problem will be minimized by future fiberglass blades being built for the recent large WECS appears an unknown at this time.

In toto, the potential environmental impacts of WECS from the systems standpoint appear negligible compared to some of the nonsolar electric energy options, as nuclear and conventional fossil plants.

4-11 Barrier To Wind-Electric Integration

By barriers we mean here those problems or issues standing in the way of WECS progress. By integration we mean here both our electrical integration and technology integration of Sec. 3-8. Including the technology integration is necessary as one looks ahead a few years to the possibility of WECS becoming a common reality in the nation's electric utility industry. We are also considering here wind-electrics as an entity, i.e. a potential whole new industry beneficial to the nation, as opposed to considering a single WECS generator located in a specific part of the country.

After surveying an appreciable fractional part of the published literature on wind-electric energy conversion systems, it is this researcher's opinion that the following items appear to be the principal 1978 barriers to WECS integration:

- National Wind Data Base, on which the siting and economic success of all future WECS depends, is suboptimal in terms of utility company or/individuals which would use the data for siting WEC plants.

- The Centralized Vs Dispersed issue raised in Sec. 1-6, i.e. a few large WECS vs many smaller ones widely dispersed, appears largely being dealt with at the national level by mostly ignoring the technology of small-scale wind-electrics and pouring most of the available wind R&D funding into ever larger WECS.

  While it is recognized there are economies of scale for WECS, it is felt that the resourceful do-it-yourselfers' of the nation--if properly guided on the technical and economics aspects of WECS design--are capable of making a significant energy contribution to the total WECS generated electric energy. There is and will be a nearly insurmountable information barrier to constructors or purchasers of these small dispersed wind-electric generators who may attempt to get reliable wind-electric technical and economic information from the government: the information shortage is particularly acute in the area of electrical integration with the utility.

- Cost/KWe of WEC plants is and will probably continue for sometime to be a barrier to both individual potential WECS users and utility companies. Both are cost conscious in an era of increasing utility bills. WECS produced electric energy cost/KWH, in the final analysis, will have to be equal to or lower than that for fossil fuel or nuclear produced electric energy insofar as the customer is concerned.

  To substantially lower the cost of WECS plants--and the electric energy produced--will be a solvable challenge for WECS system engineers of the future.

- Lack of Methodical Cost Analysis Procedures is very apparent of WECS technology, even on a cursory examination. If there are any generalized costing procedures which would be widely applicable to families of both large-scale and small-scale WECS they are well hidden; every large-scale WECS built to date seems to have been given highly individualized costing with hardly any thought being given to generalizing the cost analysis procedures to whole families of WECS machines and thereby easing the systems engineer's later comparison of systems alternatives.

  The utility industry will eventually have to have such procedures should WECS technology continue to grow.

- No Standardization of WECS Designs We now have a few KW, 100KW, 200 KW, and 1500-2000 KW machines, all experimental and highly individualized in siting, design, and construction. Must WECS technology go the future route of customizing the many thousands of machines which will be needed in the future? The conventional utility plant, to this date, has gone this route leading to high capital costs, long equipment lead times, and minimum interchangeability of parts. Wind-electric technology from the systems viewpont would appear amenable to a substantial degree of standardization of designs and hardware to the benefit of the collective future WECS industry.

  It is not too early to start thinking toward arriving at some standardized designs which would be acceptable to WECS equipment manufacturers and users alike.

- No Mass Production WECS Factories presently exist. Further, the literature appears devoid of any even rough estimating of the sizes, market potential, output required, costs, number of workers, and all the other variables involved in creating a new WECS manufacturing facility.

  Photovoltaics technology, which appears somewhat behind WECS technology in terms of commercialization, is much futher along than WECS in that realistic studies there are already being made of the needed mass production facilities.

- Building Codes in urban areas need studying and probable modifications if small-scale WECS are to see widespread use in urban areas and in small cities and towns where there is abundant wind. Such changes are long, detailed, highly political, and may have legal ramifications, as developments in the solar heating and cooling area during the past several years have shown. A preliminary examination of the legal aspects of WECS has been surveyed by Taubenfeld [113] and Mayo [114]. There will probably be many more legal ramifications of WECS technology as it grows.

  WECS technology, to some greater (for dispersed WECS) or lesser (for centralized WECS) degree will probably have to go through the building code barrier--for either cities or counties and possibly both.

- Public Reactions to WECS may be a small potential barrier, though this cannot be determined at this time. Ferber's [115] work appears to be the first in this area.

  It appears highly unlikely, however, that WECS technology will see anything like the current anti-feeling presently being experienced with nuclear power in many portions of the country.

- The System Dynamics of WECS-Utility Interface, from the electrical standpoint, appear in a very elementary and unsatisfactory state from the standpoint of the needed theoretical work. A substantial extension of the theory along the general lines of Hwang and Gilbert's work [102] reported in Sec. 3-8 appears to be what is needed.

  A utility systems engineer would be most uncomfortable with the present-art theoretical knowledge about synchronizing a large-scale WECS into a utility system on a routine basis.

- General Lack of Utility Involvement in WECS appears a barrier, as most of the current significant R&D advancements on WECS are being made outside of the utilities by government agencies and aerospace firms.

  It would seem highly desirable for the orderly growth of the WECS technology that utilities in the known high wind areas of the country become much more involved in making progress in WECS than has heretofore been the case. Systems planning engineers, in particular, should be involved more heavily in WECS R&D.

  Somewhat akin is the subtle barrier of negative attitudes taken by at least two utilities [116] [117] [118] toward integrating small-scale WECS into their system. If such anti-R&D attitudes persist and spread, they will hold back rather than aid the eventual integration of WECS into the electric energy system.

The above are the principal barriers seen in 1978. Others undoubtedly may see different barriers or express those elucidated above differently.

4-12 Future R&D Needed For Integration

After surveying an appreciable fractional part of the published literature on wind-electric energy conversion systems, it is this researcher's opinion that the items listed below--not necessarily in order of priority--are in need of additional R&D. The viewpoint is constricted to that of the systems engineer interested in integrating WECS with utility grids.[1] The areas are listed without extensive arguments and justifications.

---

[1] Naturally other research, not identified here, must be executed in the general area of the design of the WECS proper.

- Theoretical Study (transient and steady state) of synchronous generator driven by a WECS under 'noisy' wind (gusts) and 'noisy' output conditions for both horizontal axis machines and vertical axis Darrieus machines.

  The particular problem of time of closure of the output circuit breaker would be examined, along with electrical transients.

- Experimental Study of the measured effects on a utility's power vs time load curve that the addition and subtraction of a large-scale WECS generator actually has.

  The possibilities of improving the system load factor by means of WECS should be examined. The case would be most meaningful for a small utility where the WECS power was an appreciable part of the total. Fluctuating winds less than rated wind velocity, and the effect their generated power has on the power system dynamics and stability would be of particular interest.

- A Generalized WECS Reliability Study, linked back to the wind V(t) curve, needs to be made. Reliability of a generator is important to the systems engineer. There is not much in the wind literature on this important systems parameter. The question of "What does addition of a WECS generator of probable reliability to a power system of known reliability do to the overall power system reliability?" needs to be answered; also the question "How is the reliability of a WECS calculated, given a V(t) curve and the WECS transfer curve?"

- An Ordering Of Cost Analysis--over the life of the WECS--needs to be done, aimed at structuring existing chaotic cost literature and methods into a generalized procedure for costing future WECS generators.

  The results must be presented in a form quickly useful to a utility system planner. Such procedures do not now exist. The procedures would permit quick alternative design comparisons to be made and would also indicate areas where WECS cost reduction efforts would be beneficial. They would also show detailed means for calculating the WECS energy output and costing it.

- Basic Research On Siting needs to be done, picking up from previous art. The systems problems need to be addressed of:
  - Accumulating the body of knowledge presently existing on siting and site considerations, including extracting and organizing wind information from the national weather data bank. A preliminary start on this problem has been made in this report in Sec. 3-4.
  - Organizing the siting theory in some meaningful way. Almost no organization now exists other than that of our Sec. 3-4.
  - Studying wind flow (theoretical and measured) over hills, ridges, and in the vicinity of nonidealized surroundings (trees, buildings, and the like), including examining velocity vs height. This is a very complex theoretical problem in aerodynamics.

- Structuring the total knowledge into an ordered report a utility system planner can understand and reliably use in planning WECS sites in his system area.

- Wind Instrumentation Development needs work for permitting quick, reliable accurate wind velocity measurements at various heights at a test site. Easily portable digital readout instruments suitable for use by utility siting teams appear to not exist presently. Also truly inertialess instrumentation for measuring accurately short-term gusts appears to have not been developed. This later problem is technically difficult and probably will involve invention.

- WECS Instrumentation integration needs achieving. Not a single paper of the wind literature examined dealt with the important aspects of instrumentation for a large-scale complex WECS. Temperature, speed, blade angle, stress, power and other quantities are being measured, but virtually nothing is being reported in the literature on the many engineering considerations involved in such complex measurements.

  This work would integrate the important instrumentation experiences to date on the various large WECS and present it in a form useful to future utility systems engineers and planners.

- Instrumentation Research on applying modern solid-state electronic density function measurement techniques to the statistics of the wind velocity, permitting quick automatic plots to be made of the wind's density function. Wind people appear unaware of the existence of such apparatus developed by the author and a colleague for the Harry Diamond Ordnance Laboratory in 1966-68 as a part of the proximity fuse program.

  Such an instrument could greatly facilitate the wind statistical data handling on which siting is based and ultimately energy output predictions made. It would also reinforce some of the theoretical aspects of wind statistics.

  This research might also be tied in with developing an instrument for cubing the wind speed and automatically averaging it to facilitate actual direct windpower measurements. The French are reported to have developed such an instrument, but it seems relatively unknown among U.S. WECS researchers. It probably needs improvement to be useable by future utility personnel.

- WECS Lifetime data seems conspicuous by its absence from the wind literature. A sensible controlled program should be initiated--if it isn't already underway by someone--keeping detailed histories on the kinds and degrees of component failures and lifetimes on selected existing WECS. A preliminary body of information on lifetimes should be starting to be accumulated so that a utility company has a later minimum of unpleasant costly surprises with earlier than expected WECS component failures.

The successful execution of the above research would constitute positive progress toward smoothing the integration of wind-electrics into the utility industry.

REFERENCES

1. Hayes, Denis, Energy: The Solar Prospect, Worldwatch Paper II, March 1977. Available from Worldwatch Institute, 7776 Massachusetts Ave., N. W., Washington, D. C. 20036.

2. Golding, E. W., The Generation of Electricity By Wind Power, [London, England: E.&F.N. Spoon Ltd., 1955]. Reprinted: [New York, N. Y.: John Wiley & Sons, Inc., 1976], 332 p.

3. Eldridge, Frank R., Wind Machines, October 1978. Available from NTIS, Springfield, VA 22151 as No. PB-249 936.

4. Arnold, James E., "On the Correlation Between Daily Amounts of Solar and Wind Energy and Monthly Trends of the Two Energy Sources," Proceedings of the 1977 Annual Meeting, Vol. One, pp. 19-36. Available from American Section of the International Solar Energy Society as Conf. 770603-P2.

5. Putnam, Palmer Cosslett, Power From The World, [New York, N. Y.: D. Van Nostrand Co., 1948]. This pioneering work is, unfortunately, out of print. Copies obtainable from Total Environmental Action, Inc. Church Hill, Harrisville, N. H. 03450.

6. Justus, C. G., Winds and Wind System Performance, [Philadelphia, Pa.: The Franklin Institute Press, 1978].

7. Changery, Michael, "Initial Wind Energy Data Assessment Study," Proceedings of the Second Workshop on Wind Energy Conversion Systems, June 9-11 1975, pp. 326-335. Available from NTIS as NSF-RA-N-75-050. There is also a more lengthy May 1975 version by same title above available from NTIS as NSF-RA-N-75-020.

8. Thring, Hans, Energy For Man: Windmills to Nuclear Power, [Westport, Connecticut: Greenwood Press, Publishers, 1958].

9. Parker, A., World Energy Resources and Their Utilization. The Institution of Mechanical Engineers, The Thirty-Sixth Thomas Hawksley Lecture, Vol. 160, No. 4 (1949). Cited by Golding, Ch. 1.

10. Heronemus, William E., Quoted in Exotic Generation Technology Featured, Electrical World, March 1, 1974.

11. Hardy, Donald M., and John J. Walton, "Wind Energy Assessment." Presented at Miami International Conference on Alternative Energy Sources, 5-7 December 1977. To be published in the Conference Proceedings.

12. Von Arx, William S., "Energy: Natural Limits and Abundances," American Geophysical Union--EOS Transactions, Vol. 55 (9), p. 828, September 1974. Cited in Hardy.

13. Gravel, Senator, Clean Energy Via the Wind, Congressional Record-Senate, December 7 1971, S20776.

14. Kornreich, T. R., "Wind Energy Conversion Systems (WECS) For Central Station and Dispersed Power Applications," Proceedings of Condensed Papers, Miami International Conference on Alternative Energy Sources, 5-7 December 1977.

15. Krenz, Jerrold H., Energy: Conversion and Utilization, [Boston, Mass.: Allyn and Bacon, Inc., 1976].

16. Brunt, D., Physical and Dynamical Meteorology, [Cambridge, Mass.: Cambridge Univ. Press, 1934]. Cited in Putnam.

17. An Assessment of Solar Energy As a National Energy Resource. Prepared by the NSF/NASA Solar Energy Panel, December 1972. NTIS No. PB 221-659.

18. Devine, W. D., Jr., "An Energy Analysis of A Wind Energy Conversion System For Fuel Displacement," Proceedings of Condensed Papers, Miami International Conference on Alternative Energy Sources, 5-7 December 1977.

19. Gordon, A. H., "Harnessing The Wind," WMO Bulletin, July 1954, pp. 102-107.

20. Sittler, O. Dayle, "Energy Content of Winds In The High Plains Region of Southwestern U. S.," Sharing The Sun: Solar Technology in the Seventies, Vol. 7, August 15-20 1976. Available from American Section of The International Solar Energy Society.

21. Wentick, Tunis, Jr., Study of Alaskan Wind Power and Its Possible Applications, 29 February 1976. Available from NTIS as PB 253 339.

22. Corotis, Ross B., Stochastic Modelling of Site Wind Characteristics, November 1976. Available from NTIS as PB-261 178.

23. Knox, Joseph B., Status Report: Lawrence Livermore Laboratory Wind Energy Studies, June 1976. Available from NTIS as UCID-17157-1.

24. Smith, M. C., Wind Site Selection For Optimum Wind Power Systems. Presented at Miami International Conference on Alternative Energy Sources, 5-7 December 1977. To be published in Proceedings.

25. Reed, Jack W., Wind Power Climatology of The United States, June 1975. Available from NTIS as SAND 74-0348.

26. Hardy, Donald M., Wind Power Studies: Initial Regional Applications, 22 March 1976. Available from NTIS as UCRL-50034-76-2.

27. Coty, Ugo, "Wind Energy Conversion Systems Mission Analysis," Proceedings of The Second Workshop On Wind Energy Conversion Systems, June 9-11 1975. Available from NTIS as NSF-RA-N-75-050.

28. Hurlebaus, William H., "A Perspective On Wind Mission Analysis," Proceedings of The Second Workshop On Wind Energy Conversion Systems, June 9-11 1975. Available from NTIS as NSF-RA-N-75-050.

29. Meroney, Robert N., "Sites For Wind-Power Installations," Proceedings of The Second Workshop On Wind Energy Conversion Systems, June 9-11 1975. Available from NTIS as NSF-RA-N-75-050.

30. Justus, C. G., "Wind Characteristics/Site Survey," Proceedings of The Second Workshop on Wind Energy Conversion Systems, June 9-11 1975, pp 487-489. Available from NTIS as NSF-RA-N-75-050.

31. Merriam, Marshal F., Wind Energy For Human Needs, November 1974. Available from NTIS as UCID-3724.

32. Reed, Jack W., Wind Climatology, n.d. Available from NTIS as SAND 75-5531.

33. Reed, Jack W., Predicting Wind Power At Turbine Level From An Anemometer Record At Arbitrary Height, n.d. Available from NTIS as CONF-760909-1.

34. Freeman, B. E., A New Wind Site Selection Methodology, Proceedings of The Second Workshop On Wind Energy Systems, June 9-11 1975. Available from NTIS as NSF-RA-N-75-050.

35. Ramakumar, R., "Wind-Electric Conversion Utilizing Field Modulated Generator Systems," Solar Energy, Vol. 20 (2), 1978, pp. 109-117.

36. Justus, C. G., "Wind Energy Statistics For Large Arrays of Wind Turbines (New England and Central U. S. Regions)," Sharing The Sun: Solar Technology in the Seventies, Vol. 7, August 15-20 1976, pp. 268-288. Available from American Section of The International Solar Energy Society. Also available from NTIS as PB-260 679.

37. Justus, C. G., "Wind Energy Statistics For Large Arrays of Wind Turbines (New England and Central U. S. Regions)," Solar Energy, Vol. 20, No. 5, 1978, pp. 379-386.

38. Radice, Frank C., "Siting of Wind Driven Apparatus," Proceedings of the Eleventh Intersociety Energy Conversion Engineering Conference, 12-17 September 1976, pp. 1736-1740. Copy available from American Institute of Chemical Engineers, 345 East 47 St., New York, N. Y. 10017.

39. Heronemus, William E., "The United States Energy Crisis: Some Proposed Gentle Solutions," presented before a joint meeting of ASME and IEEE, West Springfield, Mass., 12 January 1972.

40. Hill, Philip G., Power Generation, [Cambridge, Mass.: The MIT Press, 1977].

41. Kolm, K., et al, Evaluation of Wind Energy Sites From Aeolian Geomorphologic Features Mapped From Landsat Imagery, 1 December 1975. Available from NTIS as ERDA/NSF/00598-75/T1.

42. Marwitz, J., and R. Marrs, "Locating Areas of High Wind Energy Potential By ERTS Observations of Aeolian Morphology," Proceedings of The Second Workshop On Wind Energy Conversion Systems, June 9-11 1975, pp. 353-355. Available from NTIS as NSF-RA-N-75-050.

43. Duchon, Claude E., "Wind Velocity As Modified By Geomorphology," Proceedings of The Second Workshop On Wind Energy Conversion Systems, June 9-11 1975, pp. 367-371. Available from NTIS as NSF-RA-N-75-050.

44. Preuss, R. D., "Two General Methods For The Unsteady Aerodynamic Analysis of Horizontal-Axis Windmills," Proceedings of the 12th Intersociety Energy Conversion Engineering Conference, 28 August - September 2 1977, Vol. 2, pp. 1618-1623.

45. Ramakumar, R., "A Review of Wind-Electric Conversion Technology." To be published in the Proceedings of Miami International Conference on Alternative Energy Sources, 5-7 December 1977.

46. Rohrback, C., Experimental And Analytical Research On The Aerodynamics of Wind Turbines, February 1976. Available from NTIS as COO-2615-76-T-1.

47. Martinez-Sanchez, M., "A Performance Comparison Between Constant RPM and Constant Velocity Ratio Operation of A Windmill," 15 February 1976. Available from NTIS as PB-256 198.

48. Wulff, Hans, Traditional Crafts Of Persia, [Cambridge, Mass. Press, 1966].

49. Stokhuyzen, Frederick, Dutch Windmills, [Bussum, Holland: Van Dishoeck, 1962].

50. DeLittle, R. J., The Windmill Yesterday and Today, [London, England: John Baker, Ltd., 1972].

51. Juul, J., "Wind Machines" in Wind and Solar Energy, UNESCO, Vol. III, Paris, 1956.

52. U. S. Windmill Industry Still Survives, Despite Battering by Winds of Change, The Wall Street Journal, Monday, April 15 1968.

53. Solar Energy As A National Energy Resource, NSF/NASA Solar Energy Panel, December 1972. Available from NTIS, Springfield, Va. 22151 as No. PB-221-659.

54. Federal Wind Energy Program, Energy Research and Development Adminstration, October 1975. Available from NTIS, Springfield, VA. 22161 as No. ERDA-84.

55. Westh, H. Claudi, "A Comparison of Wind Turbine Generators," Proceedings of The Second Workshop On Wind Energy Conversion Systems, June 9-11 1975, pp. 156-161. Available from NTIS as NSF-RA-N-75-050.

56. Linscott, Bradford S., et al, "Experimental Data and Theoretical Analysis of An Operating 100 KW Wind Turbine," Proceedings of the 12th Intersociety Engineering Conversion Engineering Conference, Vol. Two, 28 August - 2 September 1977, pp. 1633-1650.

57. Thomas, Ronald L., "Introduction To Large System Design," Proceedings of The Second Workshop On Wind Energy Conversion Systems, June 9-11 1975, pp. 13-20. Available from NTIS as NSF-RA-N-75-050.

58. Glasgow, John C., et al, Early Operation Experience On the ERDA/NASA 100 KW Wind Turbine, September 1976. Available from NTIS as N77-10640.

59. Thomas, R., et al, Plans and Status of The NASA-Lewis Research Center Wind Energy Project, October 1976. Available from NTIS as N75-21795.

60. Gilbert, Leonard J., A 100 KW Experimental Wind Turbine: Simulation of Starting, Overspeed, and Shutdown Characteristics, n.d. Available from NTIS as N76-18672.

61. Chamis, C. C. and T. L. Sullivan, Free Vibrations of the ERDA-NASA 100 KW Wind Turbine, March 1976. NASA TM X-71879. Avaialbe from NTIS as no. N76-18674.

62. Hoffman, John A., Coupled Dynamics Analysis of Wind Energy Systems, February 1977. Available from NTIS as N77-20558.

63. Linscott, Bradford S., et al, Tower and Rotor Blade Vibration Test Results for a 100 Kilowatt Wind Turbine, October 1976. Available from NTIS as N76-33628.

64. Puthoff, Richard L., Fabrication and Assembly of the ERDA/NASA 100-Kilowatt Experimental Wind Turbine, April 1976. Available from NTIS as N76-21703.

65. Torrey, Volta, Wind-Catchers, [Brattleboro, Vermont: Stephen Greene Press, 1976], 222 p.

66. Puthoff, Richard L., "Status of 100 KW Experimental Wind Turbine Generator Project," Proceedings of the Second Workshop on Wind Energy Conversion Systems, June 9-11 1975, pp. 21-36. Available from NTIS as NSF-RA-N-75-050.

67. Ramakumar, R., et al, Development and Adaptation of Field Modulated Generator Systems for Wind Energy Applications, October 1975. Available from NTIS as PB-263 604.

68. Hamilton, Roger, "Can We Harness The Wind?," National Geographic, Washington, D. C., December 1975.

69. Balcomb, Douglas, "The Department of Energy Solar Program: A Question of Balance," Solar Age, Vol. 3, No. 5, May 1978, p. 12 ff.

70. Southern California Edison To Sign Contract With Schedule For 2,700 KWE Windmill, Solar Energy Intelligence Report, Vol. 4, No. 10, March 6 1978.

71. Handbook of Homemade Power, [New York, N. Y.: Bantam Books, Inc., 1974].

72. Weisbrick, A. L., "Feature Review of Some Advanced And Innovative Concepts In Wind Energy Conversion Systems," Proceedings of Condensed Papers, Miami International Conference on Alternative Energy Sources, 5-7 Decmeber 1977. Available from Clean Energy Research Institute, University of Miami, Coral Gables, Florida.

73. Heronemus, William E., "Pollution-Free Energy From Offshore Winds," Preprints, 8th Marine Conference and Exposition, Marine Technology Society, 11-13 September 1972.

74. 200 KW Turbine Supplies Power To New Mexico Town, Electric Light & Power, Vol. 56, No. 4, April 1978, p. 6.

75. New Wind Turbine To Bring Power To N.M. Town; Dedication Jan. 28, Energy Insider, Vol. 1, No. 8, January 23 1978, p. 1.

76. Puerto Rico, Rhode Island Sites Picked to Test 200 Kilowatt Turbines Hooked To Power Plants, ERDA News, June 27, 1977, p. 5.

77. North Carolina Town to Test Record Size DOE Wind Turbine, Energy Insider, Vol. 1, No. 4, November 14, 1977.

78. GE To Build World's Largest Windmill, Electric Light And Power, January 1977, p. 37.

79. Thomas, Ronald L., Large Experimental Wind Turbines--Where We Are Now, March 1976. Available from NTIS as N76-21683.

80. Wilson, R. E., "Applied Aerodynamics of Wind Power Machines," Proceedings of the Second Workshop on Wind Energy Conversion Systems, June 9-11 1975. Available from NTIS as NSF-RA-N-75-050.

81. Weber, W., The Optimum Configuration of Rotor Blades For Horizontal Wind Energy Converters, February 1977. Trans. from German. Available from NTIS as N77 17562.

82. Jordan, Peter F., "Segmented And Self Adjusting Wind Turbine Rotors," Proceedings of the 12th Intersociety Energy Conversion Engineering Conference, Vol. 2, August 28 - September 2 1977. Available from The American Nuclear Society, Inc., 555 N. Kensington Ave., La Grange Park, Ill. 60525.

83. Dutch Researching Windmill Power, Gainesville Sun, Wednesday, February 29, 1978.

84. Windmills In Tvind, June 1976. Availabe from NTIS as ERDA-TR-230.

85. Merriam, Marshall F., "The Big Windmill at Tvind," Sunworld, Vol. 2, No. 2, May 1978, pp. 57-59.

86. New Windmill Based On Old Patent, Popular Mechanics, May 1974, p. 58.

87. Muraca, R. J. and R. J. Guillotte, Wind Tunnel Investigation Of A 14' Vertical Axis Windmill, March 1976. Available from NTIS as N76-19551.

88. Bankwitz, Hermann, et al, Development of A Vertical Axis Wind Turbine, Phase I, October 1975. In German. Abstract only translated. Available from NTIS as N77-17112.

89. Templin, R. J., Aerodynamic Performance Theory For the NRC Vertical-Axis Wind Turbine, June 1974. Available from NTIS as N76-16618.

90. Blackwell, B. F., "Some Geometrical Aspects of Troposkiens as Applied to Vertical-Axis Wind Turbines," May 1975. Available from NTIS as SAND 74-0177.

91. Blackwell, Ben F., Status of The ERDA/Sandia 17 Metre Darrieus Turbine Design, n.d. Available from NTIS as SAND-76-5683.

92. Sandia's Six-Story Tall Wind Turbine will Generate Power for Grid System, ERDA News, Vol. 1, No. 5, February 6 1976, p. 1.

93. Weingarten, Larence I., Material and Manufacturing Considerations for Vertical-Axis Wind Turbines, n.d. Availabe from NTIS as SAND 75 5512.

94. Low-Cost Materials Tested for Solar, Wind Power Uses, Electric Light and Power, January 1977.

95. Strickland, J. H., The Darrieus Turbine: A Performance Prediction Model Using Multiple Streamtubes, October 1975. Available from NTIS as SAND 75-0431.

96. Blackwell, Bennie F., Wind Tunnel Performance Data for the Darrieus Wind Turbine with NACA 0012 Blades, May 1976. Available from NTIS as SAND 76-0130.

97. Banas, J. F., Methods For Performance Evaluation of Synchronous Power Systems Utilizing the Darrieus Vertical-Axis Wind Turbine, April 1975. Available from NTIS as SAND 75-0204.

98. Smith, R. T., et al, "Operational, Cost, and Technical Study of Large Windpower Systems Integrated with Existing Electric Utility," Proceedings of the Eleventh Intersociety Energy Conversion Engineering Conference, September 12-17 1976, pp. 1754-1760. This paper is also available from NTIS as CONF 760906-8.

99. Mulcahy, Michael J., "Interface With Utilities And Users," Proceedings of the Second Workshop on Wind Energy Conversion Systems, June 9-11 1975, pp. 502-505.

100. Sorensen, Bent, "On The Fluctuating Power Generation of Large Wind Energy Converters, With and Without Storage," Solar Energy, Vol. 20, No. 4, 1978, pp. 321-331.

101. Swanson, R. K., et al, "Operational, Cost and Technical Study of Large Windpower Systems Integrated with Existing Public Utilities," Proceedings of the Second Workshop On Wind Energy Conversion Systems, June 9-11 1975, pp. 92-97.

102. Hwang, H. H., and Leonard J. Gilbert, Synchronization of the ERDA 100 KW Wind Turbine Generator With Large Utility Networks, March 1977. NASA TMX-73613. Available from NTIS as N77-19580.

103. Reitan, Daniel K., "A Progress Report on Employing A Nonsynchronous AC/DC/AC Link In A wind-Power Application," Proceedings of the Second Workshop On Wind Energy Conversion Systems, June 9-11 1975, pp. 290-297. Available from NTIS as NSF-RA-N-75-050.

104. Meyer, Hans, "Synchronous Inversion: Concept and Application," Sharing The Sun: Solar Technology In The Seventies, Vol. 7, August 15-20 1976. Available from the American Section of the International Solar Energy Society.

105. Lindsley, E. F., "New Inverter Gives Windpower Without Batteries," Popular Science, October 1975, pp. 50-52.

106. Energy Rate Initiatives. Study of the Interface Between Solar and Wind Energy Systems and Electric Utilities, March 1977. Available from NTIS as PB-265 607.

107. Bereny, James A., Survey of the Emerging Solar Energy Industry, 1977 Edition, p. 177. Available from Solar Energy Information Services, P. O. Box 204, San Mateo, Calif. 94401.

108. Meier, Richard C., "Concept Selection, Optimization, and Preliminary Design of Large Wind Generators," Proceedings of the Second Workshop On Wind Energy Conversion Systems, June 9-11 1975, pp. 46-58. Available from NTIS as NSF-RA-N-75-050.

109. Quinn, Brian, "The Consumer's Cost of Electricity from Windmills," Proceedings of Eleventh Intersociety Energy Conversion Engineering Conference, September 12-17 1976, pp. 1746-1753.

110. Solar Energy Update, a subscription publication of DOE. Available through Technical Information Center, United States Department of Energy, Oak Ridge, Tenn.

111. Rogers, S. E. et al, Evaluation of the Potential Environmental Effects of Wind Energy System Development, August 1976. Available from NTIS as ERDA/NSF/07378-75/1.

112. Solar Program Assessment: Environmental Factors. Wind Energy Conversion, March 1977. Available from NTIS as ERDA-77-47/6.

113. Taubenfeld, Rita Falk, Barriers to the Use of Wind Energy Machines: The Present Legal/Regulatory Regime and a Preliminary Assessment of Some Legal/Political/Societal Problems, July 1976. Available from NTIS as PB-263 576.

114. Mayo, Louis H., "Legal-Institutional Implications of Wind Energy Conversion Resources," Proceedings of the Second Workshop on Wind Energy Conversion Systems, June 9-11 1975, pp. 385-389.

115. Ferber, Robert, "A Pilot Study On Public Reactions To Wind Energy Devices," Proceedings of the Second Workshop On Wind Energy Conversion Systems, June 9-11 1975, pp. 380-384.

116. Utility Company Nixes Hookup To Windmill, AP Dispatch, Gainesville Sun, Sunday, June 6 1976.

117. His Windmill-Produced Electricity May be of Some Use to Industries, AP Dispatch, Gainesville Sun, Monday, April 14 1977, page 5C.

118. Con Ed to Establish Windmill Charge, AP Dispatch, Gainesville Sun, Friday, May 6 1977.

119. Windpower And Windmills, July 1975. Bibliography series No. 58 prepared by the National Aeronautical Laboratory, Bangalore, India 560017. Available from NTIS, Springfield, Va. 22151, as No. NP 20897.

120. Lotker, Michael, "Northeast Utilities' Participation In The Kaman/NASA Wind Power Program," Proceedings of the Second Workshop On Wind Energy Conversion Systems, June 9-11 1975, pp. 59-68.

121. DOE, Danish Agency To Refurbish and Test Neglected 20-Year-Old 'Gedsermill,' Solar Energy Intelligence Report, Vol. 3, No. 42, November 28, 1977.

CHAPTER 4

ENERGY STORAGE

"Storage ranks high among the uncertainties that impede the use of long-term energy sources. Although studies have been performed, none has yet established which storage systems will have an economic edge."

Denis Hayes [1, p. 61]

- Introduction
- Prior and Present Art of Energy Storage Methods
- System Integration of Energy Storers
- Economics
- Future R & D Needed for Integration of Energy Storers

4-1 Introduction

The two previous Chaps. dealt with converting solar energy directly into electricity via photovoltaics and wind-electrics. The need for energy storage, while briefly recognized there, was not treated. It is entirely appropriate to treat energy storage as a discrete in-depth topic sandwiched between the earlier basic methods of solar-electric conversion and Chap. 5 dealing with systems of solar-electrics. Thus storage completes the component building blocks needed for a reliable solar-electric system.

The need for energy storage of some kind is almost intuitively evident for a solar-electric system. An optimally designed solar-electric system will collect and convert when the insolation is available during the day. Unfortunately the time when solar energy is most available will rarely coincide exactly with the demand for electrical energy, though both tend to peak during the daylight hours. There is also the problem of clouds with photovoltaic plants, and cloud cover for several days may result in substantially lowered electrical output compared to high insolation cloud-free days. Similarly for wind-electrical plants, the wind may not blow for some time. Obviously during such days energy previously stored during high insolation times could be used to provide a continuous electrical output. Thus the addition of storage can increase the reliability of being able to deliver electrical power at an arbitrary needed time.

We conclude that energy storage in a solar-electric system may:

- Permit solar energy to be captured when insolation is highest and then later used when the need is greatest. It can thus transform a diurnal solar energy input into a more uniform desired electrical output [2, p46].

- Make it possible to deliver electrical load power demand during times when insolation or wind is below normal or nonexistent. Storage also makes it possible to deliver short peaks of power far exceeding the rated power capacity of the plant.

- Be located close to the load, thereby minimizing the need for costly transmission and distribution facilities which would otherwise be required to meet peak load demands were storage not present there.

- Improve the reliability of the solar-electric system over what it would be without storage.

- Permit a better match between the solar energy input and the electrical load demand output than would be the case without storage.

- Assist in permitting short-term maintenance to be done to the principal solar-electric generator on a planned basis without loss of load. It is initially clear that adding energy storage to a solar-electric system,should it be needed, inherently increases the plant's capital cost which must translate to an already high electrical energy cost at the busbar output of the plant.

The amount of storage needed for a given solar-electric system would be determined by:

- The size of the solar-electric generator.

- The shape, magnitude, and timing of the load demand curve.

- Whether the system is to be operated in a peak-shaving mode or a load leveling mode.

- The insolation history for that particular plant site.

- The cost per KWH of the stored energy.

- The permissable capital cost allocated to storage.

- Environmental and safety considerations.

Determining the optimal kind and size of energy storage for a given solar-electric system is thus a complex procedure dependent on a number of major factors.

The principal classes of energy storage means are shown in Fig. 4-1. We will survey the state-of-the-art of each in this chapter. They are either now used on various solar-electric systems or might be used. Our interest here is chiefly in energy storage means potentially applicable to solar-electric systems. We therefore exclude other known energy storage methods not relevant to this goal.

We define an energy storage means as that method, device, material, or apparatus which permits energy to be deliberately placed therein and to be retrieved at some later desired time. The energy may be stored in a variety of forms, e.g. as heat, chemical, mechanical, or magnetic.

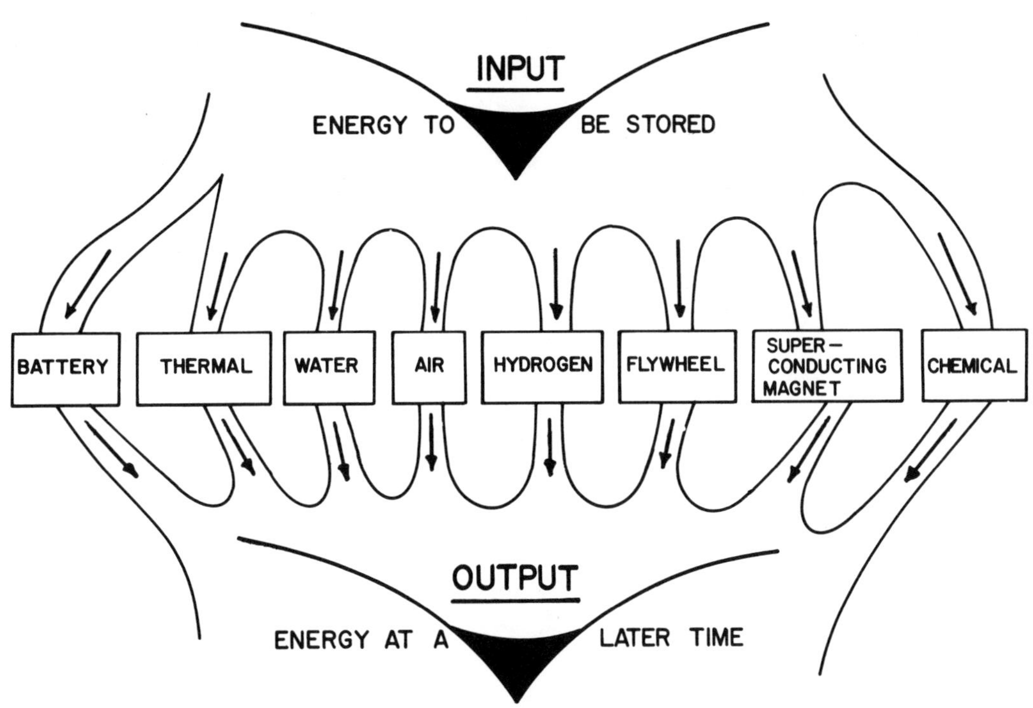

FIG. 4-1 Principal energy storage means most relevant to solar-electrics.

We should note here that energy storage, regardless of the means used, is always less than 100 percent efficient because of various loss processes involved getting the energy into storage, loss while it is in storage, and loss in retrieving it.[1]

The subject of energy storage has heretofore generally been associated with the advanced energy conversion field except for hydro-storage which has been in commercial use in conventional utility systems many decades. It is interesting to observe that with the exception of Angrist [3, Ch.2], not a single engineering textbook was uncovered in this research which treats the principal energy storage means existing in the research literature on energy storage. We should therefore not be surprised over the fact that engineers generally know little about the energy storage subject. Angrist has an excellent summary of the first six storage means of Fig. 4-1; he does not treat the latter two methods, however. His treatment of energy storage used in conjunction with a conventional electric plant [pp37-41] will be of special introductory interest to systems engineers. References are given where more information can be found. Thus Angrist's excellent Chap. 2 appears to contain the only succinct overview textbook on energy storage existing in 1978.

Other helpful sources of energy storage information include Simmons [4, pp116-129], Stoner [5, Chs. 1 and 2], Sullivan [6, Ch.4], Denton [7, Ch.8], Krumm [8], Mc Collom [9, pp813-819], and Centrell [10, p285].

We now turn to examining the state-of-the-art of each method for storing energy.

[1]Once again, we are reminded of the cascaded conversion principle of Sec. 1-7.

4-2 Prior and Present Art of Energy Storage Methods

As in previous Chaps., we define prior art as pre-1975 and present art as post 1975. Here, however, for conciseness we combine prior and present art reviews; it permits structuring the subject more logically, e.g. chronologically. By subtitles, however, we retain the prior and present art distinction.

This Sec. will inherently be lengthy because of the volume of published literature uncovered in this research. While previous state-of-the-art reviews have been made in single energy storage areas, e.g. batteries, with the exception of Angrist [3], no one appears to have attempted to integrate this entire field of energy storage. To do so we choose to partition the field after Fig. 4-1, surveying each energy storage method shown as a sub-part of this section, thus rendering a clear structure to the subject.

We limit our attention strictly to energy storage aspects throughout the treatment, even though some papers or books may show the storage in a larger systems context. We shall give only the briefest glimpses of those systems and what storage method(s) best go with which systems so far as is presently known.

Some of the energy storage methods have a long history, e.g. batteries, while other newer storage methods have a much smaller literature, e.g. superconducting magnets.

Our viewpoint continues to be principally that of the systems engineer; however excursions into the devices and materials areas are unavoidable. One cannot intelligently discuss energy storage by batteries, for example, without dealing briefly with materials and to some extent the processes involved in making the battery.

In some energy storage areas, e.g. batteries, there is no way this extensive field can be satisfactorily summarized in a single short document of this kind. For such a large field the present survey can, at best, only chart the general terrain of the territory, leaving details to the references. Thus it is absolutely essential that the serious reader search out and study the references if an in-depth knowledge is desired.

Batteries

Before surveying the battery literature in-depth, we first present a few key terms and facts about batteries in general which may be helpful to the systems engineer. Then we shall treat more detailed information about batteries.

History of Batteries is linked with the names of Faraday, Volta, Galvin, Daniell, and many other 'greats'. The historical development of the important lead-acid battery as well as other types has been ably summarized in Vinal's [11] book. His book is particularly appealing to the systems engineer interested in using batteries rather than studying the detailed electro-chemistry within them. An excellent historical little known summary of the Nickel Cadmium vs Edison batteries has been made by Berg [12]. Berg's account is of particular interest to those developing batteries for electric vehicles.

Battery Defined A battery is a combination of individual cells. A cell is the elemental combination of materials and electrolyte constituting the basic electro-chemical energy storer. A battery can also be thought of as a black box into which electrical energy is put, stored electrochemically, and later retrieved as electrical energy.

Battery Divisions

There are two principal divisions:

- Primary Batteries nonrechargeable, e.g. "dry cell" flash light batteries. In primary batteries the chemical reactions are non-reversible.
- Secondary Batteries rechargeable, e.g. a lead acid battery. There are many types of secondary batteries. The chemical reactions are reversible in secondary batteries.

Secondary batteries are of chief interest for solar electrics, and we therefore exclude here considering primary batteries.

Basic Battery Theory. A generalized cell consists of two electrodes called the anode and cathode immersed in a suitable electrolyte. When an electrical load is connected between the electrodes charge separation occurs at the interface between one electrode and the electrolyte, freeing both an electron and an ion. The electron flows through the external load and the ion through the electrolyte, recombining at the other electrode.

The elementary theory of electric cells is laid out in Vinal's book [11, Ch.4]. A more advanced treatment of the electrochemistry has been assembled in Jasinski's [13, Ch.1] book.

The polarity and magnitude of the cell terminal voltage is, in general, a function of the electrode materials, electrolyte, cell temperature, and other factors.

Many generic versions of cells exist. Each has its own unique history, and all have arisen from the electro-chemistry field.

Battery Fundamental Quantities of most interest to a systems engineer include:

- Open Circuit Voltage (OCV) The battery or cell terminal voltage with $R_{load} = \infty$, i.e. no load.
- Energy Capacity of a battery is of 3 types:
  - ▲ Energy stored (watt-hours or kilowatt-hours)
  - ▲ Energy stored/weight (watt-hours/kg. (or watt-hours/lb.))
  - ▲ Energy stored/volume (watt-hours/m^3 (or watt-hours/ft^3))
- Power Capacity is the rate at which stored energy can safely be taken out of a battery and restored.

A battery possibly could have high energy capacity but low power capacity if, for some reason, its internal resistance were large. Conversely a battery could have high power capacity for a short time but a low energy capacity.

- Specific Power is the maximum rated power output/kg.(lb.) the battery can supply. Rating various batteries by specific power permits rapid performance comparisons of different kinds of batteries.

- Energy Efficiency of a battery is, according to Vinal[11,ch.8]:

$$\text{Energy Efficiency}^{1} = \frac{\text{useful energy out (watt-hrs.)}}{\text{recharge-energy (watt-hrs.)}}$$

$$(4\text{-}1) \qquad = \frac{\int_0^{t_1} I_1 E_1 \, dt}{\int_0^{t_2} I_2 E_2 \, dt}$$

Where:

I_1 = Battery discharge current

E_1 = Battery discharge terminal voltage

I_2 = Battery charging current

E_2 = Battery charging terminal voltage

t_1 = Battery discharging time

t_2 = Battery charging time

Two special cases of Eq. 4-1 are of interest to the systems engineer. If the discharge and charge occurs at constant voltages, then Eq. 4-1 reduces to:

$$(4\text{-}2) \quad \text{Energy Efficiency} = \frac{E_1 \int_0^{t_1} I_1 \, dt}{E_2 \int_0^{t_2} I_2 \, dt} = \frac{E_1 \times \text{Amp hour output}}{E_2 \times \text{Amp hour input}}$$

The integrals might be evaluated graphically. Once found, multiplication by the respective terminal voltage gives the watt-hours of input or output. Eq. 4-2 suggest why for some types of batteries with nearly constant terminal voltage the chief interest might be in the I vs t curve for discharge or charge. The ampere-hour output of a battery is a very common way of rating , [11,ch.5].
If the discharge and charge occurs at constant currents, then Eq. 4-1 reduces to:

$$(4\text{-}3) \quad \text{Energy Efficiency} = \frac{I_1 \int_0^{t_1} E_1 \, dt}{I_2 \int_0^{t_2} E_2 \, dt}$$

1. Sometimes called the "watt-hour efficiency".

If neither current of voltage is held constant, then we must resort to Eq. 4-1 directly and either graphically or electronically evaluate the integrals to determine the energy efficiency for the battery.

- Cycle Life is the number of times the battery can be charged and discharged under specified conditions. The cycle life of a battery may vary greatly with the depth of discharge, deep discharge tending to result in short cycle life. However, deep discharge may be difficult to avoid in some battery applications, as on electric vehicles.

Battery Fundamental Characteristics in a generalized form are shown in Fig. 4-2. Those are the terminal characteristics of the battery. Clearly

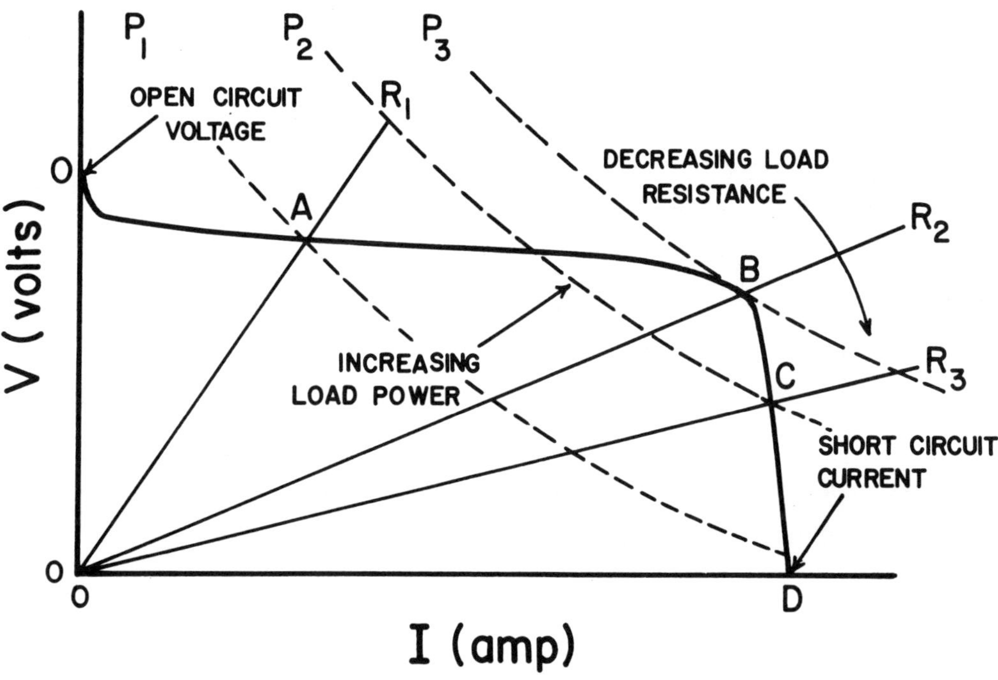

Fig. 4-2 The I-V characteristics of a generalized battery and its relationship to a DC load.

at infinite resistive load we have the open circuit voltage of point 0; as resistance decreases to R_1 the battery-load system operates at point A delivering P_1 power to the load. A further decrease in load resistance to R_2 moves the operating point to B, the maximum power output being P_3. Further decreases in load resistance to R_3 results in operation at C and a decrease in load power to P_2. If the terminals are short circuited (R=0) then we arrive at point D.

The shape of the battery's characteristic I-V curve will vary with the type and size of the battery. The 0-A-B portion of the curve is generally not flat, though for some cell types it may be approximately so. It usually will have a portion which is approximately straight. The battery's curve 0-A-B-C-D is generally found experimentally under a fixed known set of temperature, pressure, or other conditions.

Another fundamental characteristic of a battery is shown in Fig. 4-3, the voltage vs time curve under the constraint of constant discharge current.

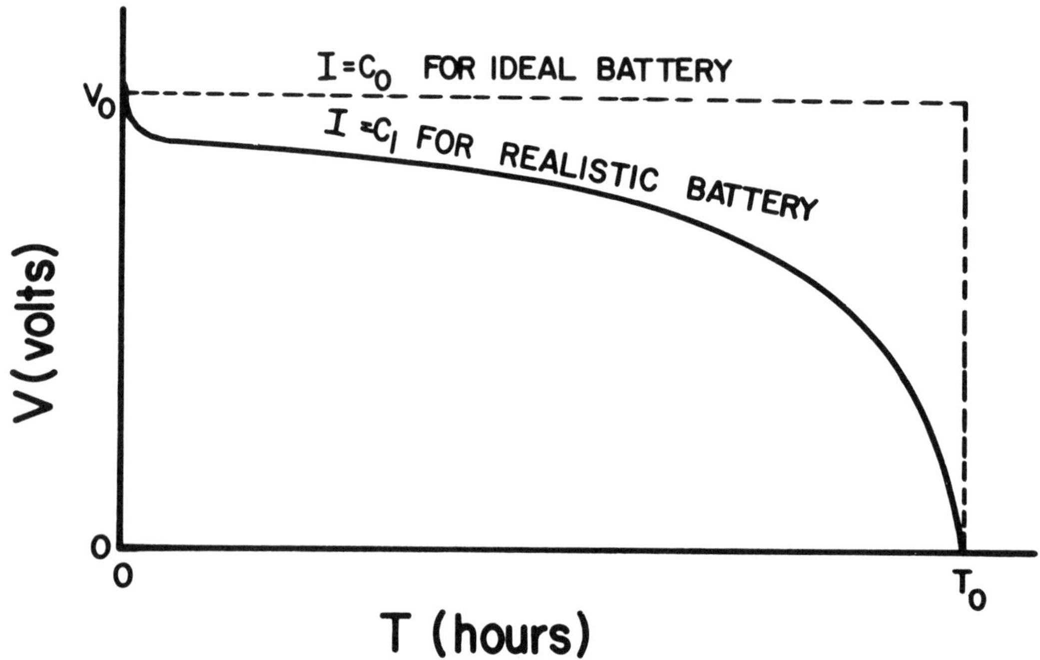

Fig. 4-3 The voltage vs time curve for a battery.

The electrical energy recovered is:

(4-4) Energy recovered $= \int_0^{T_0} I\, v(t)dt = I \int_0^{T_0} v(t)dt$

$=$ Ix area under curve (watt-hours)

For the 'ideal battery' shown maintaining a constant terminal voltage

(4-5) Energy recovered $= V_0\, I\, T_0 = V_0 \times$ "amp.-hour capacity".

In general each kind of battery has a different V(t) curve. The curves are obtained generally for a constant discharge current. The curves are generally found experimentally. There may be marked differences in these curves for different kinds of batteries, e.g. a nickel-cadmium battery is well known for its very sudden exhaustion in contrast to the gradual fall-off shown in Fig. 4-3.

Some factors which may markedly affect both the I-V characteristic curve and the V(t) curve include:

- Battery temperature. Low temperatures generally result in decreased battery performance and vice versa.
- Battery previous history.
- Whether it is used continuously or only part time.
- Battery environment, e.g. altitude, vibration, and other factors.

- Type, size, and manufacturer of battery.
- The internal resistance of the battery.

The above factors should generally be specifically identified when curves of the kinds shown in Fig.4-2 and 4-3 are being dealt with.

Battery Equivalent Circuit models are shown in Fig. 4-4. The first order model is the one most widely used if leakage effects are of no major importance.

Classes of Batteries are shown in Fig. 4-5. We shall be discussing some of these later as we review the state-of-the art. There are many different kinds of batteries which have been created and studied in the past. Each kind of battery typically has a long history, which may include failures as well as successes, and the systems engineer planning to use a battery should examine the history of each candidate battery in some detail before committing to his final choice. Otherwise he runs a serious risk of misapplying a battery.

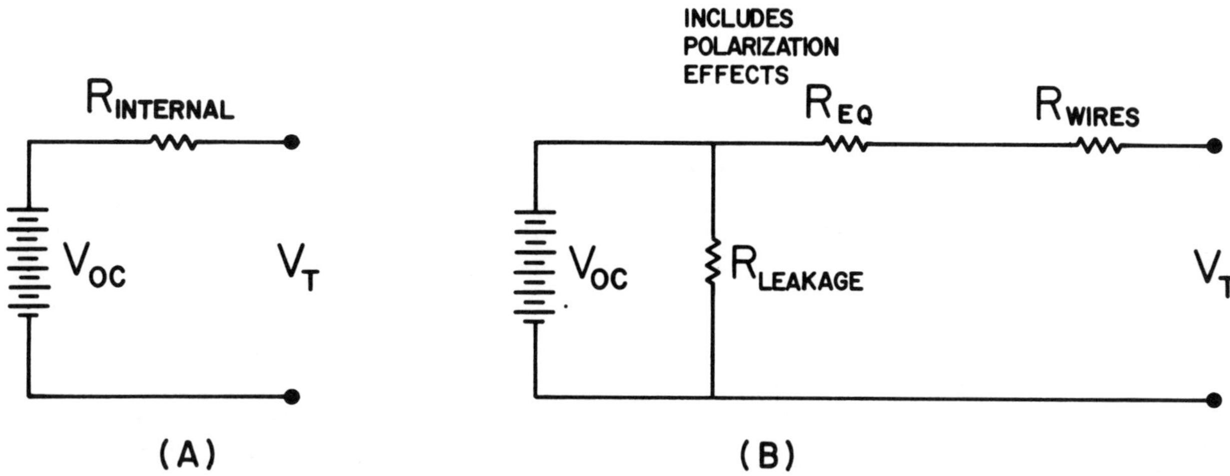

Fig. 4-4 Battery equivalent curcuit models. (A) First order model (B) Second order model showing leakage as the cause of 'run-down' and short shelf life on secondary batteries.

Conventional Batteries	Metal-Gas Batteries	Alkali Metal-High Temp Batteries
● Lead Acid	● Iron-Air	● Sodium-Sulfur
● Nickel Cadmium	● Zinc-Air	● Sodium-Chlorine
● Nickel-Iron	● Zinc-Oxygen	● Lithium-Sulfur
● Nickel-Zinc	● Zinc-Chlorine	● Lithium-Iron Sulfide
● Silver Zinc	● Nickel-Hydrogen	● Lithium-Chlorine
● Silver-Cadmium	● Cadmium-Oxygen	● Lithium-Copper
● Zinc-Bromine	● Cadmium-Air	● Lithium-Nickel halide
	● Aluminum-Air	● Aluminum-Chlorine
	● Lithium-Sulfur Dioxide	
	● Sodium-Air	
	● Magnesium-Air	

Fig. 4-5 One way of classifying secondary batteries. Other ways of classifying them also exist, and this list is by no means exhaustive of the many battery possibilities, e.g. advanced batteries developed for marine applications are not shown.

Having now acquired an elementary feel for battery terms and the general field, we now start reviewing the prior and present art of batteries.

Batteries - Prior Art (Pre 1975)

1967 Seiger and Lerner[14], Gulton Industries, Inc., published a paper treating sealed cells and auxiliary electrodes. Sealed cells avoid leaks that lead to corrosion. Auxiliary electrodes in alkaline batteries are used as charge control electrodes and for scavenging hydrogen or oxygen or both and thereby preventing pressure buildup within the cell. Their cells are small ones as might be used on space vehicles.

Lutwack[15], JPL, reported progress on a JPL program to develop heat sterilizable high impact resistant batteries for use on space craft. This program dealt with the many internal material problems inherent in battery development. Principal attention was on the silver-zinc battery because of its high energy density. While high impact is not important for terrestrial solar-electric applications, perhaps the high temperature aspects of this development may have carry-over value to terrestrial solar-electric batteries.

Bacher and Nekrasov[16], RCA, analyzed secondary batteries in a space vehicle application. Emphasis was on nickel-cadmium and silver-cadmium batteries. Since this was a systems analysis approach, portions of their work possibly may have a carry-over value to terrestrial solar-electric battery systems.

Shimotake and Cairns[17], Argonne National Laboratory, described research on sodium-bismuth and lithium-tellurium cells. They compared [their table 2] the specific energy densities (W-hr./kg.) and specific power densities (W/kg.) and other characteristics of 12 storage cells. Their lithium-tellurium cell ranks among the top of the cells examined in specific power density.

Shipps, et al. [18], General Dynamics Corp., described advanced research on secondary battery development for vehicle propulsion. Focus was on the zinc-air cell.

Weber and Kummer [19], Ford Motor Co., described their research on a sodium-sulfur high temperature (250-300°C) secondary battery. The battery "shows promise as a power source for electric vehicles by virtue of superior energy and power densities and uses cheap, light reactants." Their battery used liquid sodium as an anode and liquid sulfur as the oxidant in the cathode compartment.

1968 Woodward [20] summarized NASA'S progress on rechargeable batteries, temperature resistant batteries, metal-gas batteries, and the potential non-space application of space fuel cell technology. This work was principally orientated toward space vehicle batteries, but, as we have repeatedly emphasized in this work, much can be learned from studying the scientific and engineering work accomplished in the space program with the idea of carrying over some of the techniques and knowledge as it may apply to terrestrial solar-electrics.

During this year NASA personnel described a number of new battery developments springing from the space program:

- Baer and Ford [21] revealed a new recharge unit for optimum recharging of battery cells. It made use of a percent recharge unit, permitting "each cell of the battery to be charged to a preset percentage value of the discharge ampere-hour capacity of the cell."

- Ford, et al. [22] developed a means "to internally heat electrochemical cells by supplying power from an external source (solar array, power supply, etc.) to a resistive heater element incorporated in the cell construction."

- Gross [23] created "a new lightweight battery package design which provides for even cooling of individual alkaline (KOH) cells, constraint against cell expansion, and convenient replacement of cells."

- Shimotake, et al. [24] described a new lithium-tellurium bimetallic cell with increased voltage.

- Shimotake and Hesson [25] revealed a new experimental bimetallic EMF cell for a Na-Bi system operating at about 550°C.

- Paulkovitch and Ford [26] invented a circuit for controlling the charge of nickel-cadmium batteries by a coulometer and third electrode.

During this year nickel-cadmium aircraft batteries evidently began to come into use, for an aircraft battery maintenance manual was published by General Electric [27].

Mandel[28], Army Electronics Command, surveyed magnesium-air, cadmium-air, zinc-air, and iron-air secondary battery developments to 1968. Some of the problems such as 'fade-out', cadmium penetration, water loss, oxidation of anodes, and other problems were clearly exposed. Such batteries were recognized as being a marriage of fuel-cell and conventional battery technologies, holding the promise of high energy density electrical sources. It was recognized that "a considerable amount of additional effort and ingenuity will have to be expended" before the goal is reached.

Briggs [29],Gulton Industries, described hermetically sealed nickel-cadmium batteries and their solar array charging system for a satellite program.

Dunlop and Bounds [30], Comsat Laboratories, investigated adhydrode control of Ni-Cd battery charging for satellites. They found that pulse charging offered no significant advantages. "However, the adhydrode charge control provides a reliable means for minimizing overcharge, extending cycle life at deep depths of discharge."

Lutwack [31], JPL, updated their progress made in developing heat sterilizable Ag-Zn batteries for spacecraft. Some work was also starting on Ag-Cd and Ni-Cd battery systems.

Meredith and Uchiyama [32] analyzed the theoretical aspects of hot spot temperatures in silver-zinc batteries. The hot spot temperature is the maximum temperature increase at the center of the battery. They tabulated temperatures "for various types of rectangular battery dimensions as a function of the current drawn from the battery, the number of battery electrodes, and the thermal conductivity parallel to the electrode surfaces."

Gross [33] investigated heat generation in sealed spacecraft batteries. He studied nickel-cadmium, silver-cadmium, and silver-zinc cells. He found that "overcharge of nickel-cadmium cells is the most important factor contributing to heat generation."

Biddick and Nelson [34], Gould National Batteries, built and tested a more rugged version of Kapitza's bipolar battery intended for multisecond pulse discharge as in military systems. They achieved a power density of 1400 KW/M^3 ($39W/ft^3$). "The battery appears attractive for continuous high duty cycle pulses with charging accomplished between discharge pulses."

Arrance and Plizga [35] investigated "the problems involved in the development of long-life, high energy density thermal cells are discussed with particular emphasis on lithium anode cells with a variety of cathodes in fused LiCl-KCl eutectic electrolyte. The compatibility of a variety of inorganic separators and their characteristics in thermal cell test fixtures are discussed. The test cell data reported indicate the feasibility of this approach for high rate thermal cell development."

Shimotake and Cairns [36], Argonne National Laboratory, investigated a lithium/tin cell with an immobilized fused-salt electrolyte. They indicate "the materials problem has not yet been solved" for operation near 1200°C.

1969 Linden [37] summarized the characteristics and electrical performance of mechanically rechargeable zinc-air batteries. He points out its "energy density of 100 watt-hours per pound is twice that of the closest competing battery system."

Wilburn and Almerini [38] described new low temperature magnesium perchlorate batteries suitable for use as low as -40°F for portable military equipment.

Braeuer [39] described the development of a lithium cupric chloride reserve type organic electrolyte battery for military applications. They claim energy densities of 100-200 watt-hours per pound of battery weight on discharge.

Eisenberg [40] revealed "a new class of high energy density batteries employing lithium anodes and cupric halide cathodes with organic non-aqueous electrolytes." He claims they offer energy densities of 500-750 watt-hours/lb--about 50-150% higher than comparable silver oxide-zinc batteries.

Shimotake, et al. [41], Argonne National Laboratory, reported their research on secondary cells with lithium anodes and paste electrolytes at operating temperatures up to 375°C. Their "results suggest that properly designed batteries of these cells should store 360W-hr./kg. (6 hr. rate) and deliver power at 130W/Kg. (1hr. rate).

Moser et al. [42] did a design study on long life thermal cells of the Mg/LiCl-KCl/CuO type operating at 427°C. Their cell lasted 150 hours.

Opitz [43], Lockheed Electronics, described a seawater power supply using a magnesium-steel cell. It was intended for an undersea source of electrical power. Their results were based on laboratory tests and only a single field test cell. These were all very low power level cells.

Preusse et al. [44] did a parametric study on charge control of nickel-cadmium batteries in space vehicles. They studied capacity removed, depth of discharge, limiting voltage, charge rate, and environmental temperature.

Seiger, et al. [45], Gulton Industries, summarized the electrical characteristics of sealed nickel-cadmium batteries having a bipolar construction. Discharge characteristics, as a function of rate and temperature, as well as advantages and economics were treated.

Hartmann [46], JPL, described the application of bench tests in the development of heat sterilizable battery separators. This is a part of JPL's on-going program to develop heat sterilizable spacecraft batteries.

Teresa amd Corbett [47], Lockheed, summarized two years performance of their type X1 nickel-cadmium battery used in space vehicles. Their data permitted them to assign a failure rate of one failure per million hours of operation of this battery.

Palmer, et al. [48] summarized the high rate performance of mechanically rechargeable zinc-air batteries of small sizes for portable military radio transceivers.

1970 The IECEC Proceedings for this year was not available for review by this researcher.

1971 Kyle, et al. [49], Argonne National Laboratory, did an in-depth study of lithium/sulfur batteries for electric vehicles. The batteries operate at about 380°C. Their laboratory results estimate "ranges of about 193 to 343 Km (120 to 213 miles: for a 2087 Kg (4600lb.) automobile powered by a 488 Kg (1075lb.) lithium/sulfur battery." They identified areas needing additional work.

Witherspoon, et al. [50], General Motors, reported on progress with an experimental vehicular zinc-air battery with replaceable anodes. They concluded that "the zinc-air system provides adequate power and energy storage to give good range for small electric vehicles. Energy storage ranged from about 55 W-hr/lb. at 10W/lb. to 33 W-hr/lb. at 33 W/lb. The range was 3 to 5 times lead-acid batteries or 150-160 miles." They identified problems needing further work

Caprioglio and Weinberg [51], Gulf General Atomic Co., also reported on zinc-oxygen batteries for electric vehicles. Such a "prototype electric vehicle powered by a zinc-oxygen battery has been designed, fabricated, and successfully tested."

Breeskin and Taylor [52], Mc Donnell Douglas, described their research on third electrodes in Ni Cd batteries for spacecraft. Gaz [53] also reported on test results for the same type battery.

Liang [54], P.R. Mallory and Co., reported on the storage and discharge characteristics of a lithium solid state battery of small size with a 20 year anticipated shelf life.

Metcalfe, et al [55], Standard Oil Co., described a twelve-cell molten salt (LiCl-KCl) battery operating at 400°C. It operated for 100 cycles. The best cell built had an energy density of 36 watt-hrs./lb. They believe their results indicated it is possible to build a 15 KWH battery capable of delivering 60 W-hr./pound.

Schwartz [56] discussed the problem of extending the life of Nickel-Zinc batteries intended for the urban hybrid vehicle application. He showed how space technology could be applied to this problem, suggesting that much work has already been done or is underway on nickel-zinc batteries within the space program and that making use of this information would permit more confidence and a higher probability of success with hybrid vehicle batteries.

1972 Gratch et al. [57], Ford Motor Co., updated the progress on the sodium-sulfur battery development, claiming longevity in excess of 2000 deep charge-discharge cycles has been achieved. Minck [58] described the electrical performance of a prototype sodium-sulfur battery operating at 300° Celsius.

Silverman, et al. [59], TRW Inc., investigated the cycle-life of a sodium-sulfur cell of a laboratory type.

Levine, et al. [60], Dow Chemical Co. presented their research results on the sodium-sulfur battery using glass as the electrolyte. The cell operated at 300° Celsius.

Gay, et al. [61], Argone National Laboratory, laid out their research results on secondary cells using lithium anodes and sulfur-selenium-thallium cathodes. Their cells operate in the vicinity of 400°C. One problem identified is short lifetimes on cells with paste electrolytes. Liquid electrolytes demonstrated longer lifetimes.

Takehara [62] reported on studies on lithium alloys-chlorine battery research for experimental automobile use.

Klein [63] published a paper on metal-gas rechargeable batteries of a sealed design. He sees that "a family of secondary batteries employing rechargeable gas electrodes is emerging". Such batteries appear closely kin to fuel cell technology.

Stockel, et al. [64], Comsat Laboratories, described their research work on a nickel-hydrogen secondary cell for spacecraft, finding the system to have high stability. Also "charge and discharge voltages are comparable to those of the nickel-cadmium cells presently used in satellites".

Menard [65], Gould Inc., described a new line of high energy density nickel-cadmium batteries of D and C cell sizes. The D size cell yields an energy density of 27 W-hr./lb over the 4.5-6.7 amp-hr. capacity range. This is about a 40% improvement over the standard commercial cell. The improvement came through "improvements in the internal cell components".

Fono [66] treats an assessment of Nickel-Cadmium battery cycle life in spacecraft. "Cycle life curves for cycles of different time durations are constructed and presented." He presents cycle life to failure data on batteries made by many manufacturers.

Kerr and Stager [67] describe a 1 KW prototype battery system for use on spacecraft. It is based on Ni-Cd batteries.

Charkey [68] described Energy Research Corp's. development of nickel-zinc cells for sustained high rate discharge applications, as for the propulsion system of a torpedo target drone.

Silverman, et al. [69] researched the optimization of lead-acid batteries for high power-short pulse operation. They showed that "a lead-acid cell containing modified plates was shown to be capable of producing peak specific power in excess of 200 watts per pound and could sustain power of 188 W/lb. for twenty seconds".

Snider and Nagle [70] investigated nonleaking battery terminals on conventional alkaline batteries. Previous terminal designs had been less than optimal.

<u>1973</u> Berberick, et al. [71] described the many aspects of manufacturing a new Bell System lead-acid battery with a life of more than 30 years. The battery is the emergency source to run a telephone plant during times of power failure.

Eisenberg and Wong [72], Electochimica Corp., described their progress on stable lithium batteries. They presented "extensive data on a broad range of performance parameters such as effect of discharge rates, effect of temperature and effect of elevated temperature storage on delivered cell capacity, energy density per pound, energy density per cubic inch, and voltage regulation. Design trade-off characteristics and advantages in new power system designs are discussed."

Halberstadt [73] described a new type of power cell based on a rotating cathode shaft and alkali solid metal anode. The cathode is water fed. He claims the cell is unaffected by impurities and contaminants and that "lithium is actually quite well behaved in water." Such cells have been developed up to multiple kilowatt systems.

Symons and Carr [74] presented their research results on chlorine electrodes in the zinc-chlorine battery. They developed a battery testbed for a Vega automobile yielding 50 MPH and 152 mile range.

Kirsch [75], Sperry Rand, in a paper appearing to have significant carry-over value to solar-electrics, addressed the problem of modeling the nickel-cadmium battery in the Apollo telescope mount satellite. Two deterministic models were developed, a capacity degradation model and a cyclic charge/discharge model.

Gay, et al. [76], Argonne National Labs., described the development of lithium/sulfur cells for application to electric automobiles.

<u>1974</u> Klein and Baker [77] described the nickel-hydrogen secondary battery system, claiming it offers the advantages of long cycle life, insensitivity to overcharge and reversal, and high energy density. They estimate 18-20 W-hr./lb.

Gandel, et al. [78], Lockheed, described developments under way on nickel-hydrogen batteries for synchronous satellites. They suggest this system "should be considered for replacing Nickel-Cadmium batteries."

Littauer and Redlien [79] summarize the features of the lithium/water battery as a marine power source. "This battery has unique features which make it especially practical for marine use. Its electrolyte can be predominately seawater, and the system is, in principle, insensitive to deep ocean pressures." They examine a wide range of marine applications for such batteries. Some of this technology may have a carry-over value to land based solar-electrics, or possible storage systems with OTEC or ocean sited wind-electric plants. Chilenskas, et al. [80], Argonne National Laboratory, also investigated the same application but using their new bicell sealed cell design.

Solomon [81] described the performance of a new non-silver seawater battery for sonobuoys.

Gay, et al. [82], Argonne National Labs, updated their work by describing the development of lithium/sulfur cells for application to electric automobiles.

Seiger, et al. [83], Heliotek, summarized what was known about the positive electrode-plaque relationships in sealed Ni-Cd cells. Puglisi and Seiger published a corollary paper [84] dealing with methods of impregnating to produce nickel oxide electrodes.

Juvinall, et al [85], JPL and Gould Labs., revealed a novel negative-limited sealed nickel-cadmium cell free of the gassing problem. By adequate charge control a viable means exists for controlling the cell internal pressure.

Easter [86] compared the energy densities of nickel-hydrogen and silver-hydrogen cells. He found silver-hydrogen to yield higher energy densities in all cases considered.

Walsh, et al. [87], Argonne National Laboratories, proposed the use of lithium/sulfur cells being developed for use in large batteries for bulk energy storage on electric utility networks. A 100MW-hr. lithium-sulfur battery facility is proposed. Details of the cell design are described.

Mitoff and Bush [88], GE, described the characteristics of their sodium-sulfur cell intended for bulk energy storage.

Klein [89] dealt with the important problem in a utility context of removing heat from batteries used for load leveling.

Bush [90], GE, summarized GE progress made in the development of sodium-sulfur batteries intended for energy load levelling in a utility context. This work was sponsored by EPRI. They focussed on factors that limit the operating life of sodium-sulfur batteries. They found "life to abrupt failure was limited principally by the beta alumina solid electrolyte. Through changes in the formulation of the solid electrolyte, cell life was extended from less than 5 ampere-hours/cm^2 of electrolyte to about 70 ampere-hours/cm^2." They saw, however "no fundamental scientific or technical barriers to the development of a sodium-sulfur cell suitable for bulk energy storage--." Their report has a nicely done Appendix summarizing the principles of operation of the sodium-sulfur cell. There they also nicely summarized the major <u>strengths</u> of the sodium-sulfur cell as:

- High open circuit and operating voltage: 1.8-2.1V
- High specific energy: 100-140 WH/lb.
- High energy density: 5-8 KWH/$ft.^3$
- High power density or discharge rate: 1 to 3 hour rate
- Hermetically sealed construction
- No gas evolution
- Zero self-discharge or excellent charge retention
- High recharge efficiency: 85% at three hour rate
- Liquid reactants: No aging of active materials
- Potentially five to ten year life.

They see the <u>limitations</u> of the sodium-sulfur battery as:

- Operating temperature 285-350°C

- Possible aging of solid electrolyte: deterioration of mechanical and electrical properties on cycling
- Highly reactive components: Safety considerations, corrosion
- No overcharge capability: Need for voltage control when charging

A more detailed treatment of this work is in reference [91].

Present Art-Batteries (Post 1975)

1975 Hall, et al. [92], Rockwell International Corp., issued an EPRI report detailing their developmental progress on lithium-metal batteries for utility load leveling. They found the porous alumina separators being attacked, limiting the cell life by separator failure.

Nelson, et al. [93], Argonne National Labs, issued a report describing ANL's developmental efforts on high-performance lithium-metal sulfide batteries for both utility off-peak energy storage as well as for electric vehicle propulsion. Their best cell appears to "have negative electrodes of solid lithium-aluminum alloy and positive electrodes of iron sulfide ($Fe\ S_2$ or FeS), with an electrolyte of LiCl- KCl eutectic (mp, 352°C). The operating temperature of the cells is about 400-450°C." A follow-up of this work is in reference [94].

Bechtel Corp., under contract to Argonne National Labs, issued a technical report [95] describing a conceptual design of a battery energy test (BEST) facility so as to be ready for "any of several advanced batteries currently under development (which) may be available for large scale testing in a utility application by as early as 1980." "In its fully implemented state, the BEST facility will be capable of simultaneously testing three 10 MWH battery systems. Three inverter/converter units, each rated at 2.5 MW, will interface these battery systems to a utility network. A 1000V AC bus system, comprised of three individual buses, will enable the independent inverters through a series of ac breakers, switches and fuses. Similarly, a 13.8 KV AC system will connect the inverters to either the utility network or to a synthetic load."

Bechtel Corp. issued a report [96] summarizing battery energy storage applicable to centralized photovoltaic power plants. They succinctly summarize lead-acid, lithium, sodium, and chlorine batterv technologies. They also investigated the needed power conditioning for bulk power plants. They concluded that near-term (less than 7 yrs.) installations should use lead-acid batteries at about $50/KWH. Beyond 7 years advanced batteries now being developed are seen best.

Cooper, et al. [97] investigated the nation's requirements for lithium if lithium-water-air batteries were to become widespread in electric vehicles. They found that "the supply of lithium will have to be increased."

Electric Storage Battery, Inc. (ESB) issued a report [98] clearly laying out their progress on the sodium chloride battery development for utility load leveling. Many material-related details were dealt with. "A cell has been developed which offers considerable promise for lightweight, long life, and low cost batteries." Their extensive detailed work is difficult to summarize adequately.

Bro, et al. [99], P.R. Mallory Co., discussed the discharge process in high rate Li/SO_2 batteries. They conclude that "high rate cells can be make that are capable of operating at power levels of 460 watts/Kg while delivering 100 W-hrs./Kg. At high rates of discharge the cell temperature and pressure increase, and cells must be provided with vents to deactivate the cells well below the melting point of lithium. For high rate applications the Li/SO_2 cells should be provided with properly designed heat dissipation structures."

Holleck, et al. [100], EII Inc., described their developmental results on transition metal sulfides as cathodes for lithium batteries. They investigated niobium disulfide and trisulfide. With unoptimized battery electrodes they achieved energy densities of 250 W-hr./lb. initially, dropping to about 150 W-hr./lb. for later cycles.

Sudworth [101] summarized an extensive British effort to develop sodium-sulfur batteries for railway use in England. Their battery requirements are:

● Specific energy	150 W-hr./Kg.
● Specific volume	200 KWH/m^3
● Specific power	90-150 W/kg.
● Overall efficiency	70 %
● Cost	< 20 L/KWH
● Life	3 years (1,000 cycles)

He claims "to have produced a cell dêsign which is safe, potentially suitable for automatic assembly and which meets our requirements of specific energy, volume, and cost. Our goal of a three year life still remains to be realised, but results obtained so far lead us to believe it is attainable".

Levine [102], Dow Chemical, described Dow's progress in the development of the hollow fiber sodium-sulfur cell. The cell uses fine hollow glass fibers as the electrolyte. They found two types of failure modes of these fibers.

Marcoux, et al. [103], TRW, described the design of a sodium-sulfur test cell utilizing a β alumina flat plate separator. This test fixture appears to be highly useful in laboratory testing of various combinations of materials, etc. for laboratory cells of the sodium-sulfur kind.

Gay, et al. [104], ANL, describes their development of high-perfomance iron-sulfide electrodes with porous current collector structures. This work is aimed eventually at batteries for off-peak energy storage in utility networks' and for electric vehicle applications.

Schertz, et al. [105], ANL, described their multiplate Li-Al/FeS cell and modules as envisioned for electric vehicle propulsion. The battery has a nominal output of 43 KWH and can be designed to fit under the hood of a Mustang II.

Sudar, et al. [106] described the development of a rechargeable lithium/iron sulfide battery underway at Atomics International. It is intended for utility load-leveling application. They describe a 100 W-hr. demonstration cell and deal with electrode and ceramic separator developments.

Schaefer, et al. [107], ESB Inc., described the development of a LiAl anode and KCl-Li Cl electrolyte cell being developed for fork-lift truck applications. The energy per unit volume of 110 W-hr./m^3 exceeds that of the lead-acid cell.

Klein [108], Energy Research Corp., revealed the design of a 20 A-hrs and a 50 A-hr. nickel-hydrogen secondary battery. They have energy densities of 21 and 24 W-hr./lb. respectively. In this battery "during charge, hydrogen gas is evolved and stored within the housing and during discharge the same gas is reconsumed." It was developed for satellite application.

Miller [109], Eagle-Picher Industries, Inc., showed several designs of nickel-hydrogen batteries. He claims long life, "high energy density, long cycle life, deep depth-of-discharge capability, and an inherent insensitivity to overcharge and overdischarge." The hermetically sealed battery is intended for aerospace use as a lower cost competitor for the nickel-cadmium battery type.

Malaspina [110], ESB, Inc., analyzed lead chloride-magnesium sea water batteries for deep submergence operation in sonobuoys. He finds "the lead chloride-magnesium battery, using a non-pressure sensitive system, will operate independent of pressure from the surface to depths far in excess of the capability of present sonobuoy with little electrical noise and virtually no mechanical noise below 1,350 feet".

Yao and Birk [111], did an extremely significant and well-done state-of-the-art paper on advanced batteries for utility load leveling and electric vehicles. In fact, they brought to the battery field a much needed integration which was largely previously lacking, based on this researcher's findings of the battery field to date. Yao and Birk clearly make the case, from the utility systems view, of the need for battery energy storage, and they clearly identify the "benefits and advantages of batteries for bulk energy storage, and contrasted to those of gas turbines, can be summarized as follows:

- Mitigation of oil shortage and import problems
- Lower cost of electric energy--as the result of savings in fuel cost and improved utilization in capital equipment
- Environmental pollution abatement of air and noise, and aesthetic conservatior
- Modulability and lower capital cost of batteries
- Substantive savings in power transmission
- Shorter lead time for construction
- Reliability in power emergency and regulation."

They point out that "presently, no commercial battery can meet both the technical end economic requirements" needed of load leveling battery application.

They surveyed battery applications to electric vehicles. Their paper also included an excellent summary of the state-of-the-art of advanced battery development programs in the U.S.A., including succinctly summarizing the state-of-the-art of each of the chief battery options.

Overall, Yao and Birks' paper, in this researcher's opinion, was outstanding in its integrative quality in a field characterized by fragmentation, details, and many individual battery systems being promoted by their proponents. It was indeed refreshing to this researcher to find such a well-done paper. It is particularly recommended to systems engineers.

Halberstadt, et al. [112], Lockheed, cited the physical and economic characteristics of the lithium-water marine battery for sonobouy, torpedo target, and submersible applications. They show specific energies of 102, 224, and 404 W-hr./kg. respectively for batteries for these applications.

Charkey [113], ERC, reviewed the improvements in electrode structure, cell construction, and charging methods made on nickel-zinc batteries of ERC'S. nickel-zinc batteries are seen as competitors for lead-acid or nickel-cadmium batteries. They are available both sealed and vented in capacities from 7-100 amp-hours. Cycle life falls in the 150-200 cycle range.

McCoy [114], Rockell Int'l, revealed "a new type of Zinc-Nickel oxide cell using a revolving thin aperatured shutter-separator does not suffer from zinc electrode shape change and therefore retains its performance characteristics on repeated deep discharge". He shows an experimental cell built and tested. This battery type appears to be in an early state of development. The intended application is for "utility energy storage and all electric vehicles".

Maskalick, et al. [115], Westinghouse, in a paper highly significant to system engineers, makes the case for the lead-acid battery for utility electrical energy storage. He clearly cites the criteria any proposed electrochemical storage system must satisfy and examines nine possible batteries for peaking power use. They nicely summarized the key numbers for each and discussed a comparison of all. They conclude "the 115 year old, well-known and understood lead-acid system offers higher probability of successful application of technology than any other proposed system".

1976 Weiner [116], Ford Motor Co., as a part of an on-going program, described research results on electrodes and electrolyte for the Ford sodium-sulfur battery. He says their "cell testing program has yielded additional promising results, and more information has been accumulated concerning the reasons for cell failure. The oldest laboratory cell is now in its 18th month of operation, has passed over 1060 A-hr./cm^2 in discharge and completed more than 6400 cycles".

Chilenskas, et al. [117], ANL, studied the nation's lithium requirements for lithium-aluminim/iron sulfide batteries for load- leveling and electric vehicles. They concluded that "expansion of exploration for new sources of lithium and increases in production capacity will be needed".

Holleck, et al.[118], EIC Corp., described their developmental results on sulfur-based lithium-organic electrolyte batteries. Their goal is to develop "a rechargeable organic electrolyte lithium battery with the following specifications: An energy density approaching 100 W-hr./lb; a cycle life in excess of 500; high charge retention on stand; operating range-40 to +160° F---". This appears to be an extensive detailed program difficult to summarize adequately.

Early in 1976 Birk [119], assembled an EPRI sponsored workshop on near-term technologies for utility load leveling using batteries. Focus was on the lead-acid battery. Ten papers by various authors active in the battery field are included in the report of the workshop. This entire volume will be of much interest to systems engineers interested in the utility energy storage problem via batteries. More of a systems approach is taken here than one finds exhibited in the many detailed R&D battery papers.

Yao [120], ANL, in a paper on the applications and prospect of energy storage batteries nicely summarized the preliminary goals for battery development. These are repeated here in Table 4-1.

	Utility Load-Leveling	Electric Vehicle
Duty Cycle	5-10 hr. discharge/day	2-4 hr. discharge
	5-7 hr. charge/day	4-8 hr. charge
Energy Efficiency (Round trip)	> 70 %	> 50 %
Cycle Life	> 2500 Cycles (10 yr.)	> 1000 Cycles (3-10 yr.)
Installed Cost	$20 - 30/KWH	$20-35/KWH
Specific Energy	---	60 WH/lb.
Specific Power		
Sustaining	---	40 W/lb.
Peak	---	100 W/lb.
Typical Installation Size	200 MWH	20-30 KWH
Environmental Impact	Minimal	Minimal

TABLE 4-1 Preliminary goals for battery development. From Yao [120].

He points out the stringency of those goals and that the current effort to develop advanced batteries aims to meet these requirements.

Sudar, et al. [121], Rockwell International, reported on the development of lithium-metal sulfide batteries for load leveling. The report deals with the many details of developing practical lithium-metal sulfide batteries and culminates in a description of a 150 W-hr. sealed cell they developed.

Nelson, et al. [122], ANL, described high performance batteries being developed for off-peak energy storage and electric vehicle propulsion. Their major cells consist of FeS or FeS_2 positive electrode, molten LiCl-KCl, and negative electrodes of lithium-aluminum alloy.

Their cell operates at 400-450°C. They are trying to commercially develop this cell. The report also treats cell chemistry, advanced cell engineering, and alternative secondary cell systems.

Birk [123], Energy Development Associates, reported on the evaluation of a 1KWH zinc chloride battery. "The 1 KWH test battery -- demonstrated an average energy efficiency for the 100 cycles of 74.6% at a current density of 33 ma/cm^2. The system was operating at 27°C---." He concludes that "the results of this study demonstrate the technical feasibility of the zinc chloride battery for utility applications. --- This high energy efficiency and the stability of performance through the 100 charge and discharge cycles suggests this battery is an attractive candidate for development into utility energy storage system."

Silverman, et al. [124], TRW, described their development work on sodium-sulfur batteries. It "uses molten sodium and sulfur as the active electrode materials, employs a solid electrolyte (beta alumina), and is operated at 300-350°C. The system is particularly promising for utility application because of the intrinsically low materials cost." They see "the major challenges of developing the sodium-sulfur battery into a commercially viable energy storage system are:

- Development of corrosion resistant sulfur-electrode containers
- Design and development of a less highly stressed seal
- Fabrication of long-life electrolytes
- Design of cells capable of full utilization of the active positive-electrode material (sulfur)."

Unfortunately this work was terminated by TRW "because the lengthy development period was not consistent with TRW's present business interests in the energy management field."

Briggs, et al. [125] described engineering aspects of the silver-zinc battery for the deep submergence rescue vehicle, including especially a method for predicting battery capacity which they claim has been applied accurately in laboratory tests.

Charkey [126], ERC, described advances in component technology for nickel-zinc cells intended for electric vehicles. They are working on "(1) development of inexpensive non-sintered nickel electrodes, (2) development of long-life zinc electrodes, (3) development of stable separators, and (4) development of 300 A-hr. size cells." They have documented battery failure mechanisms. They are projecting a life of 1000 deep cycles.

Kemp, et al. [127], Lockheed, described the design and performance features of a 0.45 KW, 4 KWH lithium-water marine battery.

Marmaduke and Donaghey [128], Univ. of California, discussed the engineering evaluation of solid transition-metal sulfide electrodes for use in high energy-density batteries. They developed "simple criteria --- for screening potentially, important solid electrode materials for battery applications."

Shimotake, et al. [129], ANL, described "the development of engineering-scale uncharged Li-Al/Fe S_x cells capable of high performance."

This cell is a part of "a family of new rechargeable battery systems comprising negative electrodes of lithium or a lithium alloy, positive electrodes of sulfur or a metal sulfide, and a molten-salt electrolyte has been under development at Argonne National Laboratory." Their "future work will concentrate on long lifetimes and the refinement of cell design to obtain lower cost cells."

Gay, et al. [130], ANL, reviewed electrode designs and fabrication techniques for lithium-aluminum/iron sulfide cells intended for utility and electric vehicle applications. They recognized, correctly, that "the final choice of electrodes is based on the ease of manufacture and component costs."

Mc Coy and Heredy [131], Rockwell International, described the development status of lithium-silicon-iron sulfide load leveling batteries being developed for electric utility load leveling. These batteries "use a molten LiCl-KCl electrolyte and are operated at a temperature of 400°C or higher."

Dunning, et al.[132], General Motors, described the development of compact lithium-iron disulfide electrochemical cells for advanced electric vehicles. They see the goals for such high-temperature batteries as:

- Specific energy — 200 W-hr./kg.
- Specific power — 200 W/kg. peak
- Specific power — 50 W/kg. sustained
- Minimum cycle life — 300-1000
- Minimum lifetime — 3 yr.
- Cost — $20/KWH
- Cell capacity — 100 Amp-hr.

They identified 5 areas where improvements in performance are necessary of which improving cell performance, high cost, and durability are chief.

Bauer, et al. [133], Brown, Boveri and CIE AG, described the development of sodium sulfur cells at Brown Boveri (Germany). They are also developing these for utility load leveling and for electric vehicles, the same as in the U.S.A. They conclude that "the most urgent task in the development of sodium-sulfur batteries appears to be at present to keep the effective capacity of cells constant on a reasonably high level."

Dell, et al. [134] described the development of sodium-sulfur batteries in the United Kingdom. The intended applications are for road vehicles and for rail traction. They describe their tubular cells and conclude that the "sodium sulfur cell can be developed to meet the technical targets and the cost targets of motive power applications in the United Kingdom."

Betz, et al. [135] described the use of nickel-hydrogen storage batteries for use on a navigation satellite. These appear to be the same nickel-hydrogen batteries which were cited in an earlier year herein. These batteries have now been flight qualified.

Sparks and Scott [136], TRW, consider the engineering problems in applying nickel-cadmium batteries in deep discharge (65-85% depth of discharge) satellite applications.

Kirsch and Paschal [137] summarized the state-of-the-art of nickel-cadmium cell reliability model as a function of cumulative cyclic operation, cell temperature, and cyclic depth-of-discharge. They also presented "a simple and convenient cell failure analysis technique."

Miller [138], Eagle Picher Industries, Inc., described an advanced sealed nickel-cadmium cell design of approximately 20 A-hr. capacity. The intended application is not clear.

Kozawa and Takagaki [139] summarized new technology and recent Japanese products in lead-acid batteries. Emphasis is on automobile batteries.

Battey [140], [141], Bechtel Corp., dealt with the problem of engineering design for the chlorine storage for a zinc-chlorine utility load-leveling battery. The chlorine must be stored external to the electrochemical converter. "The storage function is a key aspect and major part of the battery system. For example, a 40MW, 5 hr. zinc-chlorine battery system will require storage capability for 150 tons of chlorine." He concluded that "the chlorine hydrate approach is a simpler and less expensive storage system than liquid chlorine."

Chatterji [142], GE, summarized a year development effort on sodium-sulfur batteries for utility bulk storage of energy. Their cell "consists of many individual beta-alumina tubes used as the separator between single sodium and sulfur compartments." A considerable ceramic development effort is involved on this program which was supported by EPRI. Their Appendix II on sodium-sulfur battery system reliability and performance analysis particularly impressed this researcher as having wide-spread applicability to all future sodium sulfur batteries. Chatterji dealt with the various ways the cells might be connected including crossover connections and analytically treated failure probabilities for the various configurations. There is a striking similarity to the configurations and analysis methods here and much earlier work cited in Chap. 2 in connection with photovoltaic array reliability, the work done by NASA researchers. Chatterji's impressive report concludes that "this year's work considerably strengthened our previous opinion that the sodium-sulfur battery will meet all technological requirements of a successful load-leveling and peaking system."

<u>1977</u> Lazennec [143], Compagnie General d'Electricite, recited their improvements of electrolyte and seal technology for sodium-sulfur and sodium antimony trichloride load leveling batteries. The electrolyte portion concentrated on the manufacturing aspects; the seal development was for alpha-alumina to aluminum junction. They also did some work on cell design and optimization, including some cost studies.

Rauh, et al. [144], EIC Corp., have investigated the rechargeability of lithium-sulfur batteries operated at ambient temperature. The "sulfur is completely dissolved in an organic electrolyte, as polysulfide, Li_2 Sn." They found that "the principal difficulty with the rechargeability of the Li/Li_2 Sn battery in its present form is the control of the self-discharge rate during extended cycling" brought on by the "deteriorating morphology on the Li electrode results in an increase in the self-discharge rate".

They saw that "a rechargeable battery--operating at ambient temperature could be useful in electrical load leveling and for vehicular propulsion--". Ambient temperature lithium-sulfur cells appear to be in a very early state of research.

Donovan and Imamura [145], Martin Marietta, detail the development of three design approaches to individual cell monitoring and control for sealed secondary battery cells of the Ni Cd and Ag Zn varieties. Their control system has substantial electronics. They wisely conclude that "the best method of providing control and protection of the rechargeable cells is dependent on the specific application."

Lear and Imamura [146], Martin Marietta, summarized their life test experiences with cycle life characteristics of sealed silver-zinc cells with inorganic separator. This work is based on spacecraft use. They conclude that "cell-level charge control and discharge protection are required on Ag-Zn batteries to achieve long cycle life and (that) the sealed Ag-Zn cells--have demonstrated their capability to cycle at a fairly high depth of discharge--."

Murie [147], General Motors, "discusses the post test analysis of two compact Li/Fe S_2 cells--that operated for 10,462 and 1,733 hrs. --at 475°C." These cells were candidates for electric vehicle power. A thorough x-ray, visual, and scanning electron microscope analysis was made of the various materials in the cells. The paper appears to be both developing the methods for post-test analysis as well as determining failure areas on these particular batteries.

Mitoff, et al., [148] summarized the GE sodium-sulfur development effort cited here earlier under 1976 progress by Chatterji.

Sudar, et al., [149], Rockwell International, recounted their development progress on lithium-silicon/iron sulfide load leveling batteries. These are "molten salt batteries using lithium-silicon negative electrodes and iron suffide positive electrodes." An energy density of 79 W-hr./Kg has been reached. "A 150 W-hr. cell has been operated for over 110 cycles in a period of more than 1 year. Scaleup of these cells has been carried out without difficulty to that of a multi-electrode cell with a capacity of 1 KWH."

Lai and Sammells [150], Rockwell International, recounted their development progress on negative electrodes for lithium-iron sulfide load leveling batteries. They found that "the electrochemical properties of electrodes using the modified alloy are superior to those obtained with negative electrodes using the lithium-silicon alloy alone." They show photomicrographs of some failed materials.

<u>1978</u> Fewer documents on 1978 battery research were available to this researcher.

A <u>Journal of Power Sources</u>, published by Elsevier in Lausanne Switzerland, has recently come into existence. It is an international journal on "the science, technology and application of non-mechanical, electrical power <u>storage</u>, generation and conversion."

A new book by Palz [151] briefly treats storage batteries in a cursory way [p. 199, 238-240]. He makes the important point [p. 239] that "if lead-acid batteries are discharged slowly and overcharge is avoided, they have a life expectancy of up to 20 years." He also points out that "Near Bordeaux, France, a system comprising a lead-acid battery and a 240W silicon solar array has been in operation for more than seven years. After five years' service the battery cells were analyzed, but no deterioration could be detected." He compactly summarizes [p238] in a single table the state of the art of lead acid, nickel-cadmium, nickel-iron, silver-zinc, zinc-air, lithium-silver chloride, lithium sulphur, and sodium sulphur cells. Unfortunately he fails to cite references for the sources of his various data, so it is not possible to acquire additional information therefrom. One necessarily must fall back in this case on a summary of the original research literature, as the present report.

Sullivan [152] has helpful practical information about batteries applicable to aerogenerator systems.

Closure On Batteries This sub-section laid out a few fundamentals of batteries and chronologically summarized the 1967-1978 research and development progress on various kinds of batteries. Unfortunately time did not permit grouping all papers relating to a single battery type; but such a grouping is easily and quickly possible, if needed, as a small step beyond the present report. It is clear that the battery field is indeed a dynamic one.

It should be noted that summarizing the state-of-the-art of battery technology is, at best, a difficult task. Like photovoltaics in Chap. 2, it was absolutely necessary to cite and summarize many detailed papers. It can safely be said that the rather large and important battery field is, at present, characterized by these many technically detailed papers and that there has been no previous attempt to integrate the entire technology within the scope of the present summary. Instead, as in other fields, battery technology appears to spawn specialists in a single type battery system rather than generalists who have perspectives on the entire field. Hopefully, perhaps the present summary has been a small step in establishing such perspectives.

The thoughtful reader will observe we have not surveyed fuel-cell technology which is closely related to battery technology. It was felt the fuel cell is principally a converter of energy rather than a storer of energy, the principal thrust of this Chapter. Surveying fuel-cell technology was simply beyond the scope of the present research.

As a final comment, it is interesting to observe that of the many various battery types surveyed, virtually all of them necessarily resort to developmental work of proposing and actually physically testing. It appears that, at the present state-of-the-art, one does not create successful long-life physical batteries via computer programs. Perhaps this says something about the state-of-the-art of the basic science involved in battery technology and the modeling thereof. Many of the complex electrochemistry processes appear simply not well understood.

Heat Storage

The next most important energy storage means for solar-electrics, especially solar-thermal-electrics (as in the Power Tower concept), is heat storage. Much of this technology overlaps with heat storage for solar heating of buildings, though the storage temperature needed for energy storage in buildings is lower than that needed for solar-thermal-electrics. Though our principal interest here is in solar-electrics, because of the overlap it is difficult to separate the heat storage literature, e.g. into low temperature and high temperature storage categories. It is therefore unavoidable in the review to follow to intermix these two objectives.

We also here slightly alter our presentation format toward an even tighter one of simply indicating the author, a reference citation, and the general problem he treated. The reader needing more information can then find it in the references. The necessity for this even tighter format is brought on by the large number of references encountered in the literature search for this Chapter on energy storage and its several sub-fields as well as the limited time available for this survey.

It should be pointed out a priori that the field of heat storage is quite old, and it is not possible in this short review to attempt an exhaustive review. Rather the references cited below are simply those of a relatively recent origin which were turned up in this research and should probably be called to the attention of those probing the heat storage field for the first time.

Heat Storage Prior Art

1972 Saunders [153], Univ. of Pa., described the design and performance of an electrically heated calorimeter used to determine approximate values for the heat of fusion of candidate thermal energy storage materials.

Dudley [154], Univ. of Pa., described his research on a thermal energy storage unit for air conditioning systems using phase change material.

Kauffman and Pan [155], Univ. of Pa., investigated thermal energy storage in sodium sulfate decahydrate mixtures.

1973 Belton [156], Univ. of Pa., studied the thermochemistry of salt hydrates. He finds that the latent heat of fusion offers the best possibility, and he concentrates on inorganic salt hydrates.

Kauffman et al. [157] [158], Univ.of Pa., researched congruently melting materials for thermal energy storage in air conditioning. A number of paraffin waxes were identified as promising thermal energy storage materials.

The idea of storing solar thermal energy at the bottom of a stratified pond, otherwise known as a solar pond, may possibly be later relevant to solar-electrics. Dake [159] has performed analytical and laboratory studies on a two layered solar pond. Jain [160] has studied the heating of solar ponds.

1974 Van Vechten [161], Bell Laboratories, proposed the use of a covalent semiconductor material, germanium sulfide, as a heat storage medium. He made preliminary cost estimates.

Lorsch, [162] examined latent heat and sensible heat storage for solar heating systems.

Heat Storage Present Art

1975 Shelton [163] researched the underground storage of heat in solar heating systems.

Borucka [164] surveyed the selection of inorganic salts for application to thermal energy storage.

Golibersuch, et al., [165], GE, investigated thermal energy storage for utility applications.

Rabl and Nielsen [166], Ohio State Univ., investigated solar ponds for space heating.

1976 Kovach [167] assembled a NATO Science Committee Conference on thermal energy storage. The report contains working group reports on high temperature and low temperature energy storage as well as heat transfer and thermal energy storage. A final report examines the impact of thermal energy storage on energy structures.

Bundy, et al., [168], GE, summarized the status of thermal energy storage. He has a nice synopsis of the fundamental principles to design a TES system and illustrates these by low temperature and high temperature applications. This report contains a most comprehensive literature review along with table of pertinent TES material properties.

Green, et al., [169], Sandia Laboratories, surveyed current technology relating to high temperature thermal energy storage. He cites an extensive bibliography.

Mitchell, et al., [170] described a new gravel and liquid storage system for solar thermal power plants.

Nemecek, et al., [171] NRL, proposed demand sensitive energy storage in molten salts based on heat-of-fusion energy storage and on-demand steam combined with heat pipes. Very large "energy storage-boiler tanks" are involved in this proposal.

Assaf [172] proposed the imaginative scheme of using the Dead Sea as a solar lake. His idea appears to be to utilize the stored heat in a power plant, though the paper is more concerned with the physics of such lakes.

Dickinson, et al. [173], Lawrence Livermore Lab., proposed the shallow solar pond energy conversion system "as an effective way to produce large-scale electric power from solar energy." "The stored hot water heats a thermodynamic fluid, probably Freon 11, which drives a turbine and an electric generator." Thus this paper clearly ties solar pond energy storage to solar-electrics.

LeFrois and Venkatasetty [174], Honeywell, Inc., investigated inorganic phase change materials for energy storage in the solar thermal program in connection with the Central Receiver Solar Thermal Power System Pilot Plant.

Riaz, et al. [175] explored "the technical feasibility of large volume packed beds using native earth or rock materials for long duration (months) storage of thermal energy at high temperatures (up to 500° C) and power levels (in the tens of MWt range) compatible with electric power generation and community heating in total energy systems context."

Birchenall and Telkes [176], Univ. of Delaware, investigated thermal storage in metals, notably Al, Cu, Mg, Si and Zn. This work is for large scale energy storage systems.

Turner [177], JPL, investigated thermal energy storage (to 500°C=930°F) using large hollow steel ingots. He estimates the installed system cost to be $13.50/KWHt.

Hall and Howerton [178] investigated thermal storage by paired ammoniated salts. It is "an energy system that accepts thermal energy at high temperatures, stores it at near ambient temperature, and recovers the energy on demand at a temperature near the original high temperature." Unfortunately only the abstract was published. Applicability is to power plant thermal storage systems.

Molz et al. [179] performed an experimental study of the subsurface transport of water and heat as related to the storage of solar energy.

Chahroudi [180] investigated thermocrete heat storage materials applications and performance specifications. "Thermocrete energy storage materials are made by combining phase change materials with open cell cements to produce low cost energy storage materials with structural and thermostatic properties." It is a patented material. It is for HVAC applications.

Edie and Melsheimer [181], Clemson Univ., did preliminary studies on an immiscible fluid-heat of fusion energy storage system.

Angrist [3, pp. 56-58]makes some generalizations about thermal energy storage, including tabulating properties of a few candidate energy storage materials using both sensible heat and latent heat.

Gresko and Glenn [182], GE, assessed the role of thermal energy storage applied to residential heating systems. They studied both technical and economic feasibility of thermal energy storage.

Nicholson and Cahn [183], Exxon, proposed a novel method of storage in oil of off-peak thermal energy from large power stations.

Venkatesetty and LeFrois [184], Honeywell, investigated thermal energy storage for large scale solar power plants. They summarize the thermodynamic properties of candidate energy storage materials.

Lior, et al. [185], Univ. of Pa., examines the problem of thermal energy storage for solar-thermal power generation. In addition to showing various energy storage candidate materials, they also developed the mathematical details relevant to transient analysis of thermal storage.

Qvale [186], Denmark, examined the theoretical and practical aspects of seasonal storage of thermal energy in underground water.

Libowitz and Blank [187], Allied Chemical Corp., evaluated the use of metal hydrides for solar thermal energy storage.

Ridgway and Dooley [188] examined the underground storage of off-peak power for a 1000 MW plant.

1977 Herrick and Golibersuch [189], GE, described the qualitative behavior of a new latent heat storage device for solar heating/cooling systems. It is based on the idea of a rolling cylinder heat storage means using Glauber's salt.

Morrison and Abdel-Khalik [190], Univ. of Wisconsin, studied the effects of phase-change energy storage on the performance of air-based and liquid-based solar heating systems.

Fong and Miller [191], Univ. of Calif., researched the heat capacity of paraffin wax as a thermal energy storage system.

Hooper and Attwater [192] examined the optimization of an annual storage solar heating system over its life cycle.

Tsang, et al. [193] have modeled underground storage in aquifers of hot water from solar power systems.

Elwell, et al. [194] have calculated and experimentally verified the stability criteria for solar ponds.

Lin [195], Concordia Univ., investigated experimentally the thermal performance of a two-compartment water storage tank.

Phillips and Pate [196], in a paper on mass and energy transfer in a hot liquid energy storage system, found that "the models which are presently being used to predict the stratification in a hot liquid storage tank are considerably in error. A new theoretical model is proposed which is found to give better agreement with experimental data than the models currently being used."

Riaz [197], Univ. of Minn., did a transient analysis of packed-bed thermal storage systems.

Clark, et al. [198] created a computer program, called ROCKBED, for analyzing rockbed thermal storage.

Eshleman, et al. [199] have analyzed and numerically simulated heat transfer in rock beds.

Edie, et al. [200], Clemson Univ., have done fundamental studies of direct contact latent heat energy storage.

Pfannkuch and Edens [201], Univ. of Minn., have carefully examined rock properties for thermal energy storage systems in the 0° to $500^{\circ}C$ range.

Chubb, et al. [202] have studied energy storage-boiler tanks. It is an energy storage device in which heat is stored as heat-of-fusion and in which heat transfer is effected by condensation and evaporation of a heat transfer fluid."

Turner and Truscello [203], JPL, studied large-scale thermal energy storage using sodium hydroxide (NaOH).

Mitchell, et al. [204], Rockwell International, summarize their work on a dual-medium thermal storage system for solar thermal power plants. Their concept is to be applied "to the 10 MWe pilot plant and to subsequent commercial plants of 100 MWe and larger." They have successfully designed, constructed and tested a 5 MWt storage subsystem and completed a preliminary design for the 10 MWe pilot plant thermal storage subsystem.

Cabelli [205], CSIRO, Australia, has made a numerical study "of a two-dimensional model of the motion which takes place in storage tanks used for solar water heating applications."

Bryant and Colbeck [206] have studied the possibilities of creating a large solar pond for London.

Tanaka, et al. [207], Japan, have performed fundamental studies on heat storage of solar energy. They analyze the unsteady case.

Lavan and Thompson [208], IIT, performed an experimental study of thermally stratified hot water storage tanks.

Givoni [209] did an overview paper on the subfield of underground long-term storage of solar energy. Different options are described.

Hill, et al. [210], NBS, propose a "test method for determining the 'effective capacity' and heat loss characteristics of thermal storage devices."

1978 Solar Age magazine [211] published a special edition on storage. Several papers, not cited in detail here, deal with solar energy storage for buildings.

Heat storage citations and abstracts of papers on this subject will also be found in Solar Energy Update [212, p. 77-78]. The current issue of this publication can be a continuing source of one's updating sources of information on heat storage.

Dean [213] wrote a small book devoted exclusively to thermal storage for buildings. Some references are cited, though the list is not extensive.

Riaz [214], Univ. of Minn., performed a transient analysis of packed-bed thermal storage systems. He used a one dimensional single-phase conductivity model.

Yoneda and Takanashi [215], Japan, briefly present the results of their research on eutectic mixtures for solar low temperature heat storage.

Hull [216], Iowa State Univ., analyzed the effects of radiation absorption on convective instability in salt-gradient solar ponds.

Wood and Platt [217], Lawrence Livermore Labs, studied the feasibility of shallow solar ponds for powering irrigation pumps.

Nemecek, et al. [218], NRL, summarize the results of their research on a 2 megawatt-hour energy storage-boiler tank. A laboratory test tank has been built and tested.

Altman, et al. [219] revealed the design, construction and test of a two stage thermal storage system to be used on DOE's Central Receiver Solar Thermal Power System.

Iannucci and Eicker [220] studied "the worth and the way in which storage should be utilized in stand-alone solar/fossil hybrid electric plant."

Nielsen [221], Ohio State Univ. examined the physics of the equilibrium thickness of the stable gradient zone in solar ponds.

We bring this review of the state-of-the-art of heat storage to a close by recognizing that thermal energy storage indeed has many facets of a physics, chemistry, thermodynamic, and engineering nature. This sub-field, as many others we've surveyed in this report, also appears to be growing in a somewhat unstructured way with its devotees to a particular thermal storage method rather than an overall systems view of the entire field which, as in the battery field, is notable by its absence.

Much work of significance to solar-electrics has been clearly done in thermal energy storage, but it is equally clear that more work toward structuring the existing body of knowledge into a more coherent whole needs to be done. The survey above is recognized to be only a first imperfect step in that direction.

Water Storage

The idea of utilizing the potential energy in stored water is very old. The idea has been extensively employed in the hydropower energy field. Extensive engineering treatises and books on hydropower engineering were written in the early part of the twentieth century. It is inappropriate and beyond our ability here to survey that literature, most of which is already known to specialized utility systems planners.

What is not generally recognized is that hydropower, and hence water storage, is basically solar fueled. The sun's insolation evaporates the earth's surface water causing it to rise in the atmosphere as clouds. When their moisture is again deposited on the earth's surface, the water may then be at a higher elevation than before. Thus it is the sun's energy which lifted the water to the storage area from whence via a dam and a turbine of some kind the water's potential energy can be converted into useful shaft power and thence into electric power.

Here we simply cite the sources of information on water energy storage which turned up in this non-exhaustive research. Time did not permit the logical ordering of this field either, but it is initially clear that one way of structuring it is via small scale and large scale storage. We are again back to the decentralized vs centralized controversy treated in Chap. I. It should be recognized a priori that on the contemporary solar-electric scene there are places for each philosophy; both small scale and large scale energy storage means are being developed in the United States to some degree. It is well known within the literature, for example, that there are currently about 50,000 small hydro dams within the U. S. which potentially could be applied to generate electric power.

Water Power-Prior Art

1974 Stoner [222] nicely summarizes both the simple basics of small water power storage and conversion along with some of the engineering considerations involved in the construction of small dams. One basic [p. 64] is the net power output from a hydro turbine available for input to an electric generator. His formula is:

$$P_T = \frac{V_W \times h \times 1{,}000}{75} \times \eta_T \qquad (4\text{-}6)$$

where:

P_T = Power output of the turbine shaft

V_W = Velocity of minimum water flow (m^3/sec.)

h = net head = gross head (m) - head losses (m)

η_T = Turbine efficiency.

Methods of measuring the flow rate are shown, as well as estimating the head losses (nomographs). Various classes of dams and turbines are shown. Unfortunately this reference fails to relate the fundamental of the total volume of water stored to the total energy stored therein.

A similar practical treatment of water storage and its conversion for small-scale use will be found in reference [223, pp. 66-123].

Water Power - Present Art

1975 A brief review of early American small-scale use of water storage and its useful conversion is in Clegg's book [224, pp. 133-152]. He also includes manufacturer's of small-scale water power apparatus.

1976 Some of the important environmental and ecological issues and problems brought on by the Aswan Dam (Egypt) are summarized in an article by Critchfield [225]. Hopefully other countries designing large-scale water energy storage facilities would learn some of the hard lessons found in the Aswan case and where possible avoid the mistakes.

Krenz [226, Ch. 7] briefly summarized hydropower in the U. S. He also briefly exposes a few analytical fundamentals of the stored energy in a tidal power dam and turbine. This is one of the very few places where water energy storage and hydropower exist in modern advanced energy textbooks.

Angrist [3, pp. 58-60], another modern DEC textbook, has a few paragraphs on pumped-storage hydroelectric plants. He points out that " in 1970 there was 3,600 MW of pumped storage, by 1980 there will be 27,000 MW and by 1990 70,000 MW. The demand reflects the fact that peak loads are believed to be expanding much faster than base loads."

Overall, it tentatively appears that a comprehensive truly integrated treatment of the subject of water energy storage appears in no modern advanced engineering textbook. Such a treatment would clearly lay out all the engineering fundamentals, cite specific references to this old art, and in general give the reader perspectives and an accurate systems view of water energy storage.

The above citations are only a small beginning to that end. Much more integrative work on water energy storage needs yet to be done before it can be said that a well-done state-of-the-art document exists. There simply was not time to achieve this goal in the present research.

Compressed Air Storage

Another alternative for possibly storing energy in future solar-electric systems is to store the energy in a volume of compressed air. A wind turbine, for example, could be created which would directly pump air into a suitable pressurized storage tank. Then later when the wind is not blowing the energy stored in the air could be utilized to drive an air turbine whose shaft would then drive a generator, thus supplying the needed electric power when the wind is not blowing. Compressed air storage may also be applicable to other solar-electric conversion systems.

Compressed air storage has been of increasing interest to utility system engineers as an alternative to pumped storage for utilizing excess generating capacity during valleys on the load demand curve and later utilizing the stored energy for electric generation during peak load times, thereby effectively tending to 'flatten' the load power curve.

Here we collect some of the literature on air storage technology. It is not an exhaustive survey--only that which was uncovered in this research.

Air Storage - Prior Art

1972 Scott [227, p18]claims that "the Swedish firm of Stal-Laval has hatched the idea for the world's first air storage plant. It's planned for Husqvarna, 180 miles south of Stockholm. In it, a 70 megawatt (MW) turbine is cleverly married to the hydraulic forces of a nearby lake to deliver 200 MW to the local generating station." He shows sketches of the basic system and briefly describes it, pointing out "the air system needs only the right combination of water and geology."

1974 Van Cleave [228 , p. 6] briefly reviews air storage in a utility context for peaking power to avoid gas turbines. He describes the key idea: "During off-peak periods, the generator acts as a motor and drives the compressor components to store air under pressure in a cavern. When power generation is needed, the compressed air is heated and used in the turbine combustion chamber." The estimate is that such storage would cost about $165/KW firm capacity. He also calls attention to the fact that at Huntorf, West Germany "two washed-out salt caverns will store air for a 210-MW compressed air plant that will be built by Brown-Boveri for Nordwestdeutsche Kraftwerke AG. The $40 million plant is scheduled to go on line in mid-1977." The same idea was described in a Brussels, Belgium news release [229].

Air Storage - Present Art

1975 Giramonti [230], United Technologies, performed a preliminary feasibility evaluation of compressed air storage power systems for peaking power. He identified prospective areas suitable for air storage in the Eastern U. S., did conceptual design studies, and specific application studies for utilities. His report appears to be highly significant to systems engineers interested in air storage.

Simmons [231, pp. 127-128] in a book on wind power briefly discusses compressed air storage. He says "it has the lowest capital cost of any storage system, and certain beneficial environmental advantages that include not using surface area."

1976 Willett [232], Acres American Corp., examined the current status of underground pumped storage to identify research priorities and encourage adoption of the concept by electric power utilities. He proposed the construction of a pilot facility. His report represents an impressive piece of engineering. Many references are cited.

Kreid [233], Batelle, performed a technical and economic feasibility analysis of the no-fuel compressed air energy storage concept. He concludes that "no insurmountable technical problems have been identified which would rule out (Compressed Air Energy Storage) implementation" and that "the no-fuel CAES concept does not appear to be economically competitive with conventional CAES systems for conditions forseeable at this time." He nicely collected the formulae defining the various component costs in several appendices.

Bush, et al. [234], GE, ERDA, ANL, summarized fuel fired compressed air energy storage. They see it as a near-term option for utility application. Sketches of various systems are shown, and areas of the U. S. where CAES site possibilities appear most sensible are identified. Design of the aquifer reservoirs and various other aspects of this technology are treated. In their references they cite significant patents and also indicate that in 1975 a Compressed Air Energy Storage Workshop was sponsored by ERDA. This appears to be a significant paper to this technology.

Angrist's textbook [3, pp. 60-62] briefly summarizes the fuel fired compressed-air energy storage system and tabulates the technical data for a 220 MW air storage power plant.

1977 An announcement was made [235] by ERDA of a program to study air energy storage for utility use. The sponsoring utilities were Potomac Electric Power Co., Middle South Services, Inc., and Public Service Co. of Indiana.

Vosburgh [236], GE, extensively studied the conceptual design for a pilot/demonstration compressed air storage facility employing a solution-mined salt cavern in southwestern Alabama. Power would be generated at 800 MW for 2000 hrs. each year. Capital costs were estimated at $196-200/KW (Mid 1976 dollars). No technical or environmental barriers were seen to constructing such plants. His report is an impressive piece of engineering.

From the above citations, it seems clear that the use of air as a storage medium for energy is a viable peaking option already being seriously examined by the electrical industry as one near-term load leveling means. Thus the use of air energy storage appears less speculative in nature than perhaps some of the other advanced energy storage methods. This state-of-the-art makes it appear all the more evident that compressed air energy storage could be successfully applied to solar-electric technology.

Energy Storage Via Hydrogen

Several places in this research we have indicated energy from various solar-electric systems can be stored in hydrogen. On wind-electric or photovoltaic systems, for example, with a DC output the power can be fed directly into an electrolyzer tank which produces hydrogen and oxygen from ordinary water. The gases may be produced either under pressure or near atmospheric pressure and then via external pumps compressed to the desired pressure. The latter approach, however, requires auxiliary energy. The hydrogen and oxygen gas produced can be stored either in gas or liquid forms for a long time. It can, when needed, be quickly and easily converted again directly into electric energy via the well known fuel cell. The system thus effectively stores the sun's energy as hydrogen and oxygen, and from this storage a smooth reliable power output may be taken--for a limited time set by the hydrogen storage capacity.

Hydrogen can also be used as a fuel to drive automobiles and other useful engines, and there appears to be a growing number of people within the technical community advocating this application of hydrogen. They have extended this idea to encompass the entire energy system of the nation, and one hears and sees frequent references to 'an all hydrogen economy'. Surveying that systems field is beyond our scope here.

Since our principal interest here is in storage, we survey only the part of the hydrogen literature uncovered in this research which bears on hydrogen storage. The topic is unavoidably intertwined with the larger hydrogen literature referred to above, however.

At best here, we can only indicate where more hydrogen energy storage information can be found in the references.

Hydrogen Storage - Prior Art

1971 Jones [237] wrote an article on liquid hydrogen as a fuel for the future. He summarized [his Table 2] the properties of liquid hydrogen and compared hydrogen to gasoline and fuel oil. Liquid hydrogen has 29,000 cal./g whereas gasoline is only 11,500--almost 3 times the energy/mass of gasoline. He points out that "as far as I know, there is no other chemical fuel which can equal hydrogen on an energy-per-unit-weight basis. Because of its very low density, hydrogen is about one-third as good a fuel as hydrocarbons on an energy-per-unit-volume basis."

1972 Reilly and Wiswall [238], BNL, studied hydrogen storage and purification systems. Focus was on metal alloy-hydrogen stystems.

The IECEC Proceedings for 1972 contains an appreciable number of papers on the many facets of hydrogen. Papers which appear to be most relevant to energy storage are those by Bartlit, et al. [239], Gregory and Wurm [240], Martin [241], and Wiswall and Reilly [242].

1974 Cummings and Powers [243] analyzed the storage of hydrogen as metal hydrides.

Reilly, et al. [244] studied the storage of hydrogen using metal hydrides. The automobile is the application.

This year a conference was held (Miami) on the hydrogen economy. Many papers on hydrogren appear in the Conference Proceedings, some of which relate to the storage issue [245].

Hydrogen Storage - Present Art

1975 Lindsey [246] wrote a popular type article on hydrogen power. He briefly dealt with the practical storage and handling problems.

Rohy, et al. [247] reported on automotive storage of hydrogen using modified magnesium hydrides.

Biederman, et al. [248], Inst. of Gas Technology, investigated for EPRI the utilization of off-peak power to produce industrial hydrogen. Hydrogen storage was dealt with in several places in this report. It also dealt with oxygen storage, a problem which many proponents of hydrogen appear to ignore. This is an impressive piece of engineering.

Srinivasan, et al. [249], BNL, discusses hydrogen production, storage, and conversion for electric utility and transportation applications. He has an interesting discussion on hydrogen storage (particularly via metal hydrides) and transportation.

Kelly, et al. [250], JPL, appraised the R & D needs for hydrogen production and use and identified and related NASA's experience to meeting these needs.

The 1975 IECEC Conference Record contained several papers on hydrogen technology of which the following appear relevant to the storage issue: Billings [251], Konopka and Gregory [252], Kissel, et al. [253], Powell [254], Voth [255], Boser and Lehrfeld [256], Lundin and Lynch [257], and Lundin and Lynch [258].

1976 Mathis' book [259] on hydrogen technology appeared. This is believed to be the first comprehensive single book on hydrogen energy which attempts to integrate the many facets of this growing new field. It is well organized and easy to follow; numerous references are cited where the reader can find more information. The author appears to have done an exceptionally good job in drawing together in a coherent form this still young field. He has an entire Chapter [pp. 135-216] dealing with the various aspects of hydrogen energy storage. Separate Chaps. on liquid hydrogen and solid hydrides also treat the storage problem.

Mathis' book is must reading for newcomers to the field desiring to obtain the needed knowledge and perspectives.

Angrist [3, p. 55] cursorily mentions energy storage in hydrogen and points out "the economics have not been fully established because of uncertainties regarding suitable containment materials." It is encouraging to see that the subject of hydrogen storage is starting to be mentioned in engineering textbooks.

Beaufrere, et al. [260] examined hydrogen storage via Iron-Tritium for a 26 MWe peaking electric plant. A preliminary plant design and cost estimate is presented for such a plant. R & D needed is identified.

Dickson, et al. [261], Stanford Res. Inst., prepared a large report on a preliminary assessment of the hydrogen economy. The storage, distribution, end use, and safety aspects are treated among many other subjects. Many references are cited. This comprehensive work cannot be justifiably and adequately summarized here. It appears to integrate this complex subject well and is recommended reading for newcomers to the field.

Salzano, et al. [262], BNL, discussed hydrogen storage, electrolysis and fuel cells for electric utility energy storage.

Reilly [263], BNL, treats metal hydrides as hydrogen storage media and describes the application of same. His report is extensively illustrated and contains many references.

Salzano, et al. [264], BNL, in a well referenced report treated hydrogen via metal hydrides for utility and automotive energy storage applications.

Ramakumar [265], Oklahoma State Univ., in a brief paper assessed hydrogen as a means to store solar energy.

1977 McRae, et al. [266], in an extensive new book on energy, briefly treat hydrogen [pp. 371-377] production, distribution, and utilization.

A brief news article describing Billings' Provo, Utah, house* which was converted to use hydrogen energy, appears in reference [267] . The hydrogen is stored in a tank about the size of a home washing machine. This 'hydrogen homestead' is described along with other uses for hydrogen in more detail in an attractive pamphlet published by Billings Energy Corp. [268].

This year an International Association for Hydrogen Energy [269] came into being.

Cox and Williamson [270] edited a new 5 volume work on hydrogen production, transmission and storage, properties, utilization, and implications. Though these books were not available to this researcher, it is evident they constitute a comprehensive work, integrating the many facets of this complex new energy field. The editors see no major barriers to hydrogen use. They claim this work is "the most timely evaluation of the subject to date."

Strickland [271], BNL, summarized the state-of-the-art of the technical problems involved in the storage of hydrogen via metal hydrides.

In December of this year an International Conference On Alternative Energy Sources was held at Miami at which 25 technical papers were presented on various aspects of hydrogen, including storage [272].

Nakamura [273], Japan, described hydrogen production from water utilizing direct thermal decomposition of water at high temperatures. The method possibly may have relevance to the solar-thermal-hydrogen-fuel cell-electric chain.

Sandrock [274], International Nickel Co., revealed a new family of hydrogen storage alloys based on the system nickel-mischmetal-calcium. This systems offers, he claims, "(1) raw material costs per unit of hydrogen storage capacity less than one-third that of present $LaNi_5$, (2) room temperature plateau pressures ranging from 30 to 0.5 atm., (3) very low hysteresis, and (4) easy activation."

Beaufrere, et al. [275], BNL, GE, EDA, assessed a hydrogen-halogen energy storage system for electric utility applications.

<u>1978</u> No 1978 information on hydrogen energy storage was available to this researcher. More current information undoubtedly exists, but for practical reasons it could not be acquired during this research.

From the above collection of citations, it is evident that a rather extensive advanced technical literature on hydrogen utilization already exists; it appears to be rapidly growing. The same can be said for the sub-field pertaining to hydrogen production and storage which are of most interest in solar-electrics.

Unfortunately time and space did not permit us to extract key illustrations from the literature and describe them here; the interested reader simply must make use of the referenced literature.

We now leave the subject of hydrogen energy storage to survey another alternative energy storage means.

---

* Interestingly, as Billings relates in another place, the house is located on a street renamed Hindenberg Lane!

Energy Storage Via Flywheels

The basic idea of flywheel energy storage, sometimes referred to as a 'superflywheel', is to accelerate a suitably designed physical rotor to a very high speed in a vacuum, as via an electric motor, at which state high energy storage densities are achieved. The energy is stored as kinetic energy most of which can be electrically retrieved when the flywheel is run as a generator. Such flywheels are presently classed as advanced energy storage means perhaps potentially applicable to the energy storage problems of solar-electric technology. The proponents of super flywheel energy storage claim this storage method has higher energy storage per pound than conventional flywheels or lead-acid storage batteries. Superflywheels are claimed to also have many other advantages [276, p. 1119].

We collect the literature here as on previous energy storage means.

Flywheel Energy Storage - Prior Art

1971 Rabenhorst [276], Johns Hopkins Univ., wrote a paper describing the potential applications for the superflywheel. He appears to have pioneered the superflywheel.

Dugger, et al. [277], Johns Hopkins Univ., examined the flywheel and flywheel/heat engine hybrid propulsion systems for low-emission vehicles.

Lawson [278], Lockheed, exposed the design and testing of high energy density flywheels for application to flywheel/heat engine hybrid vehicle drives.

1972 Jakubowski [279], IBM, described a wound rim flywheel used as an energy buffer.

1973 Post and Post [280] described various designs of flywheels and anticipated "giant flywheels for the storage of energy in electric-power systems--".

1974 Lampe [281], in a popular type article, described several of the various kinds of superflywheels being developed in the U. S.

Flywheel Energy Storage - Present Art

1975 Fullman [282], GE, wrote a paper describing "the best performance and cost effectiveness that are likely to be attainable in the storage of energy by flywheels." This work was also reported in reference [283].

A Flywheel Technology Symposium was held in Berkeley, California. A group of 36 technical papers were presented and appear in the Proceedings [284]. They cover virtually every aspect of flywheel R & D as it existed in 1975.

Gordon [285], William M. Brobeck & Associates, in a research study for EPRI, reports on the development of high-density inertial-energy storage. "Fiber-matrix composite-ring rotors of weights of 200 pounds and maximum diameters of 38 inches in vacua are described." It runs at 15,000 RPM. "No insuperable barriers to building rotors of greater size and number of rings have been encountered." This report has many interesting appendices on basic eqs. for inertial energy storage, vibrational stability of composite rotors, and other relevant topics.

Chiao and Stone [286], Lawrence Livermore Laboratory, described the LLL program for composite flywheels.

Clements [287], LLL, summarized engineering design data for composite materials for flywheels. Rinde [288], LLL, found 3 epoxy resin systems to be suitable for flywheel applications. Penn and Chiao [289], LLL, reviewed fiber composites for energy storage flywheels. Blake [290], LLL, described the mechanical design of a composite centerless flywheel. Stone, et al. [291], LLL, described a fiber composite program for flywheel applications. Penn and Chiao [292] evaluated fibers for flywheel applications.

Simpson, et al. [293], Atomic Energy of Canada Ltd., researched superflywheels as kinetic energy storage of off-peak electricity in a utility context. Several interesting appendices are included.

1976 Angrist's textbook [3, pp. 62-65] appeared, containing a brief summary of energy storage in flywheels. This appears to be the first time this technology has appeared in an engineering textbook.

Pepper [294], William M. Brobeck Associates, investigated multi-ring fiber-composite flywheels for energy storage under an EPRI contract.

Rinde, et al. [295], LLL, reported on a composite flywheel for energy storage.

Gordon [296], William M. Brobeck & Associates, concluded his earlier investigation of multi-ring fiber-composite flywheel.

1977 Huddleston, et al. [297], Union Carbide Corp., summarized their development of a composite flywheel.

1978 Schneider [298] is reported to have developed a unique rope type flywheel suitable for energy storage.

In closing this sub-section on flywheel energy storage, it is known additional R & D work not cited above is underway on flywheels. Unfortunately this project's funds were exhausted, and the more recent documents could not be procured.

It is evident that an impressive body of literature exists on flywheel energy storage. This technology appears still in the developmental phase, and no known commercially available superflywheel appears to exist as of 1978. Virtually all made to date as reported in the above literature are laboratory models.

We now leave flywheels to examine yet another energy storage means.

Energy Storage Via Superconducting Magnets

Superconducting electro magnets would be wound with superconducting wire and refrigerated to near absolute zero wire temperature. The current passing through the coil would create a magnetic field without loss of energy because of the phenomenon of superconductivity of the wire. It is believed that large energy densities can be stored in such magnets.

It is anticipated it could be developed as an off-peak storage means for conventional electric power systems. If that development proves successful, the technology might also be applicable to solar-electric systems.

This area was not extensively searched in this research but the following items turned up. It is highly probably that more literature than this has been published on energy storage via superconducting magnets.

Superconducting Magnets - Present Art

1975 ERDA announced an R & D contract on Los Alamos Scientific Laboratory to study superconductive magnetic energy storage [299]. Hassenzahl [300], LASL, recited the progress made to early 1975 in designing a 100 MJ superconducting magnetic energy storage system.

Segal and Boom [301], Univ. of Wisconsin, recited their work on conductor development for large superconductive energy storage magnets aimed at superconductive energy storage units for utility use in the 1,000-10,000 MWH range.

Overall, the state-of-the-art of energy storage via superconducting magnets appears highly physics orientated and is in the research phase. Most engineers would class this method of energy storage as highly speculative in 1978.

Energy Storage Via Chemical Means

The final energy storage option from our Fig. 4-1 is via chemical means which we now briefly survey. The chemical option mostly deals with thermal energy storage via chemical means.

Only a few items in this category were turned up in this research.

Chemical Energy Storage - Prior Art

1974 In an AP dispatch [302] Dunnelly of Readi Temp. Corp., Piqua, Ohio, is reported to have invented a heat storage can "that can store heat much the way a sponge holds water" via sterile chemicals inside. "When the chemicals are contaminated--in this case by air--a reaction takes place, releasing heat." Evidently he has also designed a way to store solar heat therein via Fresnel lenses.

Chemical Energy Storage - Present Art

1976 Prengle and Sun [303] published a paper in which they examined operational chemical storage cycles for utilization of solar energy to produce heat or electric power. They favor using an ammonium hydrogen sulfate (AHS) cycle which meets all needed engineering criteria, they say.

Schmidt and Lowe [304] examined the selection criteria and performance of thermochemical energy storage systems suitable for operating a power plant or heat buildings. The various candidate chemicals and their properties are clearly tabulated, permitting easy comparisons. All these chemicals react both exothermically and endothermically.

Offenhartz [305], EIC Corp., has examined in some detail chemical methods of storing thermal energy. He also examined thermochemically driven heat pumps.

Huxtable and Poole [306] examined thermal energy storage by the sulfuric acid-water system utilizing "solar thermal energy to evaporate water from a sulfuric acid-water reaction. The sulfuric acid and water are stored separately until heat is required. Upon recombination--heat is released---."

Baurle, et al. [307], Rockwell International, investigated a novel solar thermal energy storage by use of inorganic oxide/hydroxides for use in heating and cooling buildings.

Simmons [308] preliminarily researched the reversible oxidation of metal oxides for thermal energy storage. "Feasibility is yet to be demonstrated experimentally."

Kauffman and Lorsch [309] were developing thermal energy storage with saturated aqueous solutions. They believe such a system to have major advantages.

1977 An Australian system was reported [310] based on "using solar energy to effect a heat-absorbing chemical reaction--the breaking down of ammonia into nitrogen and hydrogen. The gases would be piped to a central recovery plant where they would be recombined to release energy in the form of heat for producing electricity."

Lawrence Berkeley Laboratory was reported [311] studying thermochemical heat storage for solar plants on a DOE/ERDA grant. They were also examining batteries, fuel cells, heat absorbing bricks, molten salts and flywheels as other possible energy storers.

Fujii and Tsuchiya [312], Japan, performed an experimental study of thermal energy storage by use of reversible chemical reactions, mostly using inorganic hydroxides.

Wentworth, et al. [313] propose and discuss "energy storage cycles based on reversible, uncatalyzed chemical reactions."

Kutal et al. [314], Univ. of Georgia, investigated the use of transition metal compounds to sensitize a photochemical energy storage reaction.

From the above citations most R & D on chemicals for energy storage has focussed on thermal energy storage. Many of these investigations have a distinct scientific flavor and fail to come to grips with engineering realities as the size and cost of the various alternatives. Also here, once again as seen so many times in this state-of-the-art research review, one clearly sees each researcher examining his subfield in depth with, in this case, few taking the larger systems view of comparing thermochemical systems.

This completes our review of the energy storage subject. We have summarized the state-of-the-art of eight major subfields therein, each potentially applicable to future solar-electrics.

We now examine some systems aspects of energy storage.

4-3 System Integration of Energy Storers

Energy storage, to be useful, must typically be integrated into a larger energy system of some kind. From this systems view an energy storage means then becomes a 'component' in the larger system. The larger energy system must then be made to work in an optimal manner. From this perspective it is evident that ultimately energy storage is not an isolated subject but is inherently interrelated with the system in which it is used.

Here we, all too briefly, indicate the nature and scope of integrating an energy storage means with the larger electric energy system. We limit our scope to energy storage integration as it has been conceived applicable to solar-electric systems perhaps integrated with a conventional utility.

The various storage means are here, unavoidably, intermixed with the various solar-electric means.

Energy Storage Integration - Prior Art

1948 Putnam's book [315] clearly envisioned hydro-storage being used with large-scale aerogenerators. "As long as water-power remains economically justified, special partnerships between wind and water will be justified" [p. 211]. The system cost and ultimately the busbar energy cost of wind-electrics is influenced by the water storage costs.

1969 Koslover [316] examined the optimization of battery subsystems for earth satellite lifetimes of greater than 5 years. He was particularly concerned with system reliability. Some of the analysis philosophy may have carry-over value to terrestrial solar-electric systems. This is only one sample of numerous system optimization studies made in the space program.

Ramakumar et al. [317], Oklahoma State University, summarized their research on wind energy storage and conversion systems for use in underdeveloped countries. Their system uses water electrolysis to generate hydrogen and oxygen which is stored and used to run a fuel cell, an internal combustion engine, or a burner. A system view prevails in this work.

1971 Kordesch [318], Union Carbide Corp., relates the system integration problems on a city car with hydrogen-air fuel cell and lead battery based on one year operating experiences.

1974 Fernandes [319], Niagara Mohawk Power Corp., studied the integration of hydrogen cycle peak-shaving for electric utilities. A system view is evidenced throughout.

Lotker [320] examined some long range systems possibilities of hydrogen use for the electric utilities. Burger and Lewis [321], BNL, and Pratt and Whitney Aircraft, examined systems aspects of energy storage for utilities via hydrogen systems based on also using 26 MWe fuel cells.

Hanneman, et al. [322], GE, described closed loop chemical systems for energy transmission, conversion and storage.

Ordin [323], LRC, in connection with hydrogen energy storage, reviewed the hydrogen accidents and incidents in NASA operations. Specific mishap descriptions and causes are tabulated. This paper is must reading for systems engineers considering hydrogen systems, permitting them to build on NASA's extensive experience with hydrogen as an energy storage material.

El-Badry and Zemoski [324], Public Service Elec. & Gas Co., studied systems aspects of the potential for rechargeable storage batteries in electric power systems. Brown and Cronin [325], Westinghouse, studied battery systems for peaking power generation.

Stoner [326] treats aspects of energy storage in a small scale wind-electric system [pp. 22-32].

Energy Storage Integration - Present Art

1975 Ackerman [327], ANL, did an assessment study of devices for the generation of electricity from stored hydrogen. Srinivasan, et al. [328], BNL, investigated hydrogen production, storage, and conversion for electric utility and transportation applications.

Bechtel Corp. did a study [329] for ERDA on energy storage and power conditioning aspects of photovoltaic solar power systems. This report is must study for a systems engineer interested in photovoltaics and associated energy storage.

Yao and Birk [111], ANL and EPRI, did an outstanding comprehensive paper on battery energy storage for utility load leveling and electric vehicles. A systems view permeates this excellent paper. Maskalick, et al. [115], Westinghouse, made an impressive case for lead-acid storage batteries for peaking systems in utilities.

1976 NASA announced [330] the initiation of a research program to combine wind and energy- storage utilizing "a special storage medium called Thermkeep, developed by Comstock and Wescott" of Cambridge, Mass. It is heated to about 900° Fahrenheit.

Public Service Electric and Gas Co., Newark, N. J., on an EPRI/ERDA contract completed an assessment of energy storage systems suitable for use by electric utilities [331].

EPRI made a compilation of workshop papers [119] on the lead-acid battery as a near-term energy storage technology for utilities. The report is from a systems view, not getting lost in the many details of battery technology.

Tison, et al. [332], Inst. Gas Technology, did an impressive engineering report on wind-powered hydrogen electric systems for farm and rural use.

Hartnett's book [333, p. 236-243] briefly treats the energy storage problem in the utility context. Thermal energy storage and chemical energy storage are the only two alternatives considered. He points up storage as a means of improving the reliability of solar-electric plants.

1977 An article in the EPRI Journal [334] deals with the technology transfer process from laboratory to application. It also says that "EPRI's efforts have also served to demonstrate that some innovations are not practical. More than a dozen such projects have been phased out, including work on the flywheel and superconducting magnetic energy storage."

A news report [335] indicates that Battelle is working on a 10 MW electrolysis-hydrogen storage-fuel cell system for utility use.

Zachmann [336] studied the feasibility of obtaining liquid hydrogen from solar energy.

Lu and Srinivasan [337] surveyed the state of development in the area of water electrolysis. The electrolyzer is an important component of the hydrogen energy storage system.

Halpin, et al. [338], Drexel Univ., examined the integration of a solar electric-utility system for a residence using load management. The system contains energy storage batteries.

In conclusion, it is evident that there is some literature on the systems integration aspects of energy storage means. This systems literature is small in comparison with the total literature on the individual energy storage means, however. There is, naturally, much overlap between these systems aspects and the individual storage means which have been surveyed earlier in this Chapter. As solar-electric technology continues to grow it appears highly probable that more systems type literature on energy storage will arise.

4-4 Economics

We collect here the papers which were turned up in this research dealing principally with economics of solar-electric storage. The subject of economics is an important one in the evolution of a new technology. Economics is interwoven throughout the technical literature already cited on energy storage, and it is therefore not possible to collect in a single place all information bearing only on economic aspects of energy storage. Thus the citations to be made here should be regarded as supplementing previous references as regards costs and cost projections. Because of space and time limitations we cannot go into any detail here on the content of the references.

We make the added observation of the difficulty of extracting from the literature reliable accurate information on costs, particularly where an energy storage concept is still in the laboratory and an accurate projection of its later mass production cost is to be made. We make no effort here to sort the economic literature by storage means.

The following references were discovered:

Economics of Storage - Prior Art

1968 Bruckner, et al. [339], LSU, presented a paper on the economic optimization of energy conversion with storage, describing "an experimental energy storage system being developed at the Oklahoma State University--to formulate a model for optimizing an energy conversion and storage system. ---It is assumed that the conversion system is a steam or hydro plant with a controlled input as opposed to a solar or wind plant with a random input, although its extension to the latter is possible."

Economics of Storage - Present Art

1975 Braun, et al. [340], BNL, examined the economic incentive for introducing electric storage devices into the national energy system. They consider electric storage devices as peaking plants operating at load factors of 0.05, 0.10, and 0.20. They see storage devices supplying peak load demands and competing against expensive gas turbines for peaking applications. "Higher breakeven capital costs are found at all conversion efficiencies for electric storage plants designed to supply peak load demand only, than for storage devices operating at the intermediate plus peaking mode."

Appleby [341], France, reported on the technical and economic aspects of the C. G. E. Zn-Air vehicle battery.

1976 Nelson and Yao [342], ANL, published a paper treating batteries for utility load leveling. Their paper has a strong economic slant, attempting to make preliminary cost estimates of a 20 MW, 200 MWH lead-acid battery plant.

EPRI published a report [119] on the near-term energy storage technologies based on the lead-acid battery. A section of the report [pp. 25-42] dealt with the economic analysis of energy storage systems. Another section [pp. 188-193] collects the pertinent formulae for making economic analyses on battery storage systems.

1977 Ferrell, et al. [343], ESB Inc., prepared a report for EPRI on the design and cost study for state-of-the-art lead-acid load leveling and peaking batteries. They "estimate the selling prices for one 2500 cycle 10 MWH load leveling battery and two 2000 cycle 20 MW peaking batteries delivering 60 and 100 MWH." They amortize in the price the 14.4 million dollar cost of a battery manufacturing plant.

EPRI prepared a thick workshop report [344] on lead-acid batteries for utility application. This volume is a compilation of workshop papers presented by many authors. A large fraction of the papers deal with economic aspects of energy storage in a utility context. This volume appears to be the most comprehensive document retrieved in this research which bears on the economics of battery energy storage in a utility system. It appears to be must reading for systems engineers interested in battery storage economics.

Li [345], N. Dakota State Univ., published a paper on the thermoeconomic analysis of thermal energy storage. He presents "a simple, but general analysis of thermal energy storage system (TESS) from a thermoeconomic viewpoint." He identifies the important parameters affecting the acceptance of TESS. He performed a sensitivity study as well.

In conclusion, it seems clear to this researcher, after examining the papers cited here and those in Sec. 4-2, that the energy storage community is sharply aware of the economic aspects of energy storage, regardless of the kind to be employed. Undoubtedly the economic issues will come into increasing importance as this overall energy storage technology continues moving toward the eventual marketplace.

4-5 Future R & D Needed For Integration of Energy Storers

We close this necessarily lengthy Chap. by opining the future R & D needed. We have already seen that the field of energy storage is a broad one, falling into eight major classes (Fig. 4-1). Each sub-field appears well aware of the major R & D problems facing them for making the necessary progress; this researcher therefore does not presume to indicate the future R & D needed for each specialized sub-field. Rather some observations are made here which possibly may be generally applicable to all the energy storage sub-fields.

In the opinion of this researcher, after having examined a significant fraction of the published literature on energy storage, the following items need additional study and research:

- Comparison of Storage Systems appears to stand out as much needing the viewpoint of the systems engineer. While, as this research has shown, there is an abundance of information on each individual storage means, there is a relative sparcity of information summarizing and carefully comparing them on all parameters important from the systems viewpoint. What are the system trade-offs between storing solar energy thermally vs doing so via batteries? And can anyone examine the vast literature on batteries and truly be satisfied that the available information has been clearly, accurately and carefully summarized (with reference citations) in a form to permit a wise choice to be made of the best battery type for a given storage application?

  Future systems engineers attempting to use and build on the energy storage literature will surely experience the frustrations of trying to summarize and compare options from such an enormous field characterized by attention to too much detail, e. g. batteries.

  This present state-of-the-art document is only a small but necessary first step in this direction.

- Safety And Environmental Aspects of the various energy storers appears hardly scratched. The battery people are developing advanced batteries using many exotic chemicals, some of which are known to be dangerous. Virtually no studies appear to have been made of the safety and potential environmental impact the future large scale deployment of such electrochemical systems might have. Also one does not walk away from the superflywheel development with the feeling that a widespread concern for personnel safety exists near such an energy storer. What happens if something goes wrong (Murphy's Law) to a wheel rotating at such high speeds? Can/will it injure people nearby? Will flying debris impale them like projectiles? Is this a desirable kind of storer to deploy on a decentralized basis?

Hydrogen storage proponents also appear to be in need of considerably more excruciation on the safety and environmental issue, particularly in light of NASA's extensive experience storing hydrogen.

One does _not_ have to wait until one or several of these advanced energy storage methods are fully developed and ready for the marketplace to suddenly discover these issues and their potential impacts.

Such a study should probably best be done by someone wholly disconnected from any of the eight storage technologies to insure objectivity. It probably should also encompass studying _all_ of them from the safety and environmental point instead of investigating only one area in depth.

To round out this Sec., we make the final comment that it may be entirely possible that the future of all solar-electrics hinges on the successful creation of practical, economical energy storers.

REFERENCES

1. Hayes, Denis, Energy: The Solar Prospect, Worldwatch Paper 11, March 1977. Available from Worldwatch Institute, 1776 Massachusetts Ave., N.W., Washington, D.C. 20036.

2. An Assessment of Solar Energy As A National Energy Resource, NSF/NASA Solar Energy Panel, December 1972. Available from NTIS as PB 221-659.

3. Angrist, Stanley W., Direct Energy Conversion, Third Edition, [Boston, Mass.: Allyn and Bacon, Inc., 1976], 518 pages.

4. Simmons, Daniel M., Wind Power, [Park Ridge, N.J.: Noyes Data Corp., 1975], 300 pages.

5. Stoner, Carol Hupping, Producing Your Own Power, [Emmaus, Pa.: Rodale Press, Inc., 1974], 322 pages.

6. Sullivan, George, Windpower For Your Home, [New York, N.Y.: Cornerstone Library, 1978], 127 pages.

7. Denton, Jesse C., "Solar-Thermal Power Systems," Chapter 8 in Hartnett, James P., Alternative Energy Sources, [New York, N.Y.: Academic Press, 1976], 328 pages.

8. Krumm, Roger V., Means and System of Energy Storage Other Than Batteries, NASA File, Bibliography No. FLA 203 1700, prepared April 28, 1972. Available from North Carolina Science and Technology Research Center, Research Triangle Park, N.C. 27709.

9. McCollom, Kenneth A., "Use of Energy Storage With Unconventional Energy Sources To Aid Developing Countries," Advances In Energy Conversion Engineering, 1967 IECEC Conference. Available from The American Society of Mechanical Engineers, 345 East 47th St., New York, N.Y. 10017.

10. Centrell, Joseph S., "Integrating Solar Energy Concepts Into Chemistry/ Science Curriculum," Sharing The Sun: Solar Technology In The Seventies, Vol. 8 of 1976 ISES Conference.

11. Vinal, George Wood, Storage Batteries, Fourth Ed., [New York, N.Y.: John Wiley & Sons, Inc., 1951], 446 pages.

12. Berg, C. Letter to Sales Dept., University of Fla., Gainesville, Fla., November 15, 1948, 2nd Edition. Unpublished. Copy available from Prof. Robert L. Bailey, EE Dept., Univ. Of Fla., Gainesville, Fla. 32611.

13. Jasinski, Raymond, High-Energy Batteries, [New York, N.Y.: Plenum Press, 1967], 313 pages.

14. Seiger, H.N., and S. Lerner, "Sealed Cells And Auxiliary Electrodes," Advances In Energy Conversion Engineering, papers presented at 1967 IECEC Conference, pp 353-356. Available from ASME, 345 East 47th St., New York, N.Y. 10017.

15. Lutwack, Ralph, "The Heat Sterilizable Battery," Advances In Energy Conversion Engineering, papers presented at 1967 IECEC Conference, pp 365-373. Available from ASME, 345 East 47th St., New York, N.Y. 10017.

16. Bacher, Joel, And Paul Nekrasov, "Analysis of the Operating Parameters of Secondary Batteries In An Orbiting Space Vehicle Application," Advances In Energy Conversion Engineering, papers presented at 1967 IECEC Conference, pp375-388. Available form ASME, 345 East 47th St., New York, N.Y. 10017.

17. Shimotake, H. and E.J. Cairns, "Bimetallic Galvanic Cells With Fused-Salt Electrolytes," Advances In Energy Conversion Engineering, papers presented at 1967 IECEC Conference, pp 951-952. Available from ASME, 345 East 47th St., New York, N.Y. 10017.

18. Shipps, P.R., et al., "Advanced Secondary-Battery Development For Vehicle Propulsion," Advances In Energy Conversion Engineering, papers presented at 1967 IECEC Conference, pp 879-886. Available from ASME, 345 East 47th St., New York, N.Y. 10017.

19. Weber, Neill, and J.T. Kummer, "A Sodium-Sulfur Secondary Battery," Advances In Energy Conversion Engineering, papers presented at 1967 IECEC Conference, pp 913-916. Available from ASME, 345 East 47th St., New York, N.Y. 10017.

20. Woodward, William H., Release of statement of William H. Woodward, Space Power and Electric Propulsion, before Subcommittee On Advanced Research and Technology, 1968. Publication source unknown.

21. Baer, D. and F.E. Ford, Recharge Unit Provides For Optimum Recharging of Battery Cells, NASA Tech Brief 68-10273.

22. Ford, F.E., et al., Electrochemical Cell Has Internal Resistive Heater Element, NASA Tech Brief 68-10325.

23. Gross, S., Battery-Package Design Provides for Cell Cooling and Constraint, NASA Tech Brief 68-10398.

24. Shimotake, H., et al., Lithium-Tellurium Cell Has Increased Voltage, AEC-NASA Tech Brief 68-10400.

25. Shimotake, H., and J.C. Hesson, "New Bimetallic EMF Cell Shows Promise In Direct Energy Conversion," AEC-NASA Tech Brief 68-10415.

26. Paulkovitch, John, and Floyd Ford, Charge Control of Nickel-Cadmium Batteries by Coulometer and Third Electrode Method, NASA Tech Brief 68-10431.

27. Aircraft Battery Core: The ABC's Of Proper Maintenance. Nickel-Cadmium Supplement to GEJ-4344 dated March 1968. Available from General Electric Co., Battery Business Section, Gainesville, Fla.

28. Mandel, Hyman J., Metal Air Cells. Available from NTIS as AD 683 461.

29. Briggs, Donald C., "Hermetically Sealed Nickel-Cadmium Batteries For The Initial Defense Communication Satellite Program/Augmentation (IDCSP/A)," pp.13-18, IECEC '68 Record/Papers. Available from IEEE.

30. Dunlop, James D., and R.W. Bounds, "Adhydrode Control of Ni-Cd Battery Charging To Evaluate Charging Methods," pp.19-24, IECEC '68 Record/Papers. Available from IEEE.

31. Lutwack, R., "Progress In Development of Heat Sterilizable AG-ZN Battery," IECEC '68 Record/Papers, pp.25-31. Available from IEEE.

32. Meredith, R.E. and A.A. Uchiyama, "Theoretical Evaluation of Hot Spot Temperature of Silver-Zinc Batteries," IECEC '68 Record/Papers, pp.32-37. Available from IEEE.

33. Gross, Sidney, "Heat Generation In Sealed Batteries," IECEC '68 Record/Papers, pp. 38-46. Available from IEEE.

34. Biddick, R.E. and R.D. Nelson, "Lead-Acid Bipolar Battery For Multisecond Pulse Discharge," IECEC '68 Record/Papers, pp. 47-51.

35. Arrance, F.C. and M.H. Plizga, "Separator Materials For Long Life, High Rate Thermal Cells," IECEC '68 Record/Papers, pp. 65-68.

36. Shimotake, H., and E.J. Cairns, "A Lithium/Tin Cell With An Immobilized Fused-Salt Electrolyte: Cell Performance And Thermal Regeneration Analysis," IECEC '68 Record/Papers, pp. 76-91.

37. Linden, David, Mechanically Rechargeable Zinc-Air Battery, August 1969. Available from NTIS as AD 693 865.

38. Wilburn, Nicholas T., and Achille L. Almerini, "New Low Temperature Batteries," Proc. Fourth IECEC, 1969, pp. 522-524.

39. Braeuer, Klaus, H.M., "Reserve Type Organic Electrolyte Batteries," Proc. Fourth IECEC, 1969, pp. 525-527.

40. Eisenberg, M., "High Energy Lithium Organic Electrolyte Batteries For Wide Temperature Range Applications," Proc. Fourth IECEC, 1969, pp. 528-537.

41. Shimotake, H., et al., "Secondary Cells With Lithium Anodes and Paste Electrolytes," Proc. Fourth IECEC, 1969, pp. 538-547.

42. Moser, J.R., et al., "Long Life Thermal Cell," Proc. Fourth IECEC, 1969, pp. 548-554.

43. Opitz, C.L., "Seawater Power Supply Using A Magnesium-Steel Cell," Proc. Fourth IECEC, 1969, pp. 699-704.

44. Preusse, K.E., et al., "Parametric Charge Studies For Aerospace Nickel-Cadmium Batteries," Proc. Fourth IECEC, 1969, pp. 705-709.

45. Seiger, H.N., et al., "Electrical Characteristics Of A Sealed Nickel-Cadmium Battery Having a Bipolar Construction," Proc. Fourth IECEC, 1969, pp. 710-714.

46. Hartmann, Werner Von, "The Application of Bench Tests In The Development of Heat Sterilizable Battery Separators," Proc. Fourth IECEC, 1969, pp. 715-720.

47. Teresa, Michael J., and Robert E. Corbett, "Two Year Flight Performance of the Lockheed Type XI Nickel-Cadmium Battery," Proc. Fourth IECEC, 1969, pp. 721-729.

48. Palmer, Nigel I., et al., "High Rate Performance of Zinc-Air Batteries," Proc. Fourth IECEC, 1969, pp. 911-919.

49. Kyle, M.L., et al., "Lithium/Sulfur Batteries For Electric Vehicle Propulsion," Conference Proceedings 1971 IECEC, pp. 80-95.

50. Witherspoon, Romeo R., et al., "An Experimental Vehicular Zinc-Air Battery With Replaceable Anodes," Conf. Proc. 1971 IECEC, pp. 96-102.

51. Caprioglio, G., and A. Weinberg, "Fabrication and Initial Testing Of An Experimental Zinc-Oxygen Battery For Electric Vehicles," Conf. Proc. 1971 IECEC, pp. 140-146.

52. Breeskin, S.D. and A.D. Taylor, "Battery and Third Electrode Performance Characteristics for Varying Charge and Discharge Rates," Conf. Proc. 1971 IECEC, pp. 181-187.

53. Gaz, Richard A., "Battery Cell With Third Electrode Performance Test Results," Conf. Proc. 1971 IECEC, pp.188-194.

54. Liang, C.C., "The Storage and Discharge Characteristics of A Lithium Solid State Battery System," Conf. Proc. 1971 IECEC, pp. 673-676.

55. Metcalfe, J.E., et al., "Characteristics of a Tellurium-Lithium/Aluminum Battery," Conf. Proc. 1971 IECEC, pp. 685-689.

56. Schwartz, Harvey J., "Space Vehicle Technology Applicable To Hybrid Vehicle Batteries," Conf. Proc. 1971 IECEC, pp. 690-694.

57. Gratch, Serge, et al., "Recent Development of The Ford Sodium-Sulfur Battery," Conf. Proc. 1972 IECEC, pp. 38-41.

58. Minck, R.W., "Electrical Performance Of A Prototype Sodium Sulfur Battery," Conf. Proc. 1972 IECEC, pp. 42-46.

59. Silverman, H.P., et al., "Cycle-Life Study of Sodium/Beta-Alumina/Sulfur Cell," Conf. Proc. 1972 IECEC, pp. 47-49.

60. Levine, C.A., et al., "The Dow Sodium-Sulfur Battery," Conf. Proc. 1972 IECEC, pp. 50-53.

61. Gay, E.C., et al., "Secondary Cells Using Lithium Anodes and Sulfur-Selenium-Thallium Cathodes," Conf. Proc. 1972 IECEC, pp. 54-62.

62. Takehara, Zenichiro, et al., "Studies on the Lithium Alloys-Chlorine Secondary Battery," Conf. Proc. 1972 IECEC, pp. 63-70.

63. Klein, M., "Metal-Gas Rechargeable Batteries," Conf. Proc. 1972 IECEC, pp. 79-85.

64. Stockel, J.F., et al., "A Nickel-Hydrogen Secondary Cell For Synchronous Orbit Application," Conf. Proc. 1972 IECEC, pp. 87-94.

65. Menard, Claude J., "High Energy Density Nickel-Cadmium Batteries," Conf. Proc. 1972 IECEC, pp. 95-97.

66. Fono, Peter, "Spacecraft Nickel-Cadmium Battery Cycle Life Assessment," Conf. Proc. 1972 IECEC, pp. 98-102.

67. Kerr, R.L., and D.N. Stager, "Kilowatt Battery Systems," Conf. Proc. 1972 IECEC, pp. 103-109.

68. Charkey, Allen, "Nickel-Zinc Cells For Sustained High Rate Discharge," Conf. Proc. 1972 IECEC, pp. 110-113.

69. Silverman, H.P., et al., "Optimization of Lead-Acid Batteries For High Power-Short Pulse Operation," Conf. Proc. 1972 IECEC, pp. 119-124.

70. Snider, William E., and William J. Nagle, "Nonleaking Battery Terminals," Conf. Proc., 1972 IECEC, pp. 125-129.

71. Berberick, Frank A., et al., "Developing for Manufacture of the Bell System Lead-Acid Battery," The Western Electric Engineer, January 1973, pp. 18-28.

72. Eisenberg, M., and K. Wong, " Stable High Energy Non-Aqueous Electrolyte Lithium Batteries," Proceedings, IECEC, 1973, pp. 58-62.

73. Halberstadt, H. J. "The Lockheed Power Cell," Proceedings, IECEC, 1973, pp. 63-66.

74. Symons, Philip C., and Peter Carr, "Chlorine Electrodes In The Zinc-Chlorine Battery System," Proceedings, IECEC, 1973, pp. 72-77.

75. Kirsch, Walter W., "Nickel-Cadmium Battery Performance Prediction Models Apollo Telescope Mount Application," Proceedings, IECEC, 1973, pp. 78-85.

76. Gay, E. C., et al., "The Development of Lithium-Sulfur Cells for Application To Electric Automobiles," Proceedings, IECEC, 1973, pp. 96-103.

77. Klein, M. and B. S. Baker, "Nickel-Hydrogen Battery System," 1974 Proceedings, IECEC, pp. 118-122.

78. Gandel, M. G., et al., "Nickel-Hydrogren Battery Development For Synchronous Satellites," 1974 Proceedings, IECEC, pp. 123-127.

79. Littauer, E. L. and J. J. Redlien, "The Lithium/Water Electrochemical Cell As A Marine Power Source," 1974 Proceedings, IECEC, pp. 615-619.

80. Chilenskas, A. A., et al., "Lithium/Sulfur Batteries for Marine Application," 1974 Proceedings, IECEC, pp. 654-659.

81. Solomon, Frank, "Performance Of A New Non-Silver Sea Water Battery," 1974 Proceedings, IECEC, pp. 660-664.

82. Gay, E. C., et al., "The Development Of Lithium/Sulfur Cells for Application to Electric Automobiles," 1974 Proceedings, IECEC, pp. 862-867.

83. Seiger, H. N., et al., "High Energy Density Sintered Plate Type Sealed Nickel Cadmium Battery Cells. Part I. The Positive Electrode/Plaque Relationships," 1974 Proceedings, IECEC, pp. 868-872.

84. Puglisi, V. J., and Seiger, "High Energy Density Sintered Plate Type Nickel-Cadium Battery Cells. Part II. Electrochemical Impregnation Methods to Produce Nickel Oxide Electrodes," 1974 Proceedings, IECEC, pp. 873-879.

85. Juvinall, G. L., et al., "A Novel Negative-Limited Sealed Nickel-Cadmium Cell," 1974 Proceedings, IECEC, pp. 881-887.

86. Easter, Robert W., "Predicted Energy Densities for Nickel-Hydrogen and Silver-Hydrogen Cells Embodying Metallic Hydrides For Hydrogen Storage," 1974 Proceedings, IECEC, pp. 888-891.

87. Walsh, W. J., "Development of Prototype Lithium/Sulfur Cells for Application To Load-Levelling Devices in Electric Utilities," 1974 Proceedings, IECEC, pp. 911-915.

88. Mitoff, S. P. and J. B. Bush, "Characteristics of A Sodium-Sulfur Cell For Bulk Energy Storage," 1974 Proceedings, IECEC, pp. 916-923.

89. Klein, I. S., "Heat Removal From Large Batteries For Load Leveling," 1974 Proceedings, IECEC, pp. 929-933.

90. Bush, J. B., Sodium-Sulfur Battery Development: EPRI Project RP128. Final Report, July 1974. Available from EPRI as EPRI 128-0-0.

91. Bush, J. B., Sodium-Sulfur Battery Development For Bulk Power Storage, September 1975. Available from NTIS as PB-246 000.

92. Hall, John, et al., Development of Lithium-Metal Sulfide Batteries for Load Leveling, June 1975. Available from EPRI as EPRI 116.

93. Nelson, P. A., et al., High Performance Batteries For Off-Peak Energy Storage And Electric-Vehicle Propulsion, July 1975. Available from NTIS as ANL 75-1.

94. Nelson, P. A., et al., High Performance Batteries for Off-Peak Energy Storage and Electric-Vehicle Propulsion, Progress report for July-December 1975. Available from NTIS as ANL-76-9.

95. A Conceptual Design of A Battery Energy Test (BEST) Facility, August 1975. Available from EPRI as EPRI 225.

96. Energy Storage And Power Conditioning Aspects of Photovoltaic Solar Power Systems, October 1975. Available from NTIS as COO/2748-75/T1.

97. Cooper, J. F., Lithium Requirements For Electric Vehicles Using Lithium-Water-Air Batteries, November 1975. Available from NTIS as UCRL 77440.

98. Sodium-Chloride Battery Development Program For Load Leveling, Interim report for the period January 1, 1975-December 31, 1975. Available from EPRI as EPRI EM-230.

99. Bro, P., et al., "High Rate Li/SO_2 Batteries: I. The Discharge Process," Record 1975 IECEC, pp. 432-436.

100. Holleck, G. L., et al., "Transition Metal Sulfides As Cathodes For Secondary Li Batteries: I. Niobium Sulfides," Record 1975 IECEC, pp. 444-448.

101. Sudworth, J. L., "Sodium/Sulfur Batteries For Rail Traction," Record 1975 IECEC, pp. 616-620.

102. Levine, Charles A., "Progress In The Development of the Hollow Fiber Sodium-Sulfur Secondary Cell," Record 1975 IECEC, pp. 621-626.

103. Marcoux, L. S., et al., "A Sodium'Sulfur Test Cell Utilizing A β-Alumina Flat-Plate Separator," Record 1975 IECEC, pp. 624-626.

104. Gay, E. C., et al., "Development of High-Performance Iron Sulfide Electrodes With Porous Current Collector Structures," Record 1975 IECEC, pp. 627-633.

105. Schertz, W. W., et al., "Battery Design and Cell Testing For Electric Vehicle Propulsion," Record 1975 IECEC, pp. 634-641.

106. Sudar, S., et al., "Rechargeable Lithium/Iron Sulfide Battery," Record 1975 IECEC, pp. 642-648.

107. Schaefer, James C., et al., "The ESB-SOHIO CARB TEK[R] Molten Salt Cell," Record 1975 IECEC, pp. 649-650.

108. Klein, Martin, "Nickel-Hydrogren Secondary Battery," Record 1975 IECEC, pp. 803-806.

109. Miller, Lee E., "Nickel-Hydrogen As An Alternative To Lead-Acid And Nickel-Cadmium Systems In Non-Space Applications," Record 1975 IECEC, pp. 807-810.

110. Malaspina, F. P., "Lead Chloride-Magnesium Sea Water Batteries For Deep Submergence Operation," Record 1975 IECEC, pp. 817-820.

111. Yao, N. P., and J. R. Birk, "Battery Energy Storage For Utility Load Leveling and Electric Vehicles: A Review of Advanced Secondary Batteries," Record 1975 IECEC, pp. 1107-1119.

112. Halberstadt, H. J., et al., "Physical And Economic Characteristics of the Lithium-Water Marine Battery," Record 1975 IECEC, pp. 1120-1125.

113. Charkey, Allen, "Evolutionary Developments In Nickel-Zinc Cell Technology," Record 1975 IECEC, pp. 1126-1130.

114. McCoy, L. R., "Zinc-Nickel Oxide Secondary Battery," Record 1975 IECEC, pp. 1131-1134.

115. Maskalick, N. J., et al., "The Case For Lead-Acid Storage Battery Peaking Systems," Record 1975 IECEC, pp. 1135-1140.

116. Weiner, Steven A., Research On Electrodes And Electrolyte For The Ford Sodium-Sulfur Battery, January 1976. Available from NTIS as PB 252 827.

117. Chilenskas, A. A., et al., "Lithium Requirements For High-Energy Lithium-Aluminum/Iron Sulfide Batteries For Load Leveling and Electric-Vehicle Applications," January 1976. Available from NTIS as CONF-760112-3.

118. Holleck, Gerhard L., et al., Sulfur-Based Lithium-Organic Electrolyte Secondary Batteries, March 1976. Available from NTIS as AD/A-023 496.

119. Birk, James R., Near-Term Energy Storage Technologies: The Lead-Acid Battery, March 1976. Available from EPRI as EPRI SR-33.

120. Yao, N. P., Applications And Prospect Of Energy Storage Batteries, April 1976. Available from NTIS as CONF 760416-2.

121. Sudar, S., et al., Development of Lithium-Metal Sulfide Batteries For Load Leveling, May 1976. Available from EPRI as EPRI EM-166.

122. Nelson, P. A., et al., High-Performance Batteries For Off-Peak Energy Storage And Electric Vehicle Propulsion, May 1976. Available from NTIS as ANL-76-35.

123. Birk, J., Evaluation of A 1 Kwh Zinc Chloride Battery System, September 1976. Available from NTIS as PB-260 683.

124. Silverman, Herbert P., Development Program For Solid Electrolyte Batteries, September 1976. Available from EPRI as EPRI EM-226.

125. Briggs, William B., et al., "Engineering The Silver-Zinc Battery For The Deep Submergence Rescue Vehicle," Proceedings 1976 IECEC, pp. 447-451.

126. Charkey, Allen, "Advances In Component Technology For Nickel-Zinc Cells," Proceedings 1976 IECEC, pp. 452-456.

127. Kemp, D. D., et al., "Design and Performance Features Of A 0.45 KW, 4Kwh Lithium-Water Marine Battery," Proceedings 1976 IECEC, pp. 462-466.

128. Marmaduke, B. A., and L. F. Donaghey, "Engineering Evaluation of Solid Transition-Metal Sulfide Electrodes For Use in High Energy-Density Batteries," Proceedings 1976 IECEC, pp. 467-470.

129. Shimotake, H., et al., "Development of Uncharged Li-Al/Fe S_x Cells," Proceedings 1976 IECEC, pp. 471-476.

130. Gay, E. C., et al., "Review of Electrode Designs and Fabrication Techniques For Lithium-Aluminum/Iron Sulfide Cells", Proceedings 1976 IECEC, pp. 477-484.

131. McCoy, L. R. and L. A. Heredy, "Development Status of Lithium-Silicon-Iron Sulfide Load-Leveling Batteries," Proceedings 1976 IECEC, pp. 485-490.

132. Dunning, J. S., et al., "Development of Compact Lithium-Iron Disulfide Electrochemical Cells," Proceedings 1976 IECEC, pp. 491-496.

133. Bauer, R., et al., "Development of Sodium/Sulfur Cells," Proceedings 1976 IECEC, pp. 497-502.

134. Dell, R. M., et al., "Sodium/Sulphur Battery Development In The United Kingdom," Proceedings 1976 IECEC, pp. 503-509.

135. Betz, F., et al., "Nickel-Hydrogen Storage Battery For Use On Navigation Technology Satellite-2," Proceedings 1976 IECEC, pp. 510-516.

136. Sparks, R. H., and W. R. Scott, "Application of Nickel-Cadmium Batteries In Deep Discharge Synchronous Orbit Applications," Proceedings 1976 IECEC, pp. 517-520.

137. Kirsch, W. W., and L. E. Paschal, "Nickel Cadmium Cell Composite Reliability Model," Proceedings 1976 IECEC, pp. 521-527.

138. Miller, Lee E., "An Advanced Sealed Nickel-Cadmium Cell Design," Proceedings 1976 IECEC, pp. 528-531.

139. Kozawa, A., and T. Takagaki, "Japanese Lead-Acid Batteries: New Technology and Recent Products," Proceedings 1976 IECEC, pp. 532-536.

140. Battey, R. F., Engineering Design and Cost Analysis of Chlorine Storage Concepts for a Zinc-Chlorine Load-Leveling Battery, November 1976. Available from NTIS as PB-262 016.

141. Battey, R. F., Engineering Design And Cost Analysis of Chlorine Storage Concepts For A Zinc-Chloride Load-Leveling Battery, November 1976. Available from EPRI as EPRI EM-259.

142. Chatterji, D., Development of Sodium Sulfur Batteries For Utility Application, December 1976. Available from EPRI as EPRI EM-266. Also available from NTIS as EPRI EM-266.

143. Lazennec, Yvon, Improvements Of Electrolyte And Seal Technology For Sodium-Sulfur And Sodium Antimony Trichloride Load Leveling Batteries June 1977. Available from EPRI as EPRI EM-413.

144. Rauh, R. D., et al., "Rechargeability Studies of Ambient Temperature Lithium/Sulfur Batteries," Proceedings 1977 IECEC, Vol. I, pp. 283-287.

145. Donovan, R. L., and M. S. Imamura, "Cell-Level Battery Charge/Discharge Protection System," Proceedings 1977 IECEC, Vol. I, pp. 302-310.

146. Lear, John W., and Matthew S. Imamura, "Cycle Life Characteristics of Sealed Silver-Zinc Cell With Inorganic Separator," Proceedings 1977 IECEC, Vol. I, pp. 311-318.

147. Murie, Richard A., "Post-Test Analysis of Li/Fe S_2 Compact Cells," Proceedings 1977 IECEC, Vol. I, pp. 349-358.

148. Mitoff, S. P., et al., "Recent Progress In Development of Sodium-Sulfur Battery For Utility Application," Proceedings 1977 IECEC, Vol. I, pp. 359-367.

149. Sudar, S., et al., "Development Status Of Lithium-Silicon/Iron Sulfide Load Leveling Batteries," Proceedings 1977 IECEC, Vol. I, pp. 368-374.

150. Lai, San-Cheng, and Anthony F. Sammells, "Improved Negative Electrodes For Lithium/Iron Sulfide Batteries," Proceedings 1977 IECEC, Vol. I, pp. 375-379.

151. Palz, Wolfgang, Solar Electricity, [Woburn, Mass.: Butterworth (Publishers), Inc., 1978], 292 pages.

152. Sullivan, George, Windpower For Your Home [New York, N. Y.: Cornerstone Library, 1978] 127 pages.

153. Sauders, Alan P., Design and Performance of An Electrically Heated Calorimeter, July 1972. Available from NTIS as NSF/RANN/SE/GI 27976/TR 72/

154. Dudley, James C., Thermal Energy Storage Unit For Air Conditioning Systems Using Phase Change Material, August 1972. Available from NTIS as NSF/RANN/SE/GI 27976/TR 72/8.

155. Kauffman, Kenneth, and Yen-Chi Pan, Thermal Energy Storage In Sodium Sulfate Decahydrate Mixtures, December 1972. Available from NTIS as NSF/RANN/SE/GI/27976/TR 72/11.

156. Belton, Geoffrey, and Fouad Ajami, Thermochemistry of Salt Hydrates, May 1973. Available from NTIS as NSF/RANN/SE/GI 27976/TR/73/4.

157. Kauffman, Kenneth, and Yen-Chi Pan, Congruently Melting Materials For Thermal Energy Storage In Air Conditioning, June 1973. Available from NTIS as NSF/RANN/SE/GI 27976/TR 73/5.

158. Kauffman, Kennth, and Irving Gruntfest, Congruently Melting Materials For Thermal Energy Storage, Report No. NCEMP-20, November 1973. Available from National Center for Energy Management & Power, Univ. of Pa., Philadelphia, Pa. 19174.

159. Dake, Jonas M. K., "The Solar Pond: Analytical and Laboratory Studies," The Sun in the Service of Mankind, International Congress held at UNESCO house, Paris July 1973, paper E17.

160. Jain, G. C., "Heating of Solar Pond," The Sun In The Service of Mankind, International Congress held at UNESCO house, Paris, July 1973, paper EH 61.

161. Van Vechten, J. A., Latent-Heat Energy Storage is Feasible, Electrical World, August 15, 1974, p. 41.

162. Lorsch, Harold G., Latent Heat And Sensible Heat Storage For Solar Heating Systems, May 1974. Available from NTIS as NSF/RANN/SE/GI 27976/R 72/20.

163. Shelton, Jay, "Underground Stor age of Heat In Solar Heating Systems," Solar Energy, Vol. 17, No. 2, May 1975, pp. 137-143.

164. Borucka, Alina, Survey and Selection of Inorganic Salts For Application To Thermal Energy Storage, June 1975. Available from NTIS as ERDA-59.

165. Golibersuch, D. C., et al., Thermal Energy Storage For Utility Applications, General Electric Co., Technical Information Series, Report No. 75CRD256, December 1975. Available from GE Corporate Research and Development Center, Schenectady, N. Y.

166. Rabl, Ari, and Carl E. Nielsen, "Solar Ponds For Space Heating," Solar Energy, Vol. 17, No. 1, April 1975, pp. 1-12.

167. Kovach, Eugene G., Ed., Thermal Energy Storage, The report of a NATO Science Committee Conference held at Turnberry, Scotland, 1-5 March 1976. Available from NTIS by title.

168. Bundy, F. P., et al., The Status Of Thermal Energy Storage, General Electric Co., Technical Information Series, Report 76 CRD 041, April 1976. Available from Corporate Research and Development, P. O. Box 43, Bldg. 5, Schenectady, N. Y. 12301.

169. Green, R. M., et al., "High Temperature Thermal Energy Storage," Sharing The Sun, ISES Conference, August 1976, Vol. 8, pp. 4-47.

170. Mitchell, R. C., et al., "Gravel and Liquid Storage System For Solar Thermal Power Plants," Sharing The Sun, ISES Conference, August 1976, Vol. 8, pp. 84-94.

171. Nemecek, J. J., et al., "Demand Sensitive Energy Storage In Molten Salts," Sharing The Sun, ISES Conference, August 1976, Vol. 8, pp. 95-106.

172. Assaf, Gad, "The Dead Sea: A Scheme For A Solar Lake," Solar Energy, Vol. 18, No. 4, 1976, pp. 293-299.

173. Dickinson, W. C. et al., "The Shallow Solar Pond Energy Conversion System," Solar Energy, Vol. 18, No. 1, 1976, pp. 3-10.

174. LeFrois, R. T., and H. V. Venkatasetty, "Inorganic Phase Change Materials For Energy Storage In Solar Thermal Program," Sharing The Sun, ISES Conference, August 1976, Vol. 8, pp. 107-122.

175. Riaz, M., et al., "High Temperature Energy Storage In Native Rocks," Sharing The Sun, ISES Conference, August 1976, Vol. 8, pp. 123-137.

176. Birchenall, C. E., and M. Telkes, "Thermal Storage In Metals," Sharing The Sun, ISES Conference, August 1976, Vol. 8, pp. 138-154.

177. Turner, Robert H., "Thermal Energy Storage Using Large Hollow Steel Ingots," Sharing The Sun, ISES Conference, August 1976, Vol. 8, pp. 155-162.

178. Hall, Charles A., and M. Howerton, "Thermal Storage By Ammoniated Salts," Sharing The Sun, ISES Conference, August 1976, Vol. 8, pp. 176-177.

179. Molz, Fred J., et al., "Experimental Study of Subsurface Transport of Water And Heat As Related To The Storage of Solar Energy," Sharing The Sun, ISES Conference, August 1976, Vol. 8, pp. 238-244.

180. Chahroudi, Day, "Thermocrete Heat Storage Materials: Applications and Performance Specifications," Sharing The Sun, ISES Conference, August 1976, Vol. 8, pp. 245-261.

181. Edie, D. D., and S. S. Melsheimer, "An Immiscible Fluid-Heat of Fusion Energy Storage System," Sharing The Sun, ISES Conference, August 1976, Volume 8, pp. 262-272.

182. Gresko, T. M., and D. R. Glenn, "Thermal Energy Storage Applied To Residential Heating Systems," Proceedings 1976 IECEC, Vol. I, pp. 591-597.

183. Nicholson, Edward W., and Robert P. Cahn, "Storage In Oil of Off-Peak Thermal Energy From Large Power Stations," Proceedings 1976 IECEC, Vol. I, pp. 598-605.

184. Venkatesetty, H. V., and R. T. LeFrois, "Thermal Energy Storage For Solar Power Plants," Proceedings 1976 IECEC, Vol. I, pp. 606-612.

185. Lior, N., et al., "Thermal Energy Storage Considerations For Solar-Thermal Power Generation," Proceedings 1976 IECEC, Vol. I, pp. 613-622.

186. Qvale, E. B., "Seasonal Storage Of Thermal Energy In Water In The Underground," Proceedings 1976 IECEC, Vol. I, pp. 628-635.

187. Libowitz, G. E., and Z. Blank, "An Evaluation of the Use of Metal Hydrides for Solar Thermal Energy Storage," Proceedings 1976 IECEC, Vol. I, pp. 673-680.

188. Ridgway, S. L., and J. L. Dooley, "Underground Storage of Off-Peak Power," Proceedings 1976 IECEC, Vol. I, pp. 586-590.

189. Herrick, C. S., and D. C. Golibersuch, Qualitative Behavior Of A New Latent Heat Storage Device For Solar Heating/Cooling Systems, General Electric Co., Report 77CRD006, March 1977.

190. Morrison, D. J., and S. I. Abdel-Khalik, "Effects of Phase-Change Energy Storage On The Performance of Air-Based and Liquid-Based Solar Heating Systems," Proceedings 1977 ISES Annual Meeting, Vol. I, pp. 16-1 through 16-5.

191. Fong, A. D., and C. W. Miller, "Efficiency of Paraffin Wax As A Thermal Energy Storage System," Proceedings 1977 ISES Annual Meeting, Vol. 1, pp. 16-6 through 16-10.

192. Hooper, F. C., and C. R. Attwater, "Optimization of an Annual Storage Solar Heating System Over Its Life Cycle," Proceedings 1977 ISES Annual Meeting, Vol. I, pp. 16-11 through 16-15.

193. Tsang, Chin Fu, et al., "Modeling Underground Storage In Aquifers of Hot Water From Solar Power Systems," Proceedings 1977 ISES Annual Meeting, Vol. I, pp. 16-20 through 16-24.

194. Elwell, D. L., et al., "Stability Criteria For Solar (Thermal-Saline) Ponds," Proceedings 1977 ISES Annual Meeting, Vol. I, pp. 16-29 through 16-33.

195. Lin, Sui, "An Experimental Investigation Of The Thermal Performance Of A Two-Compartment Water Storage Tank," Proceedings 1977 ISES Annual Meeting, Vol. I, pp. 17-1 through 17-5.

196. Phillips, Warren F., and Robert A. Pate, "Mass and Energy Transfer In A Hot Liquid Energy System," Proceedings 1977 ISES Annual Meeting, Vol. I, pp. 17-6 through 17-10.

197. Riaz, M., "Transient Analysis of Packed-Bed Thermal Storage Systems," Proceedings 1977 ISES Annual Meeting, Vol. I, pp. 17-11 through 17-16.

198. Clark, John A., et al., "Rockbed: A Computer Program For Thermal Storage," Proceedings 1977 ISES Annual Meeting, Vol. I, pp. 17-17 through 17-20.

199. Eshleman, W. D., et al., "A Numerical Simulation of Heat Transfer In Rock Beds," Proceedings 1977 Annual ISES Meeting, Vol. I, pp. 17-21 through 17-25.

200. Edie, D. D., et al., "Fundamental Studies of Direct Contact Latent Heat Energy Storage," Proceedings 1977 ISES Annual Meeting, Vol. I, pp. 17-26 through 17-30.

201. Pfannkuch, H. O., and M. H. Edens, "Rock Properties For Thermal Energy Storage Systems In The 0° To 500°C Range," Proceedings 1977 ISES Annual Meeting, Vol. I, pp. 18-5 through 18-9.

202. Chubb, T. A., et al., "Energy Storage-Boiler Tank: Boiler Design, Materials Compatibility Tests," Proceedings 1977 ISES Annual Meeting, Vol. I, pp. 18-10 through 18-11.

203. Turner, Robert H., and Vincent C. Truscello, "Large-Scale Thermal Energy Storage Using Sodium Hydroxide (Na OH)," Proceedings ISES Annual Meeting, Vol. 1, pp. 18-12 through 18-15.

204. Mitchell, Rex C., et al., "Dual-Medium Thermal Storage System For Solar Thermal Power Plants," Proceedings 1977 ISES Annual Meeting, Vol. 1, pp. 18-16 through 18-20.

205. Cabelli, A., "Storage Tanks - A Numerical Experiment," Solar Energy, Vol. 19, No. 1, 1977, pp. 45-54.

206. Bryant, H. C., and Ian Colbeck, "A Solar Pond for London?", Solar Energy, Vol. 19, No. 3, 1977, pp. 321-322.

207. Tanaka, T., et al., "Fundamental Studies On Heat Storage of Solar Energy," Solar Energy, Vol. 19, No. 4, 1977, pp. 415-419.

208. Lavan, Zalman, and James Thompson, "Experimental Study of Thermally Stratified Hot Water Storage Tanks," Solar Energy, Vol. 19, No. 5, 1977, pp. 519-524.

209. Givoni, B., "Underground Longterm Storage of Solar Energy-An Overview," Solar Energy, Vol. 19, No. 6, 1977, pp. 617-623.

210. Hill, James E., et al., "A Method of Testing For Rating Thermal Storage Devices Based On Thermal Performance," Solar Energy, Vol. 19, No. 6, 1977, pp. 721-732.

211. Solar Age, April 1978.

212. Solar Energy Update, July 1978, SEU-78/7. Available from Technical Information Center, United States Department of Energy, Oak Ridge, Tennessee.

213. Dean, T. S., Thermal Storage, [Philadelphia, Pa.: The Franklin Institute Press, 1978], 61 pages.

214. Riaz, M., "Transient Analysis Of Packed-Bed Thermal Storage Systems," Solar Energy, Vol. 21, No. 2, 1978, pp. 123-128.

215. Yoneda, N., and S. Takanashi, "Eutectic Mixtures for Solar Heat Storage," Solar Energy, Vol. 21, No. 1, 1978, pp. 61-63.

216. Hull, John R., "The Effects of Radiation Absorption on Convective Instability in Salt-Gradient Solar Ponds," Proceedings ISES 1978 Annual Meeting, Vol. 2.1, pp. 37-40.

217. Wood, R. L., and E. A. Platt, "Shallow Solar Pond Powered Irrigation Pumping A Feasibility Study," Proceedings ISES 1978 Annual Meeting, Vol. 2.1, pp. 37-40.

218. Nemecek, J. J., et al., "2 Megawatt Hour Energy Storage-Boiler Tank," Proceedings ISES 1978 Annual Meeting, Vol. 2.1, pp. 895-896.

219. Altman, Ralph F., et al., "Two Stage Thermal Storage System For A Solar Electric Power Plant," Proceedings ISES 1978 Annual Meeting, Vol. 2.1, pp. 897-900.

220. Iannucci, J. J., and P. J. Eicker, "Central Solar/Fossil Hybrid Electrical Generation: Storage Impacts," Proceedings ISES 1978 Annual Meeting, Vol. 2.1, pp. 904-912.

221. Nielsen, Carl E., "Equilibrium Thickness Of The Stable Gradient Zone In Solar Ponds," Proceedings ISES 1978 Annual Meeting, Vol. 2.1, pp. 932-935.

222. Stoner, Carol Hupping, Producing Your Own Power, [Emmaus, Pa.: Rodale Press, Inc., 1974], 322 pages.

223. Handbook of Homemade Power, [New York, N. Y.: Bantam Books, 1974], 374 pages.

224. Clegg, Peter, Energy For The Home, [Charlotte, Vermont: Garden Way Publishing, 1975], 252 pages.

225. Critchfield, Richard, "A Melancholy Song On The Nile," International Wildlife, Vol. 6, No. 3, May-June 1976, pp. 28-32.

226. Krenz, Jerrold H., Energy: Conversion And Utilization, [Boston, Mass.: Allyn and Bacon, Inc., 1976], 359 pages.

227. Scott, David, "Compressed Air 'Stores' Electricity," Popular Science, November 1972.

228. Van Cleave, David, "Compressed Air Storage: The Answer To Your Peaking Problem?", Electric Light And Power, October 1974.

229. Suggestions Given On Ways To Beat Energy Shortage, AP dispatch from Brussels, Belgium, Gainesville Sun, Saturday, November 16 1974, p. 6B.

230. Giramonti, Albert J., Preliminary Feasibility Evaluation of Compressed Air Storage Power Systems, December 1975. Available from NTIS as PB-260 496.

231. Simmons, Daniel M., Wind Power, [Park Ridge, N.J.: Noyes Data Corp., 1975], 300 pages.

232. Willett, David C., Underground Pumped Storage Research Priorities, April 1976. Available from EPRI as EPRI AF-182.

233. Kreid, D.K., Technical And Economic Feasibility Analysis of the No-Fuel Compressed Air Energy Storage Concept, May 1976. Available from NTIS as BNWL-2065.

234. Bush, J.B., et al, "Compressed Air Energy Storage - A Near Term Option For Utility Application," Proceedings, 1976 IECEC, Vol. 1, pp. 578-585.

235. Compressed Air Stored In Underground Caverns May Drive Generators, ERDA News, September 19 1977.

236. Vosburgh, Kirby G., Conceptual Design For A Pilot/Demonstration Compressed Air Storage Facility Employing A Solution-Mined Salt Cavern, June 1977. Available from EPRI as EPRI EM-391.

237. Jones, Lawrence W., "Liquid Hydrogen As A Fuel for the Future," Science, Vol. 174, 22 October 1971, pp. 367-370.

238. Reilly, J.J. and R.H. Wiswall, Jr. Hydrogen Storage and Purification Systems, August 1972. Available from NTIS as BNL 17136.

239. Bartlit, J.R., et al., "Experience In Handling, Transport and Storage of Liquid Hydrogen -- The Recyclable Fuel," 1972 IECEC Conference Proceedings, pp. 1312-1315.

240. Gregory, Derek P., and Jaroslav Wurm,"Production and Distribution of Hydrogen As a Universal Fuel," 1972 IECEC Conference Proceedings, pp. 1329-1334.

241. Martin, F.A., "The Safe Distribution And Handling of Hydrogen For Commercial Application," 1972 IECEC Conference Proceedings, pp. 1335-1341.

242. Wiswall, R.H., and J.J. Reilly, "Metal Hydrides For Energy Storage," 1972 IECEC Conference Proceedings, pp. 1342-1348.

243. Cummings, Daniel L., and Gary J. Powers, "The Storage of Hydrogen as Metal Hydrides," Ind. Eng. Chem., Process Des. Develop., Vol. 13, No.2 1974, pp. 182-192.

244. Reilly, J.J., et al., "Motor Vehicle Storage of Hydrogen Using Metal Hydrides," October 1974. Available from NTIS as TEC-75/001.

245. Veziroglu, T. Nejat, Ed., Conference Proceedings. The Hydrogen Economy, 18-20 March 1974.

246. Lindsey, E.F., "Hydrogen Power," Popular Science, March 1975, pp. 88ff.

247. Rohy, D.A., et al., Automotive Storage of Hydrogen Using Modified Magnesium Hydrides, July 1975. Available from NTIS as TEC-75/002.

248. Biederman, Nicholas, et al., Utilization of Off-Peak Power To Produce Industrial Hydrogen, August 1975. Available from EPRI as EPRI 320-1.

249. Srinivasan, S., et al., Hydrogen Production, Storage, and Conversion for Electric Utility and Transportation Applications, October 1975. Available from NTIS as BNL-20590.

250. Kelly, James H., et al., Hydrogen Tomorrow, Demands and Technology Requirements December 1975. Available from NTIS as T76-18654-668.

251. Billings, Roger E., "Hydrogen's Potential As A Vehicular Fuel For Transportati 1975 IECEC Record, pp. 1165-1175.

252. Konopka, Alex J., and Derek P. Gregory, "Hydrogen Production By Electrolysis: Present and Future," 1975 IECEC Record, pp. 1184-1193.

253. Kissel, G., et al. "Hydrogen Production By Water Electrolysis-Methods For Approaching Ideal Efficiencies," 1975 IECEC Record, pp. 1194-1198.

254. Powell, J.R., et al., "High Efficiency Power Conversion Cycles Using Hydrogen Compressed By Absorption On Metal Hydrides," 1975 IECEC Record, pp. 1339-1347.

255. Voth, R.O., and D.E. Daney, "H_2 Liquefaction Effects of Component Efficiencies 1975 EICEC Record, pp. 1356-1362.

256. Boser, O., and D. Lehrfeld, "The Rate Limiting Processes For The Sorption of Hydrogen In $LaNi_5$," 1975 IECEC Record, pp. 1363-1369.

257. Lundin, C.E., and F.E. Lynch, "A Detailed Analysis of the Hybriding Characteri of La Ni_5," 1975 IECEC Record, pp. 1380-1385.

258. Lundin, C.E., and F.E. Lynch, "The Safety Characteristics of Fe Ti Hydride," 1975 IECEC Record, pp. 1386-1390.

259. Mathis, David A., Hydrogen Technology For Energy, [Park Ridge, N.J.: Noyes Data Corp., 1976], 285 pages.

260. Beaufrere, A.H., et al., "Hydrogen Storage Via Iron-Titunium For A 26mwe Peaking Electric Plant," n.d. (circa 1976). Available from NTIS as BNL 20902.

261. Dickson, E.M., et al., Hydrogen Economy: A Preliminary Technology Assessment, Final Report, February 1976. Available from NTIS as PB-266 607.

262. Salzano, F.J., et al., Hydrogen Storage, Water Electrolysis and Fuel Cells For Electric Energy Storage, April 1976. Available from NTIS as CONF-760482-1.

263. Reilly, J.J., Metal Hydrides As Hydrogen Storage Media And Their Applications, July 1976. Available from NTIS as BNL 21648.

264. Salzano, F.J., et al., Hydrogen Storage Via Metal Hydrides For Utility And Automotive Energy Storage Applications, August 1976. Available from NTIS as CONF-761044-1.

265. Ramakumar, R., "An Assessment Of Hydrogen As A Means To Store Solar Energy," Sharing the Sun, 1976 ISES Conference, Vol. 8, pp. 163-175.

266. McRae, Alexander, Ed., et al., The Energy Source Book, [Germantown, Md.: Aspen Systems Corp., 1977], 724 pages.

267. The Hydrogen Way of Life, Popular Mechanics, May 1978, p. 58.

268. Hydrogen Homestead. A fall quarter 1977 issue of Hydrogen Progress published by Billings Energy Corp., 2000 East Billings Ave., P.O. Box 555, Provo, Utah 84601.

269. International Association For Hydrogen Energy, P.O. Box 248266, Coral Bables, Florida 33124.

270. Cox, Kenneth E., and Kenneth D., Eds. Williamson, Hydrogen, 5 volumes, [Cleveland, Ohio: CRC Press, Inc., 1977].

271. Strickland, G., "State-of-The-Art Summary of the Technical Problems Involved In The Storage of Hydrogen Via Metal Hydrides," Preprint, Miami International Conference On Alternative Energy Sources, 5-7 December 1977.

272. Veziroglu, T. Nejat, Ed., Proceedings Of Condensed Papers, Miami International Conference On Alternative Energy Sources, 5-7 December 1977. Available from Clean Energy Research Institute, School of Engineering and Environmental Design, University of Miami, Coral Gables, Florida.

273. Nakamura, T., "Hydrogen Production From Water Utilizing Solar Heat At High Temperatures," Solar Energy, Vol. 19, No.5, 1977, pp. 467-475.

274. Sandrock, G.D., "A New Family of Hydrogen Storage Alloys Based on The System Nickel-Mischmetal-Calcium," Proceedings 1977 IECEC, Vol.I, pp. 951-958.

275. Beaufrere, A., et al., "A Hydrogen-Halogen Energy Storage System For Electric Utility Applications," Proceedings 1977 IECEC, Vol.I, pp. 959-963.

276. Rabenhorst, David W., "Potential Applications For The Superflywheel," Conference Proceedings 1971 IECEC, pp. 1118-1125.

277. Dugger, et al., "Flywheel and Flywheel/Heat Engine Hybrid Propulsion Systems For Low-Emission Vehicles," Conference Proceedings 1971 IECEC, pp. 1126-1141.

278. Lawson, Louis J., "Design and Testing of High Energy Density Flywheels For Application To Flywheel/Heat Engine Hybrid Vehicle Drives," Conference Proceedings 1971 IECEC, pp. 1142-1150.

279. Jakubowski, M. "Flywheel Energy Buffer," Conference Proceedings, 1972 IECEC, pp. 1141-1145.

280. Post, Richard F., and Stephen F. Post, "Flywheels," Scientific American, December 1973, Vol. 229, No. 6, pp. 17-23.

281. Lampe, David, "Superflywheel: The 'Battery' That Spins," Popular Mechanics, November 1974, p. 124ff.

282. Fullman, R.L., Energy Storage By Flywheels, Technical Information Series, Report 75 CRD 051, April 1975. Available from General Electric Co., Corporate Research and Development, Schenectady, N.Y.

283. Fullman, R.L., "Energy Storage By Flywheels," Record 1975 IECEC, pp. 91-100.

284. Chang, G.C. and T.T. Chiao, Eds., Proceedings of the 1975 Flywheel Technology Symposium. Available from NTIS as ERDA 76-85.

285. Gordon, Hayden S., Development of High-Density Inertial-Energy Storage, Final Report, July 1975. Available from EPRI as EPRI 269-1.

286. Chiao, T.T. and R.G. Stone, LLL Program For Composite Flywheel, December 1975. Available from NTIS as CONF-75 1133-4.

287. Clements, Linda L., Engineering Design Data For Composite Materials, December 1975. Available from NTIS as CONF-75 1133-5.

288. Rinde, J.A., Epoxy Resins For Flywheel Applications, December 1975. Available from NTIS as CONF - 751133-2.

289. Penn, Lynn, and T.T. Chiao, Fiber Composites For Energy Storage Flywheels, October 1975. Available from NTIS as UCRL 76891.

290. Blake, Alexander, Mechanical Design of A Composite Centerless Flywheel, December 1975. Available from NTIS as UCRL-51978.

291. Stone, R.G., et al., Fiber Composite Program For Flywheel Applications, November 1975. Available from NTIS as UCRL-50033-75.

292. Penn, Lynn, and T.T. Chiao, Fiber Evaluation For Flywheel Applications, December 1975. Available from NTIS as CONF-75113-3.

293. Simpson, L.A., et al., Kinetic Energy Storage of Off-Peak Electricity, September 1975. Available from NTIS as AECl 5116.

294. Pepper, J.W., Investigation of Multi-Ring Fiber-Composite Flywheels for Energy Storage, Final Report, September 1976. Available from NTIS as PB-260 718.

295. Rinde, J.A., et al., Composite Fiber Flywheel For Energy Storage, June 1976. Available from NTIS as CONF 761018-1.

296. Gordon, Hayden S., Investigation of Multi-Ring Fiber-Composite Flywheels For Energy Storage, September 1976. Available from EPRI as EPRI EM-227.

297. Huddleston, R.L., et al., Composite Flywheel Development, January 1977. Available from NTIS as Y-2072.

298. Energy-Saving Flywheel Developed by UF Prof., Gainesville Sun, Wednesday February 15, 1978.

299. Superconductive Magnetic Energy Storage Fact Sheet, Information From ERDA, Circa May 1975.

300. Hassenzahl, W.V., Superconducting Magnetic Energy Storage Project at LASL, January 1-March 31, 1975. Los Alamos Scientific Laboratory report LA-6117-PR, October 1975.

301. Segal, H.R., and R.W. Boom, "Conductor Development For Large Superconductive Energy Storage Magnets," Record 1975 IECEC, pp. 101-103.

302. Sunshine In A Can May Heat Your Home, Journal-News, Hamilton, Onio, Sunday August 18, 1974, p. A9.

303. Prengle, H. William and Chi-Hua Sun, "Operational Chemical Storage Cycles For Utilization of Solar Energy To Produce Heat or Electric Power," Solar Energy, Vol. 18, No. 6, 1976, pp. 561-567.

304. Schmidt, Eckert W., and Philip A. Lowe, "Thermochemical Energy Storage Systems," Proceedings 1976 IECEC, Vol. I, pp. 665-672.

305. Offenhartz, Peter O'D., "Chemical Methods of Storing Thermal Energy," Sharing The Sun, 1976 ISES Confernece, Vol. 8, pp.48-72.

306. Huxtable, Douglas D., and Donald R. Poole, "Thermal Energy Storage By The Sulfuric Acid-Water System," Sharing The Sun 1976 ISES Conference, Vol.8, pp. 178-191.

307. Baurle, G., et al., "Storage of Solar Energy By Inorganic Oxide/Hydroxides," Sharing The Sun, 1976 ISES Conference, Vol. 8, pp. 192-218.

308. Simmons, John A., "Reversible Oxidation Of Metal Oxides For Thermal Energy Storage," Sharing The Sun, 1976 ISES Conference, Vol. 8, pp. 219-225.

309. Kauffman, Kenneth W., and Harold G. Lorsch, "Thermal Energy Storage With Saturated Aqueous Solutions," Sharing The Sun. 1976 ISES Conference, Vol. 8, pp. 227-237.

310. New Way To Store Solar Energy, Popular Mechanics, November 1977, p.21.

311. Lawrence Berkeley Lab to try Chemicals to Store Heat Generated By Solar Plants, Energy Insider, October 17, 1977, p.3.

312. Fujii, I., and K. Tsuchiya, "Experimental Sutdy of Thermal Energy Storage By Use of Reversible Chemical Reactions," Preprint. Miami International Conference on Alternative Energy Sources, 5-7 December 1977.

313. Wentworth, W.E., et al., "An Ionic Model For The Systematic Selection of Chemical Decomposition Reactions For Energy Storage," Proceedings 1977 ISES Meeting, pp. 18-1 through 18-4.

314. Kutal, Charles, et al., "Use of Transition Metal Compounds to Sensitize A Photo-Chemical Energy Storage Reaction," Solar Energy, Vol. 19, No.6, 1977, pp. 651-655.

315. Putnam, Palmer Cosslett, Power From The Wind, [New York, N.Y.: Van Nostrand Reinhold Co., 1948], 224 pages.

316. Koslover, Monty, "Optimization of Battery Subsystems For Earth Satellite Lifetimes of Greater Than 5 years," Proceedings 1969 IECEC, pp. 61-73.

317. Ramakumar, R., et al., "A Wind Energy Storage and Conversion System For Use in Underdeveloped Countries," Proceedings 1969 IECEC, pp. 606-613.

318. Kordesch, K.V., "City Car With H_2-Air Fuel Cell/Lead Battery (One Year Operating Experiences)", Proceedings 1971 IECEC, pp. 103-111.

319. Fernandes, R.A., "Hydrogen Cycle Peak-Shaving For Electrical Utilities", Proceedings 1974 IECEC, pp. 413-422.

320. Lotker, Michael, "Hydrogen For The Electrical Utilities-Long Range Possibilities", Proceedings 1974 IECEC, pp. 423-427.

321. Burger, James M. and Peter A. Lewis, "Energy Storage For Utilities Via Hydrogen Systems", Proceedings 1974 IECEC, pp. 428-434.

322. Hanneman, R.E., et al., "Closed Loop Chemical Systems For Energy Transmission Conversion and Storage", Proceedings 1974 IECEC, pp. 435-441.

323. Ordin, Paul M., "Review of Hydrogen Accidents And Incidents In NASA Operations", Proceedings 1974 IECEC, pp. 442-453.

324. El-Badry, Y.Z., and J. Zemoski, "The Potential For Rechargeable Storage Batteries In Electric Power Systems", Proceedings 1974 IECEC, pp. 896-902.

325. Brown, Jack T., and John H. Cronin, "Battery Systems For Peaking Power Generation", Proceedings 1974 IECEC, pp. 903-910.

326. Stoner, Carol Hupping, Producing Your Own Power, [Emmaus, Pa.: Rodale Press, Inc., 1974], 322 pages.

327. Ackerman, John P., Assessment Study of Devices For The Generation of Electricit From Stored Hydrogen, December 1975. Available from NTIS as ANL-75-71.

328. Srinivasan, S., et al., Hydrogen Production, Storage, and Converison For Electric Utility and Transportation Applications, October 1975. Available from NTIS as BNL-20590.

329. Energy Storage and Power Conditioning Aspects of Photovoltaic Solar Power Systems, October 1975. Available from NTIS as COO/2748-75/TI.

330. NASA Combines Wind and Energy Storage, Electric Light & Power , August 16, 1976, p. 56.

331. An Assessment of Energy Storage Systems Suitable for Use By Electric Utilities, July 1976. Three volumes. Available from EPRI as EPRI EM-264 (Vol. I, II, III).

332. Tison, R.R., et al., Wind-Powered Hydrogen Electric Systems for Farm and Rural Use, April 1976. Available from NTIS as PB-259 318.

333. Hartnett, James P., Alternative Energy Sources, [New York, N.Y.: Academic Press, 1976], 328 pages.

334. From Bench To Busbar, EPRI Journal, October 1977, pp. 18-19.

335. Battelle Eyes Bulk Energy Storage Via Electrolysis, Electric Light & Power, July 1977, p. 12.

336. Zachmann, Howard C., Liquid Hydrogen From Solar Energy Now, Preprint. Miami International Conference on Alternate Energy Sources, 5-7 December 1977.

337. Lu, P.W.T. and S. Srinivasan, "State of Development In The Area of Water Electrolysis (Near Term)", preprint. Miami International Conference on Alternate Energy Sources, 5-7 December 1977.

338. Halpin, T.E., et al., "Integrated Solar Electric-Utility System For A Residence Using Load Management", Proceedings 1978 ISES Meeting, Vol. 2.2, pp. 351-358.

339. Bruckner, Arthur, II, et al., "Economic Optimization of Energy Conversion With Storage", presented at IEEE Winter Power Meeting, New York, N.Y., January 28-February 2, 1968. Paper 68 CP 134-PWR.

340. Braun, C., et al., "The Economic Incentive For Introducing Electric Storage Devices Into The National Energy System", Record 1975 IECEC, pp. 82-90.

341. Appleby, A.J., et al., "Economic And Technical Aspects of the C.G.E. Zn-Air Vehicle Battery", Record 1975 IECEC, pp. 811-816.

342. Nelson, P.A., and N.P. Yao, Batteries For Utility Load Leveling, prepared for 38th American Power Conference, April 20-22 1976. Available from NTIS (no number).

343. Ferrell, D.T., Jr., et al., Design and Cost Study For State-Of-The-Art Lead Acid Load Leveling and Peaking Batteries, Final report, February 1977. Available from EPRI as EPRI EM-375.

344. Lead-Acid Batteries for Utility Application: Workshop II A Compilation of Workshop Papers Presented On December 9-10, 1975. Available from EPRI as EPRI EM-399-SR.

345. Li, Kam W., "A Thermoeconomic Analysis of Thermal Energy Storage", preprint. Miami International Conference on Alternate Energy Sources, 5-7 December 1977.

CHAPTER 5

SYSTEMS OF SOLAR-ELECTRICS

"Let it be emphasized that solar energy technology may have a tremendous impact on load patterns now experienced by electric utilities."

Robert L. Sullivan [1, p22]

- Systems of Direct Solar-Electrics
- Systems of Wind-Electrics
- Systems of Solar-Electric Vehicles
- Integration of Solar-Electrics Into Existing Energy Systems
- Economics
- Barriers To Solar-Electric Integration
- Future R & D Needed for Systems Integration

In Chaps. 2 and 3 we examined a single solar-electric system, i. e. the apparatus needed to create one generator of solarly supplied electricity; there major emphasis was placed on integrating that single wind-electric or photovoltaic generator with the electric grid. Such a generator typically was located at a single geographic site.

In contrast, in this Chap. we briefly explore the state-of-the-art of systems of solar-electrics, e.g. a larger electric system composed of multiple geographically distributed solar-electric generators stochastically feeding power into an electric grid dominantly supplied by conventional generating plants in steady state. The viewpoint here is largely that of 'a system of systems', the latter possibly being in various mixes of solar-electrics and unconventional loads, as solar-electric vehicles.

The reader is forewarned that this entire Chap. is necessarily of a preliminary and exploratory nature, as relatively little technical work in this general area of integration appears in the pre-1978 solar-electric literature. The reason so little research in the area of 'a system of solar-electric systems' has been heretofore done is simply because of the many and varied technical problems previously requiring R & D, for example, to make a single photovoltaic system a reality; these problems have consumed researchers efforts and attention. Relatively few researchers have therefore had time, energy, and expertise to explore beyond to 'a system of systems' level. Such a systems viewpoint will, eventually as solar-electric technology matures, be essential to explore, understand, and predict, if solar-electrics are ever to be successfully integrated with conventional electric power systems.

It is anticipated that conventional power system planning engineers with positive attitudes have much to offer in this general area toward bridging the present gap between the newer solar-electric systems emerging and the older conventional electric systems. Such bridging appears to have not yet been initiated on any scale because solar-electrics technology had not heretofore grown sufficiently to be ready for such integration; we now appear poised for such an R & D leap of systems integration in 1978.

We first examine systems of photovoltaics.

5-1 Systems Of Direct Solar-Electrics

The conceptual idea of aggregations of photovoltaic systems--both decentralized and centralized--were envisioned by the 1972 Solar Energy Panel [2], though perhaps not as clearly and forcefully as might have been done. Systems of extremely large space stations beaming photovoltaicly generated power to earth via microwaves, rectifying it, and feeding it into the nation's utility grids were also envisioned, though it was stated such space-earth satellite systems had a low probability of success for the entire system.

The pre-1978 photovoltaic literature was surveyed in Chapter 2. In that survey of over 300 references--a substantial fractional part of the published photovoltaic literature-- not a single technical paper, article, or report was turned up dealing with 'a system of multiple photovoltaic systems spatially distributed and stochastically feeding a conventional power system.'

Only Goldstein and Case [3] of Sandia Laboratories came anywhere close to treating the above kind of 'system of systems' problem. They developed a computer program "to accurately simulate the performance of a photovoltaic system." They illustrate the use of their program via a load demand curve over a 24 hour span, peaking at 1.4 KW, in conjunction with an assumed insolation curve containing 'noise' because of cloud passages. The program permitted them to plot battery power delivered vs time and battery state of charge vs time curves. They also illustrate another example showing calculated energy surplus/deficit vs time over a 24 hour span.

Though Goldstein and Case studied only a single small photovoltaic system, it is not difficult to see the possibility of extending their analysis-simulation methods to the more complex study of multiple spatially distributed photovoltaic systems with stochastic outputs at much higher power levels feeding an otherwise steady state utility grid. This latter important theoretical problem has not been dealt with in the photovoltaic literature. There is the remote possibility it may have been treated by someone in the conventional electric energy systems literature; exploration of that literature on this point was beyond the scope of this research.

From the above it is abundantly clear that the general 'system of photovoltaic systems' problem has not been previously addressed within the photovoltaic community. This is an important systems problem which must be addressed initially at the theoretical level. While the generalized theory of such a system does not now exist, enough knowledge now exists about single photovoltaic systems and certainly about conventional electric power systems that this important problem could be solved.

We now leave photovoltaics and explore the wind-electric field from the above 'system of systems' viewpoint.

5-2 Systems of Wind-Electrics

The 1972 Solar Energy Panel members had in mind that eventually large numbers of aerogenerators would feed power into the electric networks of the nation; but the chief emphasis of the report [2] was on the R & D of a single wind-electric system--a topic we surveyed in Chap. 3. The larger systems problem, while recognized in that report, was avoided. "The effect of large numbers of closely spaced windmills has not been assessed" [2, p68].

Ramakumar, et al., Oklahoma State University, in a most significant paper [4] appears to have been the first to clearly and explicitly show a sketch [His Fig. 3, p110] for an envisioned system consisting of multiple spatially distributed wind generators feeding an existing 3 phase power grid with no intermediate storage. He had in mind that the wind generators would use his field-modulated 3 phase generator. He says [p110]:

> "The output of the wind energy system can also be tied to an existing conventional power line to supplement the generating capacity of the system as a whole. Since the output electronics of the generator will allow power flow only in one direction--from the machine to the power network--no two field-modulated generator rotors can interchange power between themselves like the rotors of two conventional alternators delivering power to the same system. Either the machine is pumping power to the network or it is cut off from it. The field-modulated generator cannot 'motor'. The implication of this is highly beneficial--a large number of relatively small generators can be operated to augment the generating capacity of an existing power system without introducing stability problems of the type described above."

It is thus clear Ramakumar both recognized the general systems problem and preliminarily formulated it; he did not solve it, however, as that was not the purpose of his review paper.

Also in 1974 Dambolena, et al. [5], Univ. of Mass., created a computer-based planning model "to evaluate the cost and simulate the performance of alternative wind-power systems" off the New England coast east of Cape Cod. Their offshore wind system included "a potentially large number of electric generators[1] powered by wind driven propellers" at geographically different sites in the Atlantic ocean. Their model either supplied electric power at a single site via inverters or alternatively via an electrolysis plant to produce stored hydrogen and oxygen, thence--at a later time--into fuel cells and an inverter and then to the grid. In their computer model they used for the system load "A fluctuating but deterministic hourly time series for electric demand was generated using as a base a typical weekly load curve and adjusting it by monthly seasonal factors obtained from average weekly demand data for New England compiled by the New England Planning Committee." The flow chart of their simulation model was included in their paper, but the program details were not. They used wind data obtained from measurement in the area. Their complete program let them simulate the operation of such a wind-electric system over a four year period. They felt that "This model and extensions and/or modifications of it should prove to be valuable tools for this further analysis."

---

[1] Three 2 MWe generators--6 MWe/site--were proposed mounted on an anchored floating structure at each site.

Davitian, Brookhaven National Laboratory, in a 1977 paper [6] on the use of wind power by electric utilities clearly envisioned 'a system of wind-electric systems' when he said, "The simplicity of the wind generator in comparison to these huge plants is counterbalanced by the need for large numbers of machines; over one thousand 1.5 MW units, operating at a capacity factor of roughly 0.35, would be needed to provide the energy produced in one such coal or nuclear plant."

Finally, in 1978 Baker [7] in a presentation at the August Denver, Colorado ISES conference described a computer simulation wherein measured wind velocity statistics at various sites throughout the state of Washington were fed into the transfer function of aerogenerators which were spatially distributed across the state with each feeding the power grid. He claimed he could thus input several years history of anemometry data along with known power load curves for the network throughout the state and, in essence, determine power and energy contributions by WECS generating systems had they been actually there. Their computer program even showed the effects on WECS generated power flows as frontal systems moved across the state, typically out of the northwest. This work appears highly significant to the 'system of WEC systems' problem; unfortunately no printed paper was passed out at the meeting nor was one in the Proceedings of the conference. The work would undoubtedly be of much interest to systems engineers.

One concludes that, in contrast to photovoltaics, wind researchers have at least recognized the larger 'system of systems' problem and some preliminary work has been done on it by a small number of researchers. Only the surface of this important complex 'system of systems' problem, typically in a utility context, appears to have been scratched. It would appear that it merits much more R & D, particularly by power systems engineers.

5-3 Systems of Solar-Electric Vehicles

Actual systems of solar-electric vehicles do not exist in 1978 in the United States or anywhere else in the world so far as is known from the open literature. The eventual emergence of such systems has been conceptually envisioned, however, rather strongly by the general solar community, apparently weakly by the photovoltaic community, and almost not at all by the wind-electric community insofar as can be told from examining the literature.

The reason that so few have been seriously concerned with 'a system of systems'[1] of solar-electric vehicles is because it is too early yet. Attention must first be focussed on creating A system of solar-electric vehicles which is about optimal for the users. This in turn hinges on creating a suitable electric automobile which development is currently slowed by the lack of a truly satisfactory battery for electric cars. Once a satisfactory battery appears on the contemporary scene, electric cars will follow. They will first be charged from the conventional electric utility lines. Then some years later after such electric vehicles and solar-electrics come into widespread mass use, then it will be appropriate and timely to speak of 'a system of systems' of solar-electric vehicles. However, we should recognize that the basic technological knowledge and hardware exists, in 1978 to, in principle, create 'a system of systems' of solar electric vehicles, though it would be far from economically optimum.

---

[1] 'A system of systems' here implies multiple systems of electric vehicles whose electrical energy would be dominantly supplied from some kind of solar-electric means.

Thus the arrival of systems of solar-electric vehicles is dependent upon the simultaneous arrival of a suitable electric car and suitable solar electrics. Both are obviously some years in the future. Though such a system is in the future, it is perhaps not too early to, in a preliminary way, survey here a few key notions which will be of interest to the systems engineer. It is beyond our scope here, and quite inappropriate to this larger study, for us to completely survey the rather large detailed literature on electric cars and batteries; many engineers and scientists all over the world are presently working and publishing in these areas, aiming eventually at making a practical electric car a reality.

A Quick Review Of Electric Vehicles has been assembled in an interesting, informative, and highly readable paper by Feder, et al. [8], Gould Inc., who point out that "The electric vehicle is an idea whose time came, went, and has come again." The idea springs back to the late 1880's when it was thought most future vehicles would be electric.

Many of the elements of electric vehicles, the vehicle itself, along with novel ideas about electric vehicles will be found in Wakefield's [9] informative little book.

More current information can be found in the trade magazine exclusively keyed to electric vehicles [10].

Summaries of the very considerable foreign R & D on electric vehicles will be found in references [11] and [12].

The Main Incentives For Developing Electric Vehicles are:

- Air Pollution Minimizers compared to the gasoline engine [8].
- Low Operating Cost to the user, though the literature is somewhat in conflict on this point. One can find numerous papers and articles claiming much lower operating costs than for a comparable gasoline automobile.
- Oil Import Saver, as electric cars could be powered by coal or nuclear or, eventually, solar-electric plants. Electric vehicles, in principle, could offer a transportation mode significantly less dependent on imported petroleum than the present auto. For the U. S. it is estimated that 18 million electric cars by the year 2000 would result in a cumulative savings of petroleum of 1.3 billion barrels [8].
- Quiet Operation characterizes all electric vehicles when compared with conventional gasoline automobiles.

This combination of incentives has motivated many engineers and scientists to work toward achieving practical vehicles which will permit realization of these unique incentives potentially offered by electric vehicles.

Effects of Electric Vehicles on Utility Load Curves is of interest to systems engineers. The universal vision within the electric vehicle community and some utilities is one of thousands of electric vehicles charging batteries at night during the off-peak 'valley' in the utility load curve. Thus the electric energy for charging could be supplied by the baseload power, the cheapest electric energy for the customer. Such a system of electric vehicles could have, from the utility standpoint, the major desirable benefit of tending to level out the power load curve on the system.

A feel for the dramatic load leveling effect electric vehicles potentially could have has been forecast by Jerabek [13], General Electric Co., and some of his load curves are reproduced in Fig. 5-1.

The improvement in the load factor[1] by the addition of electric vehicle charging in the valleys is evident. He also showed a similar case for Philadelphia. It is clear from these curves, as Jerabek says, "the personal automobile will dominate any load-levelling effect a city might experience."

It is also worthy of noting that, if the chargings of electric vehicles are properly timed, then a conventional utility can add such a system load with no increase in plant capacity, a very important point Adler [14] makes for Chicago. He also makes the point that "According to one study by the FPC, 38 million electric vehicles could be on U. S. roads by 1990. Their annual recharging energy requirements would be about 65 million megawatt-hours per year, equivalent to about 1% of the total electrical energy production projected for 1990."

In summary, the advent of electric vehicles to conventional utility systems would appear to have only beneficial effects on the utility system.

<u>Present Major Barriers To Electric Vehicles</u> appear to be, according to Feder, et al. [8]:

- <u>Limited Vehicle Range</u> compared to the internal combustion engine vehicles.
- <u>Six to Eight Hours Recharging</u> time required--or more.
- <u>Subpar Acceleration</u> which prohibits freeway driving.

The key is seen as the development of a better battery. His paper prognosticates probable future battery developments and also points out that "there is a general consensus that for optimum efficiency the electric vehicle must be designed from the wheels up." The electric vehicle community also appears well aware of the need to reduce vehicle costs to be competitive with the conventional auto.

Very many high level scientists and some engineers are currently researching the area of improved batteries. This is a large field beyond our scope here; but we should recognize that many feel the eventual successful creation of a 300-400 mile range electric automobile hinges on the successful outcome of this battery research.

---

[1] Load factor = lowest power/peak power on a given load curve during a specified time interval.

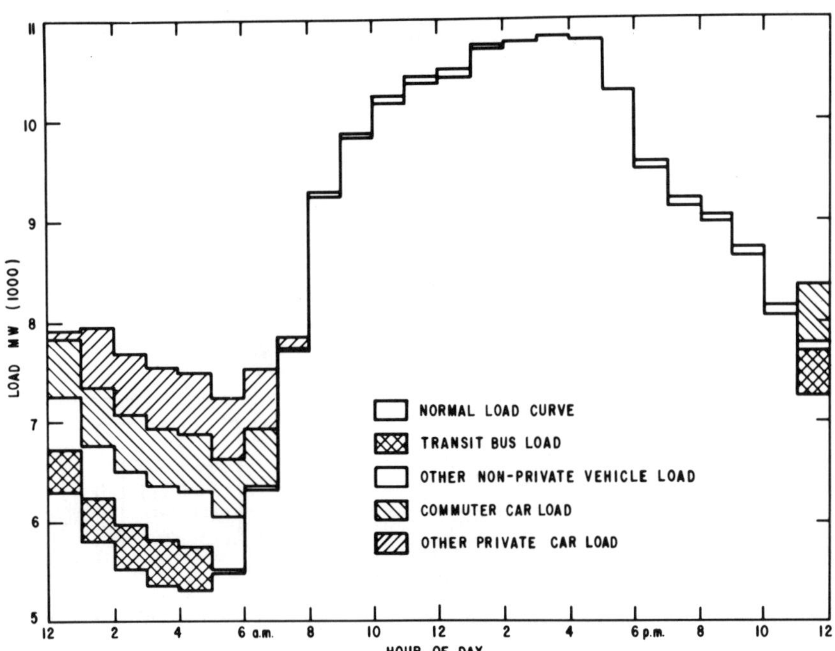

(A) Summer load curve with electric vehicle additions.

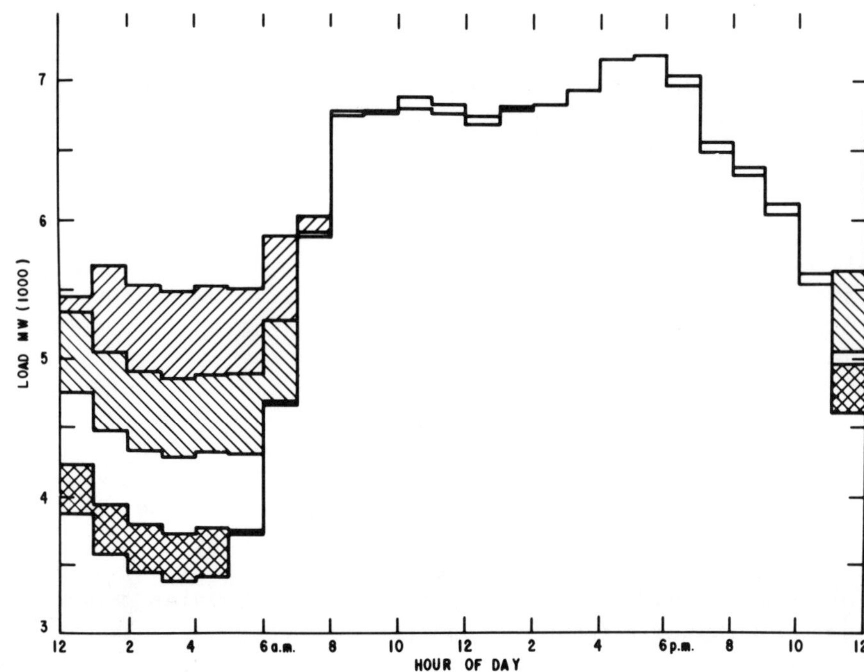

(B) Winter load curve with electric vehicle additions.

FIG. 5-1 Forecast effects on New York City load curves of adding electric vehicles. From Jerabek [13].

The Solarly Supplied Electric Car is at least as old as the 1950's and probably older. The 1972 Solar Energy Panel clearly envisioned electric cars being recharged solarly at residences, as shown in Fig. 5-2. It was principally envisioned that the solar collector would be a photovoltaic array furnishing DC power directly to the charge control, thence into the vehicle battery.

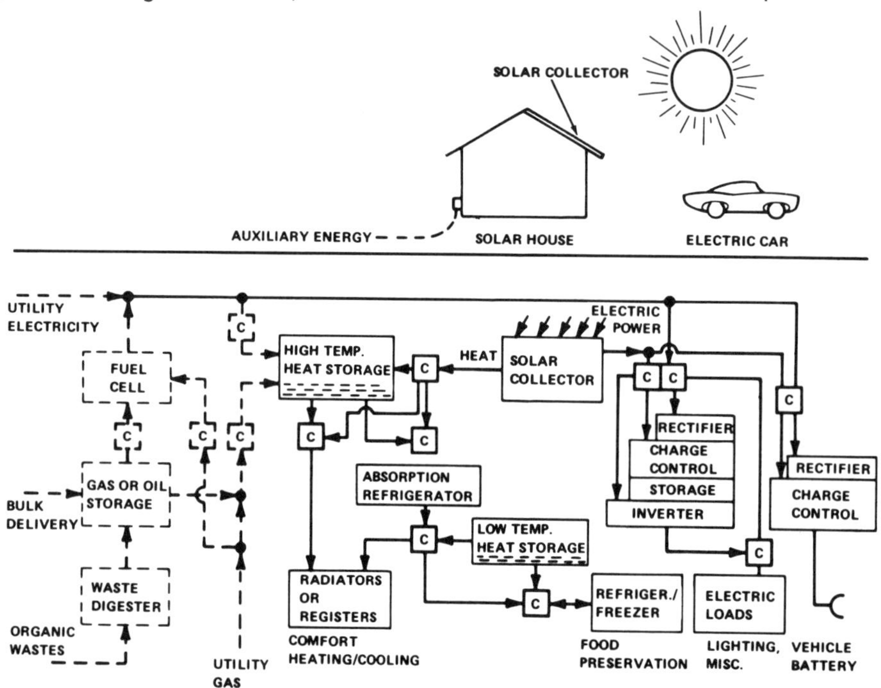

FIG. 5-2 Schematic of solarly supplied electric car for residential buildings as envisioned by the 1972 Solar Energy Panel [2, p57].

Note that the proposed solar-electric-vehicle system in Fig. 5-2 has the solar-electrics off-board the vehicle. While some experimental solar-electric cars have been built with solar cells covering some fraction of the car, it can be shown by an elementary order-of-magnitude analysis that there is not enough area available on a car--even with 100 percent efficient solar cells--to supply any significant fractional portion of the vehicle's daily energy needs. Wakefield [9, p112] has made such a calculation, showing that "such a two-square-yard surface exposed for 1 hour, assuming 100% storage efficiency and 100% efficiency of transferring electric energy into mechanical energy, a 6000-pound van would be moved only 1 mile at 30 MPH." Wakefield also investigated the use of on-board windmills and quickly saw they too are not viable.

Some readers may remember that in the 1950's the International Rectifier Corp. mounted a 'Solar King' solar cell panel on the roof of an antique 1912 Baker automobile to recharge the batteries solarly. After sufficient exposure to insolation the car could be driven several miles on solar power. This vehicle was widely exhibited around the United States.

Varde [15], Univ. of Michigan, has more recently (1977) investigated the possible on-board use of solar cells on a small commuter electric car. He finds "that the cells can supply a maximum of 5 to 70% of energy consumed during travel, depending on the mode of the driving cycle. However, the system seems to be extremely costly when compared to the conventional battery charging system."

It is clear now that the successful solarly supplied electric car of the future will have the solar-electrics off-board from the vehicle, though some on-board photovoltaic cells might be used for trickle charge.

Schaeper and Farber [16], University of Florida, appear to have been the first to actually construct circa 1973 and test a modern electric automobile whose batteries were recharged solarly. They converted a General Motors Corvair into an electric traction test-bed vehicle which could go 88 km at 40 km/hr on a level road. Battery recharging was accomplished by 145-1000 watt output Sterling hot air engine driving generators to produce electricity. In 1973, according to Farber, "the price of solar cells would have been about 28 times that of the Sterling engine systems." The solar cell cost situation has now changed dramatically, however, since 1973, as we indicated in Chapter 2.

In 1973 Bailey had his students [17] doing preliminary visionary creative engineering thinking about a new system of solar-electric vehicles. The concept was based on an off-board solar array sited in certain city or company parking lots. The array would be near, or perhaps over, a parking space. It would be so designed that the electric vehicle would automatically connect to the array power output when the car was parked. A user coin would pay for both parking and energy fees. The electric energy to recharge the vehicle batteries would be entirely supplied by the sun while the owner is working his 8 or more hours. It was argued that the nation's parking lots constitute a large solar collector area and that such a solar-electric vehicle system as proposed would use the relatively long time during high insolation daylight hours to recharge the vehicle while its user is at work. Supplementary recharging from conventional utility lines could also be accomplished once the vehicle got home; alternatively, if the array were sized correctly enough charge could have been placed in the vehicle to also get it back to the parking lot the next morning.

The above solar-electric vehicle concept has been dormant waiting for solar cell costs to decrease. The idea still appears to have much merit; a system analysis of the concept has not been done but could be. One does not have to wait until solar cell costs reach $500/KW to determine the feasibility of the concept--at least from the systems standpoint.

In conclusion it is evident from the above prior art that even one optimal solar-electric vehicle system with universal appeal to urban users has not yet been created; but the necessary pieces for such an envisioned system are currently being worked on in R & D. It is not hard to envision them converging in a few years to practical solar-electric vehicles; but much engineering and scientific work must first be accomplished.

5-4 Integration of Solar-Electrics Into Existing Energy Systems

Recall that in Sec. 3-8 we showed integration has two main facets:

- Electrical Integration
- Technology Integration

At the present state-of-the-art of both photovoltaics and wind-electrics the major present concern from the larger system view should be only electrical integration.

One 'System of Systems' Ramakumar [4, p109] of Oklahoma State University has proposed one way of electrically integrating photovoltaic arrays, wind generators, solar sea power plants, and conventional solar heating plants to result in DC and AC electrical outputs, hydrogen fuel, hydrocarbon fuel (methane, methanol, etc.) and steam outputs. The total system proposed is quite complex and need not be reproduced here. The solar-electric portion involved an array, high pressure electrolysis, and an inverter to get to AC output. The wind-electric portion was a straightforward field modulated generator directly coupled to the AC line. Though it is not explicitly stated in his paper, from the context one deduces he envisions many such systems being created at various sites, thus achieving the 'system of systems' referred to earlier. Construction of complete prototype systems was underway in 1974 at OSU.

Margin And Backup Capacity Of considerable interest to power systems engineers concerned with integrating solar-electric systems into the utility grid is a 1976 study done by McKoy [18] [19] of The Aerospace Corporation for EPRI. He studied the impact on margin requirements for utility systems of large-scale utilization of solar power plants. "A major objective of this study was to determine the amount of installed conventional capacity which is required to backup a given amount of installed solar capacity in order to satisfy a loss-of-load criterion and a scheduled maintenance requirement. Installed solar capacities up to approximately 30% of peak annual electrical demands were considered. Another objective was to determine the extent to which this backup requirement can be relaxed for a given amount of installed solar capacity when the solar plants in the system are dispersed geographically."

McKoy defines for a power plant:

$$\text{(5-1)}\quad \text{Margin}^{1} = \left(\frac{\text{Total Capacity} - \text{Peak load}}{\text{Peak load}}\right)$$

$$\text{(5-2)}\quad \text{Backup} = \left(\begin{matrix}\text{Total Capacity}\\ \text{with solar}\end{matrix} - \begin{matrix}\text{Total Capacity}\\ \text{conventional only}\end{matrix}\right)$$

and he points out that "If solar power plants comprise part of the network, some backup capacity (normally assumed to consist of conventional plant capacity) must be provided as insurance against forced outage, just as with conventional plants. Additional backup may also be required for the solar plants to account for periods of low or zero insolation."

Though his analysis contained many more details too numerous to include here, a summary of McKoy's findings is presented in Figs. 5-3 and 5-4 for single and multiple sites respectively. Note that, according to McKoy, if a solar plant at a single site were installed with 30% capacity of conventional power a backup of 55% capacity would be required. Then from Fig. 5-4 we learn that properly chosen multiple solar plant sites can significantly reduce the required backup capacity. He points out, however, that "significant reduction is achieved, however, only when weather events (and hence insolation outages) are uncorrelated between sites. Moreover, it is necessary to properly distribute the total solar capacity between the selected sites. At present the only known method of selecting this distribution is by performing system simulations---."

[1] Margins, in conventional plants, typically run 10-20% of peak load.

Summary – Single Solar Plant Sites

- REQUIRED BACKUP CAPACITY INCREASES DRAMATICALLY WITH PENETRATION

SOLAR CAPACITY AT INYOKERN % OF TOTAL	BACKUP % OF SOLAR CAPACITY
10%	23%
20%	42%
30%	55%

- ANALYSIS RESULTS ARE SITE-DEPENDENT AND SENSITIVE TO THE CORRELATION OF SPECIFIC INSOLATION AND PEAK DEMAND EVENTS
  - AT 30% PENETRATION, SITE-TO-SITE VARIATION IN REQUIRED BACKUP CAN BE AS LARGE AS 40%
  - BACKUP DEPENDS ON THOSE DAYS (≤ 5% of time) WHEN INSOLATION IS LOW
- SHAPING THE SCHEDULED MAINTENANCE PROFILE FOR THE CONVENTIONAL PLANTS REDUCED BACKUP 5% TO 10%

FIG. 5-3 Summary of backup required for <u>single</u> solar plant sites. From McKoy [18].

Summary – Multiple Solar Plant Sites

- GEOGRAPHIC DIVERSITY REDUCES REQUIRED BACKUP WHEN:
  - INSOLATION OUTAGES OCCUR FREQUENTLY
  - WEATHER IS UNCORRELATED BETWEEN SITES
- TRIPLE-SITE MIX (3/5 at Yuma, 1/5 each at Inyokern and Santa Maria) CAN REDUCE BACKUP AT 30% SOLAR CAPACITY BY ONE-SIXTH OR MORE
- DUAL-SITE MIX (1/2 at Inyokern, 1/2 at Santa Maria) REDUCES BACKUP AT 30% SOLAR CAPACITY BY ONE-FOURTH OR MORE
- LARGEST CAPACITY SHOULD BE LOCATED AT SITE HAVING BEST INSOLATION, IF THERE IS A SIGNIFICANT DIFFERENCE BETWEEN SITES

FIG. 5-4 Summary of backup required for <u>multiple</u> solar plant sites. From McKoy [18].

One of McKoy's main results appears to be that installation of a solar-electric plant in conjunction with conventional generation results in the need for significant conventional backup capacity, a notion foreign to the photovoltaic and wind-electric fields in 1978, both of which, as was pointed out in Secs. 5-1 and 5-2, have hardly probed this systems area. The sensitivity of his results to site location also underscores the earlier importance given to siting considerations in Chaps. 2 and 3.

Plastiras [20], Sandia Laboratories, is attempting to "quantify the feeling that there is a positive relationship between weather conditions at sites separated by as much as several hundred miles, the significance of this to solar plants being that the probability of two nearby plants suffering outages at the same time is considerably higher than would be expected of conventional plants. Thus, there may be advantages to a utility to geographically disperse its solar plants."

<u>Capacity Displacement</u> Plastiras [20] also is researching the capacity displacement for solar plants, presumably solar-thermal, but many of the ideas of capacity displacement are also applicable to photovoltaic and wind-electric plants. Plastiras explains the idea of capacity credit thusly:

> "The capacity credit of a solar plant is a measure of the amount of conventional generating power it replaces. To be specific, suppose that a utility comprised of a specified mix of conventional plants is able to satisfy a load with a given reliability. If a 200 MWe conventional unit is replaced by a 200 MWe solar plant, there exists a smallest generating capacity (in MWe between 0 and 200) which must be added to the altered system in order that the load may be satisfied subject to the same reliability criterion. If, for example, a 75 MWe conventional plant is sufficient to back up the solar plant, then the solar plant is assigned a capacity credit of 125 MWe for the conventional generating power it replaces. However, it must be emphasized that capacity credit does not reflect the capital or operating costs of the backup plant and that these must be known before the value of the solar plant to the utility can be determined. Furthermore, the definition of capacity credit can only sensibly be applied to a specific solar plant for which the following are also specified: the standard of reliability[1] imposed by the utility, the load profile, the insolation data, the mix of generating units in the original system, and the dispatch strategy for the operation of the solar plant and more generally for the interaction of all of the plants in the grid."

It is thus evident from the previous sub-sections that to integrate a solar-electric plant with a utility grid requires an assessment of both backup capacity and displacement credit to meet a system reliability criterion. Both of these notions undoubtedly will come as shocks to researchers in the photovoltaic and wind electric fields.

---

[1]Plastiras states that a typical reliability requirement "is that the utility be able to satisfy the load for all but at most one hour in 20 years."

Power Dispatching and Load Management. The state-of-the-art of solar-electrics is not yet sufficiently advanced at the systems level for optimal power dispatching philosophies to have been clearly enunciated. This is another important systems area deserving of more study. What should be the dispatching philosophy of a utility with solar-electric plants integrated with it? Should the philosophy be fundamentally different from that for a conventional plant? How does wind and insolation affect the philosophy? Is there an optimal philosophy, especially when mixtures of solar-electrics are dispatched into a conventional power grid? How is the solar dispatching philosophy related to the total system load profile? What relationships exist between the dispatching philosophy and the system reliability? These are only a few of the unanswered questions begging for answers as the new art of solar-electric systems engineering arises. The answers appear presently unknown. Perhaps some answers may be forthcoming from the 100kW aerogenerator at Plum Brook, Ohio.

The subject of utility load management and solar energy has been examined in a preliminary way by Davitan, et al. [21] of Brookhaven National Laboratory. Though his analysis is directed at electrically assisted solar hot water systems, much of the philosophy appears also applicable to solar-electric systems in a general way. Their "analysis indicated that if about 20-40% of the residential customers used (night time precharge electric only hot water heating) appliances in a load managed mode, the 10 PM-8 AM valley in the utility load curve would be filled. For combined electric heating and hot water, the corresponding fraction is 6-12%. It is estimated that in each case, roughly twice the number of customers could be accommodated in the valley if solar/electric (hot water systems combined) were used instead." The subject is more complex, however, than Davitan indicates when one is considering mixes of solar and conventional generation, for to fill the night time valley in the load curve using solar-electric power would necessarily imply storage at added cost. The simplest case, from the systems viewpoint, for solar-electric generation would appear to be as peak shavers during the daylight hours when insolation is available, thereby minimizing the need for costly energy storage.

The whole subject of load management via alternative rate schedules, load limiting devices, conservation, and public education for a solar-electric system feeding into a conventional power grid appears to have hardly been scratched in the solar-electric literature, though such subjects are beginning to be dealt with in 1978 by the solar-heating community in consort with various utilities. One example of the latter is a paper by Wirtshafter, et al. [22].

A Generalized Integrated Solar-Electric Systems Theory appears to not be in existence in the rather extensive solar-electric literature surveyed in this research. However, a very significant June 1978 paper by Pollard [23], Oak Ridge Associated Universities, appears to be the first major step in the direction of generating such a theory. His abstract reads:

> "The analysis relates the annual electrical output of any type of solar-electric facility directly to the effective annual insolation received on a unit area of its solar collectors. General expressions are derived for the capacity factor (in terms of demand limits, down time, and storage loss), the solar availability factor (ratio of annual solar-electric output to conventional fuel fired output at full capacity for both), and the solar fraction. The analysis takes full

> account of the daily and seasonal cycles of solar radiation and its intermittent, stochastic character. All results are given for a unit area of solar collector and are therefore independent of the size of the facility.
>
> The capital cost of solar-electric facilities is expressed in dollars for each kilowatt-hour per year of electrical output rather than dollars per kilowatt of installed capacity as is customary for conventional plants. Capital investment is divided among three components: solar-electric generation, non-solar auxiliary power, and storage. A general expression is derived in terms of actual or estimated component costs, and the results for solar generation and storage are shown graphically."

Pollard's paper is of much interest to solar-electric systems engineers, and it merits close study.

Need For "A System of Systems' Operating Experience is obvious in view of all the above very tenuous preliminary studies. While more such paper studies of a systems nature will surely be executed in the reasonably near future as the number of single solar-electric systems continues to grow, it is well to recognize that as of 1978 not a single 'system of systems' of solar-electrics integrated with a conventional power grid is known to be in existence. While several single site wind-electric plants have been successfully integrated with electric utilities, as detailed in Chap. 3, and several photovoltaic installations have likewise been integrated as cited in Chap. 2, spatially aggregated such systems have not been assembled and experimentally studied. Such a carefully thought-out and executed piece of systems work will surely have to be done before large-scale solar-electric generation appears fully integrated with the conventional power grid.

The need for actual operating experience with the above system supplying a substantial portion of its off-peak load to charging large numbers of electric vehicles is also obvious, though it presently appears it will be many years before such data will be available, as both the solar-electrics and a workable electric vehicle must yet be created.

Here, again, is the pressing need for the unique expertise of the power systems engineer. Such systems studies appear not likely to be forthcoming from the strictly solar-electric devices orientated community.

Combining Wind and Solar Systems Up to now we have in this report been considering integrating a large number of photovoltaic plants or a large number of wind-electric plants into a conventional power grid.

The possibility also exists of combining wind and photovoltaics in some optimal way to feed a conventional grid, perhaps with some energy storage. It is clear the justification is that the sun may shine when the wind doesn't blow and vice versa. Thus in principle the strengths of one type converter could compensate for the weaknesses of the other. Andrews [24], Southampton College of Long Island University, has investigated this general problem via a Monte Carlo computer simulation model, using varying proportions of wind and solar power on different runs. He concludes that there are advantages to a mixed system. His histograms indicate that about 50% wind and 50% solar results in the minimum number of energy storage hours needed.

This whole problem area of the possible optimal mixture of wind and solar systems appears to merit further R & D because of its potential significance from the systems viewpoint.

5-5 Economics

Section 5-1 through 5-4, taken collectively, indicated that the overall state-of-the-art of systems of solar-electrics is in a most preliminary state. It should therefore come as no surprise to learn that information on the economic aspects of such 'system of systems' is, essentially, non-existent in 1978. It is simply too early in the development of solar-electrics for such a body of knowledge to have arisen.

Instead, by narrowing the viewpoint somewhat to concern for a single solar-electric system perhaps integrated with an electric utility, then one finds a few papers in the literature treating economic aspects of integrated systems.

There is therefore a strong coupling between this Sec. and earlier treatments of economic aspects in Secs. 2-6 and 3-9 for photovoltaics and WECS respectively. Here, however, we look more toward economic aspects of the complete integrated system composed of conventional plus solar-electric generation.

A Generalized Costing Methodology looking at solar-electrics from the generalized systems viewpoint appears to have been formulated by Pollard [23]. He advances generalized formulae and graphs for estimating (a) the generating cost of solar electricity and (b) the cost of storage and auxiliary power. He advances the notion that it is better to express the capital cost of solar-electric facilities in dollars per kilowatt-hour per year of electrical output rather than dollars per kilowatt of installed capacity as is customary for conventional electric generating plants. He also normalizes his costs to that of a square meter basis of collector area, thus making the results independent of the size of the facility.

The systems engineer interested in the economics of solar-electrics would do well to study Pollard's paper and his notions carefully.

Pricing of Back-Up Systems based on the peak load has been dealt with by Feldman and Anderson [25]. Though this philosophically-orientated paper is aimed at the utility system as a back-up for solar heated buildings, some of the ideas contained therein appear germane to the issue of peak load pricing of solar-electric generation systems, an issue that will undoubtedly arise as solar-electrics mature.

Pricing of Hybrid Solar-Electric Plants has been preliminarily investigated by Brune [26], Sandia Laboratories. He defines "the term 'hybrid solar plant' in its broadest sense includes not only the use of solar technology with another source of energy to produce heat or electricity, but also the collection of insolation by two or more solar technologies. A hybrid system could use any of the available technologies--." His study focusses on "the use of a solar with a non-solar energy source to produce electricity for utility application." He also estimates the cost of repowering existing power plants. His paper shows estimated cost curves for the value of solar hybrid equipment as well as repowering solar equipment value.

Economic Models of WECS in the Cape Cod area have been investigated by Johanson and Goldenblatt [27], JBF Scientific Corp. Their economic model allows "determination of the life cycle value of a WECS installation to a utility. The model is designed so that it is compatible with existing utility production cost models and planning techniques. It allows the utility to evaluate the economic effect a stochastic source, such as a WECS, would have if used as a fuel saver."

They "found that the utility could spend a maximum of $792/kW for a WECS complete installation and $425 for a WECS unit, in 1984 dollars." They also found the interesting result that each successive WECS has less value since it replaces less expensive fuels."

Davitian[6], Brookhaven National Laboratory, in a paper on the use of wind power by electric utilities, philosophies briefly about analyzing the economics of wind power. He says it is a complex subject and that "the complexity of the subject is compounded by the inappropriateness of conventional utility analytical tools for studying wind power."

We conclude this brief treatment of economics of systems of solar-electrics as we began it by reiterating that economic information on a 'system of systems' of solar-electrics is essentially non-existent; there is limited information, however, on economics of a single solar-electric method integrated with a conventional utility grid, and such meager economic information as exists at the present state-of-the-art on the latter case has been cited for the interested reader.

5-6 Barriers To Solar-Electric Systems Integration

After surveying an appreciable fractional part of the published literature on solar-electric 'system of systems', it is this researcher's opinion that the following items appear to be the principal 1978 barriers to solar-electric systems integration:

- Lack of Systems Orientation within some sub-fields, particularly within the photovoltaic community. There, as has been pointed out earlier in this Chap., the important role of the systems engineer in creating a 'system of systems' of photovoltaics appears almost nonexistent. Systems aspects and system potentialities of photovoltaics is very young as was evident particularly at the June 1978 13th Photovoltaics Specialist Conference, Washington, D. C. where by far most of the emphasis was device orientated.

  As photovoltaics grows it will be increasingly essential for the field to shed some of its previous 'device orientation' and acquire more 'systems orientation.'

  This appears to be a potentially fertile field for utility systems planners who could make significant contributions.

- Need For Added Growth and maturity to occur yet within the photovoltaic and wind energy conversion communities before multiple 'system of systems' of solar-electrics are engineered and constructed. Such large systems necessarily represent substantial capital outlays; before such outlays can be made on multiple systems it is absolutely essential that the technology for a single solar-electric system be reasonably mature, including being cost effective.

This needed growth and maturity appears to be close for the wind-electric field and somewhat further in the future for photovoltaics.

- Lack of a Technical Body of Knowledge in the photovoltaic field specifically addressing the complex problems likely to occur when large numbers of smaller systems of photovoltaic generators are integrated into the power grid. This body of knowledge for multiple photovoltaic systems has been shown in this research to not exist in 1978. It is an obvious barrier to progress in the eventual integration of photovoltaic systems into the electric power networks. The barrier is believed surmountable, but it will require technical people knowledgeable in depth of both photovoltaics and conventional power systems.

  The wind-electric community appears to already have the beginnings of such a body of systems knowledge, and to this extent as a technical community it appears ahead of the photovoltaic community. However, much systems orientated work yet needs to be done within the wind community.

- Lack of Actual Operating Experience with a 'system of systems' of solar-electrics integrated with an operating utility system. No installation exists anywhere containing multiple wind generators or multiple photovoltaic generators--all spatially distributed over a large territory--all feeding a utility system under realistic wind, insolation, weather, and load conditions. Previous U. S. experience has concentrated on integrating a single wind generator or a single photovoltaic plant into the utility grid.

  In a similar vein no one knows, based on actual real-world operating experience, what an actual system load curve looks like when a large number of electric vehicles are actually used on the system. Only computer simulations have been made to date. The lack of such vital load experience and information could serve as a later barrier to the development of solar-electric vehicle systems.

- Lack of A High Energy Density Battery of low cost is an obvious block to evolvement of electric vehicle systems. Such a battery, on a larger scale, is also a barrier to the utilities now developing localized energy storage at neighborhood sub-stations. The development of such batteries could potentially hasten the arrival of complete solar-electric integration with the utilities, thereby helping alleviate the obvious energy storage problem with all solar-electrics.

- The Need For Back-up and Margin Power Capacity of solar-electric systems to meet power system reliability requirements is relatively unknown within the photovoltaic and wind technical communities.

The need for these requirements, assuming the need is corroborated and validated by other systems engineers, may be perceived by both technical communities to be a barrier to the development and integration of solar-electrics unless this need is more clearly presented and more widely disseminated than it is in 1978.

The above are the principal barriers to system integration seen in 1978, based on the present research. It is expected that as the field of solar-electric systems engineering develops other barriers can then be seen more clearly.

5-7 Future R & D Needed For Systems Integration

After surveying an appreciable fractional part of the published literature on photovoltaic and wind energy systems, and particularly the systems aspects of each, it is this researcher's opinion that the items listed below--not necessarily in order of priority--are in need of additional R & D. The viewpoint is constricted to that of the systems engineer interested in integrating 'system of systems' of photovoltaics and wind-electrics into electric utility grids. These areas are listed without extensive arguments and justifications, leaving them for future researchers.

- A Literature Search needs to be executed of the conventional electric energy systems literature to determine if anyone has examined the general theory of geographically distributed stochastic solar-electric generators driving a conventional electric power grid which is in a quasi steady state.

  We now know from the present research that within the solar-electric advanced energy field only the work of Pollard [123] gets close to addressing the above important and technically sophisticated problem.

- Develop The Generalized Theory for geographically distributed stochastic solar energy generators driving a conventional electric power grid in a quasi steady state first without storage and then with storage, assuming the above literature search indicates this important problem has not been researched by conventional electric systems engineers. This theory would furnish the theoretical base for both decentralized solar-electrics and centralized solar-electric additions to the electric power grid.

  The orderly and logical development of this theory will be essential to the future growth of solar-electric systems engineering.

  A sub-part of the problem and one expected to be very complex technically should address the problem of the transient case for the system so that system dynamics would be known and important time constants identified. It is already known that the time constant of the total rotating system on a single wind-electric system is important in the transient integration of wind-electrics. What is important when thousands of such systems are simultaneously stochastically feeding power into a utility grid? It also can be seen already that transients to the electrical system will exist when fast moving clouds pass by a photovoltaic system. What time constants are involved? If thousands of such plants feed into a utility grid what are the critical design parameters to insure a stable electric power system?

- Develop Practical System Planning Models "which address questions concerning the most effective ways to integrate windpower and other methods of producing energy." [5, p287]. The literature and art in this field needs organization and extension by some practical minded, straight thinking systems engineer; the aim should be to produce generalized planning models and methods applicable to most utilities so that each does not have to, at great expense, develop their own separate planning models.

- A Power Dispatching Philosophy for future solar-electric plants needs to be creatively thought about and shaped up into a clear coherent philosophy. It should be in a utility context and with a system of conventional electric generation using solar generated power as an important auxiliary power. The research should also address the relationship of the power load profiles to the power capabilities of both conventional and solar plants as well as how the solar plant should be economically dispatched, given a probable load curve from history.

- The Potentialities In Wind-Solar Mixtures should be investigated deeper beyond the preliminary work of Andrews [24]. From the overall systems viewpoint do such systems offer reliability improvement possibilities? Will they be cost effective as power supplementers in a conventional power system? Is, on balance, the creation of such dual systems a desirable direction for solar-electric systems to proceed in the future? Should the idea be dropped entirely?

- Experimentally Studying 'A System of Systems' of solar-electrics integrated with a conventional utility grid. The systems would have to be created within some well-defined geographic area, say, of a few counties, and the utility system would, ideally for such an experiment, be small. It might consist of a few tens of WECS, of photovoltaic plants, supplying power to the grid. The chief idea would be to conduct a well conceived small-scale experiment on 'a system of systems' from which maximum technical information could be extracted about the probable operation of much larger such electrical systems.

  It is recognized even on a small scale this proposed experiment will be a large, costly, undertaking.

  Much could be learned about solar-electric system operation from such an experiment. It could also help stimulate the photovoltaic and wind communities to prepare the hardware needed to run the experiment. It is envisioned such an experiment would require several years duration to reach a conclusive state.

- A Photovoltaic-Electric Vehicle Feasibility study based around the concept advanced by Bailey in Sec. 5-3 needs to be initiated as a preparation to the eventual arrival of an urban electric vehicle whose energy would be solarly supplied. One does not have to wait until $500/kW photovoltaics arrive to study the feasibility and economics of the overall system.

  Such a system could eventually have significant beneficial effects on the American public, the growing electric automobile industry, the oil import problem, the growing photovoltaic industry, and utilities.

The successful execution of the above R & D over the next several years would be substantial steps toward the eventual integration of solar-electrics into the nation's electric energy networks, thereby alleviating the rapid consumption of our fossil fuels and placing the nation closer to using our chief energy income, the sun.

EPILOGUE

This overall research has attempted to integrate large and important technical advanced fields around the common theme of solar-electrics. Unfortunately, and reluctantly the important sub-fields of solar-thermal electrics, solar-water-electrics (including OTEC), biological-electrics, and novel solar-electrics had to be pruned in this research study. These fields also need summarizing and integrating by one competent individual. The task was simply beyond the scope of achievability of one investigator for this short fourteen month research contract.

Perhaps these important advanced energy conversion fields can be assessed and integrated in a similar fashion to this report at some future time.

REFERENCES

1. Sullivan, Robert L., Power System Planning, [New York, N. Y.: McGraw-Hill International Book Co., 1977], 324 pages.

2. An Assessment of Solar Energy As A National Energy Resource. Prepared by the NSF/NASA Solar Energy Panel, December 1972. Available from NTIS as No. PB 221-659.

3. Goldstein, Lawrence H., and Glenn R. Case, "PVSS-A Photovoltaic System Simulation Program," Solar Energy, Vol. 21, No. 1, 1978, pp. 37-43.

4. Ramakumar, R., et al., "Prospects For Tapping Solar Energy On A Large Scale," Solar Energy, Vol. 16, No. 2, October 1974, pp. 107-115.

5. Dambolena, I. G., et al., "A Planning Methodology For The Analysis And Design of Wind-Power Systems," 1974 Proceedings, 9th Intersociety Energy Conversion Engineering Conference, pp. 281-287. Available from The American Society of Mechanical Engineers, 345 East 47th St., New York, N. Y. 10017.

6. Davitian, Harry, "The Use of Wind Power By Electric Utilities," Proceedings of the 1977 Annual Meeting, American Section of ISES, Vol. one, June 6-10 1977, p. 19-1 through 19-5.

7. Baker, Robert W., "Seasonal Wind Flow Patterns Over Pacific Northwest As Related To Wind Potential," presented at August ISES meeting, Denver, Colorado. Not in published Proceedings.

8. Feder, Robert J., et al., "The Future For Electric Vehicles," Electric Perspectives, 77/4, pp. 2-11.

9. Wakefield, Ernest H., The Consumer's Electric Car, [Ann Arbor, Mich.: Ann Arbor Science Publishers, Inc., 1977], 136 pages.

10. Electric Vehicle News, a magazine sponsored by the Electric Vehicle Council and published by Electric Vehicle News, P. O. Box 533, Westport, Conn. 06880.

11. Busi, J. D., Electric Vehicle Research Development and Technology-Foreign, April 1976. Available from NTIS as AD-A036 458.

12. Busi, James D., Electric Vehicle Research, Development, and Technology-Foreign (U), January 1977. Available from NTIS as AD/A-040 526.

13. Jerabek, Elihu C., "Load Leveling With Electric Vehicles In The Urban Environment," Proceedings Eleventh IECEC, Sept. 12-17, 1976, pp. 382-389.

14. Adler, Cy A., Plenty of Power For Electric Cars, The Wall Street Journal, Monday, August 28, 1978.

15. Varde, K. S., "Partial Energy Supply To Electric Vehicles Through Solar Cell System," preprint. Miami International Conference on Alternative Energy Sources, December 5-7, 1977.

16. Schaeper, H. R. A. and E. A. Farber, "The Solar-Electric Car--Urban Vehicle Performance," International Congress: The Sun in the Service of Mankind, Unesco House, Paris, France, 2-6 July 1973, paper E 8.

17. Hooten, Damon, Solar Powered Electric Car, Final Report in EE491 Creative Problem Solving Course, December 6, 1973. Unpublished. Copy available from Prof. Bailey, Dept. of Elec. Eng., University of Florida, Gainesville, Fla. 32611.

18. McKoy, George C., "A Study of the Impact on Margin Requirements For Utility Systems of Large-Scale Utilization of Solar Power Plants," March 1976, pp. 15-1 through 15-12 in Proceedings of First semiannual EPRI Solar Program Review Meeting and Workshop. Available from EPRI as EPRI ER-283-SR, Vol. II, May 8-12, 1976.

19. McKoy, George C., Penetration Analysis And Margin Requirements Associated With Large-Scale Utilization of Solar Power Plants, Final Report, August 1976. Available from EPRI as EPRI ER-198.

20. Plastiras, Joan K., "Capacity Displacement For Solar Plants," Proceedings of the 1978 Annual Meeting, American Section of ISES, Vol. 2.2, pp. 562-563.

21. Davitan, H., et al., "Utility Load Management And Solar Energy," Proceedings of the 1978 Annual Meeting, American Section of ISES, Vol. 2.2, pp. 576-579.

22. Wirtshafter, R., "The Impact of Solar and Energy Conservation Building Designs on Electric Utilities," Proceedings of the 1978 Annual Meeting, American Section of ISES, Vol. 2.2, pp. 571-573.

23. Pollard, William G., Analysis of Systems for the Generation of Electricity from Solar Radiation, June 1978, 47 pp. Available from Oak Ridge Associated Universities, P. O. Box 117, Oak Ridge, Tennessee 37830, as ORAU/IEA-78-12 (R).

24. Andrews, John W., "Energy Storage Requirements Reduced In Coupled Wind-Solar Generating Systems," Solar Energy, Vol. 18, No. 1, 1976, pp. 73-74.

25 Feldman, Stephen L., and Bruce Anderson, "Financial Incentives For The Adoption of Solar Energy Design: Peak-Load Pricing of Back-Up Systems," Solar Energy, Vol. 17, No. 6, 1975, pp. 339-343.

26. Brune, Joan M., "Hybrid and Repowered Solar Electric Plants," Proceedings 1978 ISES Annual Meeting, Vol. 2.2, pp. 564-570.

27. Johanson, E. E. and M. K. Goldenblatt, "An Economic Model To Establish the Value of WECS To A Utility System," Proceedings 1978 ISES Annual Meeting, Vol. 2.2, pp. 580-587.

INDEX

Of Subjects